CONTROL OF POWER INVERTERS IN RENEWABLE ENERGY AND SMART GRID INTEGRATION

CONTROL OF POWER INVERTERS IN RENEWABLE ENERGY AND SMART GRID INTEGRATION

Qing-Chang Zhong
The University of Sheffield, UK

Tomas Hornik
Turbo Power Systems Ltd., UK

A John Wiley & Sons, Ltd., Publication

IEEE PRESS

This edition first published 2013
© 2013 John Wiley & Sons, Ltd

Registered office
John Wiley & Sons Ltd, The Atrium, Southern Gate, Chichester, West Sussex, PO19 8SQ, United Kingdom

For details of our global editorial offices, for customer services and for information about how to apply for permission to reuse the copyright material in this book please see our website at www.wiley.com.

The right of the author to be identified as the author of this work has been asserted in accordance with the Copyright, Designs and Patents Act 1988.

All rights reserved. No part of this publication may be reproduced, stored in a retrieval system, or transmitted, in any form or by any means, electronic, mechanical, photocopying, recording or otherwise, except as permitted by the UK Copyright, Designs and Patents Act 1988, without the prior permission of the publisher.

Wiley also publishes its books in a variety of electronic formats. Some content that appears in print may not be available in electronic books.

Designations used by companies to distinguish their products are often claimed as trademarks. All brand names and product names used in this book are trade names, service marks, trademarks or registered trademarks of their respective owners. The publisher is not associated with any product or vendor mentioned in this book. This publication is designed to provide accurate and authoritative information in regard to the subject matter covered. It is sold on the understanding that the publisher is not engaged in rendering professional services. If professional advice or other expert assistance is required, the services of a competent professional should be sought.

MATLAB® and Simulink® are trademarks of The MathWorks, Inc. and are used with permission. The MathWorks does not warrant the accuracy of the text or exercises in this book. This book's use or discussion of the MATLAB® and Simulink® softwares or related products does not constitute endorsement or sponsorship by The MathWorks of a particular pedagogical approach or particular use of the MATLAB® software.

DISCLAIMER

The contents of this book are meant to supply information on the control of power inverters. The book is not meant to be the sole resource used in any design project. The examples and solutions presented are not to be construed as complete engineered design solutions for any particular problem or project. The authors and publisher are not attempting to render any type of engineering or other professional services. Should these services be required, an appropriate professional engineer should be consulted. The authors and publisher assume no liability or responsibility for any uses made of the material contained and described herein. The authors and publisher are not offering legal advice or endorsing any products or services that may be identified in this book.

Library of Congress Cataloging-in-Publication Data applied for.

Hardback ISBN: 978-0-470-66709-5

A catalogue record for this book is available from the British Library.

Typeset in 10/12pt Times by Aptara Inc., New Delhi, India

To those who have taught us in one way or another

Contents

Preface	xvii
Acknowledgments	xix
About the Authors	xxi
List of Abbreviations	xxiii

1	**Introduction**		**1**
1.1	Outline of the Book		1
1.2	Basics of Power Processing		4
	1.2.1	AC-DC Conversion	4
	1.2.2	DC-DC Conversion	14
	1.2.3	DC-AC Conversion	18
	1.2.4	AC-AC Conversion	21
1.3	Hardware Issues		24
	1.3.1	Isolation	25
	1.3.2	Power Stages	26
	1.3.3	Output Filters	33
	1.3.4	Voltage and Current Sensing	35
	1.3.5	Signal Conditioning	36
	1.3.6	Protection	38
	1.3.7	Central Controller	38
	1.3.8	Test Equipment	42
1.4	Wind Power Systems		44
	1.4.1	Basics of Wind Power Generation	44
	1.4.2	Wind Turbines	45
	1.4.3	Generators and Topologies	48
	1.4.4	Control of Wind Power Systems	51
1.5	Solar Power Systems		53
	1.5.1	Introduction to Solar Power	53
	1.5.2	Processing of Solar Power	54
1.6	Smart Grid Integration		55
	1.6.1	Operation Paradigms of Power Systems	55
	1.6.2	Introduction to Smart Grids	56
	1.6.3	Requirements for Smart Grid Integration	59

2	**Preliminaries**	63
2.1	Power Quality Issues	63
	2.1.1 Introduction	63
	2.1.2 Degradation Mechanisms of Voltage Quality	65
	2.1.3 Role of Inverter Output Impedance	66
2.2	Repetitive Control	67
	2.2.1 Basic Principles	67
	2.2.2 Poles of the Internal Model $M(s)$	68
	2.2.3 Selection of the Delay in the Internal Model	70
2.3	Reference Frames	71
	2.3.1 Natural (abc) Frame	71
	2.3.2 Stationary Reference ($\alpha\beta$) Frame	72
	2.3.3 Synchronously Rotating Reference (dq) Frame	74
	2.3.4 The Case with Phase Sequence acb	76

PART I   POWER QUALITY CONTROL

3	**Current H^∞ Repetitive Control**	81
3.1	System Description	81
3.2	Controller Design	82
	3.2.1 State-space Model of the Control Plant P	83
	3.2.2 Formulation of the Standard H^∞ Problem	84
	3.2.3 Evaluation of the System Stability	86
3.3	Design Example	87
3.4	Experimental Results	88
	3.4.1 Synchronisation Process	88
	3.4.2 Steady-state Performance	88
	3.4.3 Transient Response (without a Load)	91
3.5	Summary	91

4	**Voltage and Current H^∞ Repetitive Control**	93
4.1	System Description	93
4.2	Modelling of an Inverter	94
4.3	Controller Design	96
	4.3.1 Formulation of the H^∞ Control Problem	96
	4.3.2 Realisation of the Generalised Plant	98
	4.3.3 State-space Realisation of $\mathbf{T}_{ew}$	99
	4.3.4 State-space Realisation of $\mathbf{T}_{ba}$	99
4.4	Design Example	100
4.5	Simulation Results	102
	4.5.1 Nominal Responses	103
	4.5.2 Response to Load Changes	104
	4.5.3 Response to Grid Distortions	104
4.6	Summary	107

5 Voltage H^∞ Repetitive Control with a Frequency-adaptive Mechanism — 109
5.1 System Description — 109
5.2 Controller Design — 110
    5.2.1 State-space Model of the Control Plant P — 111
    5.2.2 Frequency-adaptive Internal Model M — 112
    5.2.3 Formulation of the Standard H^∞ Problem — 113
    5.2.4 Evaluation of System Stability — 115
5.3 Design Example — 116
5.4 Experimental Results — 117
    5.4.1 Steady-state Performance in the Stand-alone Mode — 117
    5.4.2 Steady-state Performance in the Grid-connected Mode — 119
    5.4.3 Transient Response: without a Local Load — 120
    5.4.4 Response to Variations of the Grid Frequency — 120
5.5 Summary — 126

6 Cascaded Current-Voltage H^∞ Repetitive Control — 127
6.1 Operation Modes in Microgrids — 127
6.2 Control Scheme — 129
6.3 Design of the Voltage Controller — 131
    6.3.1 State-space Model of the Plant P_u — 131
    6.3.2 Formulation of the Standard H^∞ Problem — 132
6.4 Design of the Current Controller — 133
    6.4.1 State-space Model of the Plant P_i — 133
    6.4.2 Formulation of the Standard H^∞ Problem — 134
6.5 Design Example — 134
    6.5.1 Design of the H^∞ Voltage Controller — 135
    6.5.2 Design of the H^∞ Current Controller — 136
6.6 Experimental Results — 136
    6.6.1 Steady-state Performance in the Stand-alone Mode — 136
    6.6.2 Steady-state Performance in the Grid-connected Mode — 138
    6.6.3 Transient Performance — 144
    6.6.4 Seamless Transfer of the Operation Mode — 145
6.7 Summary — 147

7 Control of Inverter Output Impedance — 149
7.1 Inverters with Inductive Output Impedances (L-inverters) — 149
7.2 Inverters with Resistive Output Impedances (R-inverters) — 150
    7.2.1 Controller Design — 150
    7.2.2 Stability Analysis — 151
7.3 Inverters with Capacitive Output Impedances (C-inverters) — 152
7.4 Design of C-inverters to Improve the Voltage THD — 153
    7.4.1 General Case — 153
    7.4.2 Special Case I: to Minimise the 3rd and 5th Harmonic Components — 155
    7.4.3 Special Case II: to Minimise the 3rd Harmonic Component — 156
    7.4.4 Special Case III: to Minimise the 5th Harmonic Component — 157

7.5	Simulation Results for R-, L- and C-inverters	157
	7.5.1 The Case with $L = 2.35\ mH$	158
	7.5.2 The Case with $L = 0.25\ mH$	158
7.6	Experimental Results for R-, L- and C-inverters	159
	7.6.1 The Case with $L = 2.35\ mH$	160
	7.6.2 The Case with $L = 0.25\ mH$	161
7.7	Impact of the Filter Capacitor	162
7.8	Summary	163
8	**Bypassing Harmonic Current Components**	**165**
8.1	Controller Design	165
8.2	Physical Interpretation of the Controller	167
8.3	Stability Analysis	169
	8.3.1 Without Consideration of the Sampling Effect	169
	8.3.2 With Consideration of the Sampling Effect	170
8.4	Experimental Results	171
8.5	Summary	172
9	**Power Quality Issues in Traction Power Systems**	**173**
9.1	Introduction	173
9.2	Description of the Topology	175
9.3	Compensation of Negative-sequence Currents, Reactive Power and Harmonic Currents	175
	9.3.1 Grid-side Currents before Compensation	175
	9.3.2 Compensation of Active and Reactive Power	178
	9.3.3 Compensation of Harmonic Currents	179
	9.3.4 Regulation of the DC-bus Voltage	179
	9.3.5 Implementation of the Compensation Strategy	179
9.4	Special Case: $\cos\theta = 1$	180
9.5	Simulation Results	181
	9.5.1 The Case when $\cos\theta \neq 1$	181
	9.5.2 The Case when $\cos\theta = 1$	181
9.6	Summary	184

PART II   NEUTRAL LINE PROVISION

10	**Topology of a Neutral Leg**	**187**
10.1	Introduction	187
10.2	Split DC Link	188
10.3	Conventional Neutral Leg	189
10.4	Independently-controlled Neutral Leg	190
10.5	Summary	191
11	**Classical Control of a Neutral Leg**	**193**
11.1	Mathematical Modelling	193
11.2	Controller Design	195

	11.2.1	Design of the Current Controller K_i	196
	11.2.2	Design of the Voltage Controller K_v	196
11.3	Performance Evaluation	199
11.4	Selection of the Components	201
	11.4.1	Capacitor C_N	201
	11.4.2	Inductor L_N	201
11.5	Simulation Results	202
	11.5.1	With $i_N = 0$	202
	11.5.2	With a 50 Hz Neutral Current	203
	11.5.3	With a 150 Hz Neutral Current	204
	11.5.4	With a DC Neutral Current	205
11.6	Summary	205

12 H^∞ Voltage-Current Control of a Neutral Leg — 207

12.1	Mathematical Modelling	207
12.2	Controller Design	210
	12.2.1 State-space Realisation of **P**	211
	12.2.2 State-space Realisation of the Closed-loop Transfer Function	213
12.3	Selection of Weighting Functions	214
12.4	Design Example	215
12.5	Simulation Results	216
12.6	Summary	217

13 Parallel PI Voltage-H^∞ Current Control of a Neutral Leg — 219

13.1	Description of the Neutral Leg	219
13.2	Design of an H^∞ Current Controller	221
	13.2.1 Controller Description	221
	13.2.2 Formulation as a Standard H^∞ Problem	221
	13.2.3 State-space Realisation of the Plant **P**	222
	13.2.4 State-space Realisation of the Generalised Plant $\tilde{P}$	223
	13.2.5 Design Example	224
13.3	Addition of a Voltage Control Loop	226
13.4	Experimental Results	226
	13.4.1 Steady-state Performance	227
	13.4.2 Transient Response to Changes in the Neutral Current	230
13.5	Summary	230

14 Applications in Single-phase to Three-phase Conversion — 233

14.1	Introduction	233
14.2	The Topology under Consideration	236
14.3	Basic Analysis	237
14.4	Controller Design	239
	14.4.1 Synchronisation Unit	239
	14.4.2 Control of the Rectifier Leg	241
	14.4.3 Control of the Neutral Leg	241
	14.4.4 Control of the Phase Legs	242

14.5	Simulation Results	244
	14.5.1 With Three-phase Linear Balanced Loads	244
	14.5.2 With Three-phase Non-linear Unbalanced Loads	246
14.6	Summary	248

PART III POWER FLOW CONTROL

15	**Current Proportional–Integral Control**	**251**
15.1	Control Structure	251
	15.1.1 In the Synchronously Rotating Reference (dq) Frame	251
	15.1.2 Equivalent Structure in the Natural (abc) Frame	253
15.2	Controller Implementation	254
15.3	Experimental Results	254
	15.3.1 Steady-state Performance	254
	15.3.2 Transient Performance	257
15.4	Summary	258
16	**Current Proportional-Resonant Control**	**259**
16.1	Proportional-resonant Controller	259
16.2	Control Structure	260
	16.2.1 In the Stationary Reference ($\alpha\beta$) Frame	260
	16.2.2 Equivalent Controller in the abc Frame	261
16.3	Controller Design	261
	16.3.1 Model of the Plant	261
	16.3.2 Design Example	262
16.4	Experimental Results	263
	16.4.1 Steady-state Performance	263
	16.4.2 Transient Performance	266
16.5	Summary	268
17	**Current Deadbeat Predictive Control**	**269**
17.1	Control Structure	269
17.2	Controller Design	269
17.3	Experimental Results	271
	17.3.1 Steady-state Performance	272
	17.3.2 Transient Performance	275
17.4	Summary	275
18	**Synchronverters: Grid-friendly Inverters that Mimic Synchronous Generators**	**277**
18.1	Mathematical Model of Synchronous Generators	278
	18.1.1 Electrical Part	278
	18.1.2 Mechanical Part	280
	18.1.3 Presence of a Neutral Line	281

18.2	Implementation of a Synchronverter		282
	18.2.1	Power Part	282
	18.2.2	Electronic Part	283
18.3	Operation of a Synchronverter		284
	18.3.1	Regulation of Real Power and Frequency Droop Control	284
	18.3.2	Regulation of Reactive Power and Voltage Droop Control	286
18.4	Simulation Results		287
	18.4.1	Under Different Grid Frequencies	288
	18.4.2	Under Different Load Conditions	288
18.5	Experimental Results		290
	18.5.1	Performance of Power Flow Control	290
	18.5.2	Loading Performance in the Stand-alone Mode	291
	18.5.3	Loading Performance in the Grid-connected Mode	294
18.6	Summary		296
19	**Parallel Operation of Inverters**		**297**
19.1	Introduction		297
19.2	Problem Description		299
19.3	Power Delivered to a Voltage Source		300
19.4	Conventional Droop Control		301
	19.4.1	For R-inverters	301
	19.4.2	For L-inverters	302
	19.4.3	For C-inverters	303
	19.4.4	Experimental Results with R-inverters	304
19.5	Inherent Limitations of Conventional Droop Control		304
	19.5.1	Real Power Sharing	307
	19.5.2	Reactive Power Sharing	308
19.6	Robust Droop Control of R-inverters		309
	19.6.1	Control Strategy	309
	19.6.2	Error Due to Inaccurate Voltage Measurements	311
	19.6.3	Voltage Regulation	311
	19.6.4	Error Due to the Global Settings for E^* and ω^*	312
	19.6.5	Experimental Results	313
19.7	Robust Droop Control of C-inverters		319
	19.7.1	Control Strategy	319
	19.7.2	Simulation Results	320
	19.7.3	Experimental Results	321
19.8	Robust Droop Control of L-inverters		326
	19.8.1	Control Strategy	326
	19.8.2	Simulation Results	327
	19.8.3	Experimental Results	330
19.9	Summary		330
20	**Robust Droop Control with Improved Voltage Quality**		**335**
20.1	Control Strategy		335
20.2	Experimental Results		337

	20.2.1　1 : 1 *Power Sharing*	337
	20.2.2　2 : 1 *Power Sharing*	340
20.3	Summary	346

21　Harmonic Droop Controller to Improve Voltage Quality　347
21.1　Model of an Inverter System　347
21.2　Power Delivered to a Current Source　349
21.3　Reduction of Harmonics in the Output Voltage　351
21.4　Simulation Results　353
21.5　Experimental Results　355
21.6　Summary　358

PART IV　SYNCHRONISATION

22　Conventional Synchronisation Techniques　361
22.1　Introduction　361
22.2　Zero-crossing Method　362
22.3　Basic Phase-locked Loops (PLL)　363
22.4　PLL in the Synchronously Rotating Reference Frame (SRF-PLL)　364
22.5　Second-order Generalised Integrator-based PLL (SOGI-PLL)　366
22.6　Sinusoidal Tracking Algorithm (STA)　368
22.7　Simulation Results with SOGI-PLL and STA　369
　　　22.7.1　*With a Noisy Distorted Signal having a Variable Frequency*　369
　　　22.7.2　*With a Noisy Distorted Square Wave*　372
22.8　Experimental Results with SOGI-PLL and STA　372
　　　22.8.1　*With a Voltage Taken from the Grid*　372
　　　22.8.2　*With a Noisy Distorted Signal having a Variable Frequency*　375
　　　22.8.3　*With a Noisy Distorted Square Wave*　375
22.9　Summary　378

23　Sinusoid-locked Loops　379
23.1　Single-phase Synchronous Machine (SSM) Connected to the Grid　379
23.2　Structure of a Sinusoid-locked Loop (SLL)　380
23.3　Tracking of the Frequency and the Phase　382
23.4　Tracking of the Voltage Amplitude　382
23.5　Tuning of the Parameters　382
23.6　Equivalent Structure　383
23.7　Simulation Results　384
　　　23.7.1　*With a Noisy Distorted Signal having a Variable Frequency*　384
　　　23.7.2　*With a Noisy Distorted Square Wave*　386
23.8　Experimental Results　386
　　　23.8.1　*With a Voltage Taken from the Grid*　386

	23.8.2	*With a Noisy Distorted Signal having a Variable Frequency*	389
	23.8.3	*With a Noisy Distorted Square Wave*	389
23.9	Summary		390

References   393

Index   407

Preface

It has been a long journey since the lead author had the idea of writing a research monograph on the control of power inverters, which are the common key devices used to integrate renewable energy and distributed generation into smart grids. Soon after he was appointed at Imperial College London, UK, in 2001 to work on a project on the control of power inverters, he realised that there were many challenging problems in this area and that research activities on renewable energy and grid integration would become very important worldwide in the near future. Four kinds of problem were identified at that time, that is, power quality issues, the provision of a neutral line, power flow control, and synchronisation. Over the past 10 years, he has kept working on these problems, sometimes with his collaborators and PhD students, although he has experienced several job moves. The co-author joined his team as a PhD student in 2007 and soon built up a test rig, which facilitated the research and made the idea of writing the book more concrete. The award of a Senior Research Fellowship from the Royal Academy of Engineering, UK, and the Leverhulme Trust in 2009 has considerably accelerated the process of the book project. Finally, the award of a one-year International Collaboration Sabbatical from the Engineering and Physical Sciences Research Council (EPSRC), UK, in 2011 has made the book a reality.

Energy and sustainability are two major problems the world faces today. Renewable energy has been seen as a promising means to address these problems while smart grids are being developed to improve energy efficiency, security and resilience of power systems. Integrating renewable energy and other distributed energy sources into smart grids, often via inverters, is arguably the greatest "new frontier" for smart grid advancements. Inverters should be controlled properly so that their integration does not jeopardise the stability and performance of power systems and a solid technical backbone is formed for the other functions and services of smart grids.

There are several important control problems associated with inverters. For example, how to make sure that the quality of the power fed into the grid is high, even when there are nonlinear loads and/or the grid voltage is distorted; how to make sure that a balanced neutral line is provided for applications where a neutral line is needed, e.g. when three-phase loads are not balanced; how to make sure that inverters can be operated in the grid-connected mode or the stand-alone mode and how to minimise the transient dynamics when the operation mode is changed; how to synchronise inverters with the grid so that they can be connected to the grid when needed; how to make sure that parallel-operated inverters share power proportionally according to their power ratings to avoid damage; how to operate grid-connected inverters in a grid-friendly manner so that the impact on the grid is minimised, etc. Many original ideas and control strategies, which have been developed to address these problems over the past 10 years, are presented in this book. These include different strategies to improve the power quality in smart (and/or micro) grids, inverters with capacitive output impedances (C-inverters), the provision of a neutral line for inverters, grid-friendly integration using inverters that mimic synchronous generators (synchronverters), parallel operation of inverters with robust droop controllers to share power in proportion to their ratings, harmonic droop controllers to improve power quality, sinusoid-locked loops to lock the frequency and the amplitude, in addition to the phase, of the grid voltage, etc. These advanced control strategies are expected to considerably facilitate the large-scale utilisation of renewable energy and smart grid integration.

The book consists of one introductory chapter (Chapter 1), one preliminary chapter (Chapter 2) and four parts: Power Quality Control (Chapters 3–9), Neutral Line Provision (Chapters 10–14), Power Flow Control (Chapters 15–21) and Synchronisation (Chapters 22–23). In the introductory chapter, some basics about power processing, hardware issues about inverters, and brief descriptions of wind power, solar power and smart grid integration are presented. In the preliminary chapter, some common knowledge of power quality issues, repetitive control and reference frames is introduced. In Part I, several control strategies according to different mechanisms are presented to improve the quality of the inverter voltage, and the current exchanged with the grid. In Part II, the topologies to provide a neutral line are discussed and several control strategies are presented to maintain a balanced stable neutral line, which facilitates the independent operation of phases. In Part III, both current-controlled and voltage-controlled strategies are presented to control the power flow between an inverter and the grid. Innovative concepts such as synchronverters (inverters that mimic synchronous generators), robust droop controllers, harmonic droop controllers, etc., are presented. In Part IV, conventional synchronisation methods are presented at first, followed by a sinusoid-locked loop developed according to the principles of a synchronous generator that does not exchange power with the grid.

This book is written for control engineers who are moving into the area of power electronics, renewable energy and distributed generation, smart grids, flexible AC transmission systems, power systems for more-electric aircraft and all-electric ships, etc, and researchers and practitioners working in these areas who are eager to see what benefits advanced control algorithms can bring. It systematically explores the fundamental and challenging problems with respect to control of power inverters and fully demonstrates the beauty of the integration of control and power electronics. Most of the artful control strategies presented in this book are accompanied by extensive experimental results and, hence, this book is also very useful for practitioners in this area to see how advanced control strategies could improve system performance and work in practice. This book also provides an excellent opportunity for graduate students and researchers who work in the area to become familiar with the latest developments. It can be adopted as a textbook for graduate programmes on advanced control engineering, power electronics, microgrids, renewable energy and smart grid integration.

Acknowledgments

This research monograph systematically summarises the research I, together with my collaborators and PhD students, have carried out over the past 10 years in the area of control of power inverters in renewable energy and smart grid integration. It is impossible to list everyone who has made contributions to this book as co-authors but their contributions are greatly appreciated. George Weiss, my advisor and collaborator, currently at Tel Aviv University, is the one who started this all. He opened this whole new world for me, which has paved the way for my career. Tomas Hornik, my past PhD student, has made significant direct contributions to this book and deserves to be a co-author, not just for himself but also on behalf of others. I am grateful to Long Nguyen, Zhenyu Ma, Shamsul A. Zulkifli, Wen-Long Ming, Yu Zeng, Zhi Hou, Xiaolin Wang and Xin Cao for their contributions. I am also grateful to Frede Blaabjerg, Chunmei Feng, Tim Green, Joseph M. Guerrero, Leslie Hobson, Marcel G. Jayne, Miroslav Krstic, Jun Liang and George Weiss for the collaborative work we have done.

I wish to thank the Engineering and Physical Sciences Research Council, UK, for their constant support for my research over the years. The grants which have direct contributions to this book include GR/N38190/01, EP/E055877/1, EP/H004351/1, EP/J001333/1, EP/I038586/1, EP/J01558X/1, EP/J001333/2, two grants under the Dorothy Hodgkin Postgraduate Awards (DHPA) scheme, two grants under the Knowledge Transfer Accounts (KTA) scheme and one grant under the Doctoral Training Accounts (DTA) scheme.

I also wish to thank the Royal Academy of Engineering, UK, and and the Leverhulme Trust for the award of a Senior Research Fellowship during 2009–2010. This offered me valuable time to concentrate on my research and many of the results are included in this book.

The links with, and support from, industrial partners have always facilitated our research. I am particularly grateful to Phill Cartwright and Kevin Daffey (Rolls-Royce Plc), Roger Critchley and Fainan Hassan (ALSTOM Grid), Damien Culley (National Grid, UK), Brett Dowen (add2 Ltd), Nordine Haddjeri (Nheolis, France), Tony Lakin (Turbo Power Systems), Robert Owen (Texas Instruments), Graham Chapman (Power Systems Warehouse) and David Doherty (Yokogawa Measurement Technologies Ltd) for their valuable direct support to our research.

It has been a great pleasure to work with the colleagues of John Wiley & Sons, Ltd and IEEE Press. The support and help from Liz Wingett (the Project Editor), Laura Bell (the Assistant Editor) and Peter Mitchell (the Publisher) are greatly appreciated.

Last but not least, I would like to formally thank the L$_Y$X team for having developed the L$_Y$X document processor, which has made my writing experience in the past 12 years so enjoyable and has saved me a lot of time. L$_Y$X functions as a front-end to the L$_A$T$_E$X typesetting system and is designed for authors who want professional output with a minimum of effort and without becoming specialists in typesetting. L$_Y$X is available in multiple languages for various operating systems, including Windows, Mac OS X, Linux, UNIX, OS/2 and Haiku, and can be downloaded, redistributed and modified for free from http://www.lyx.org under the terms of the GNU General Public License.

Qing-Chang Zhong
Chair in Control and Systems Engineering
Department of Automatic Control and Systems Engineering
The University of Sheffield, UK

Q.Zhong@Sheffield.ac.uk
http://zhongqc.staff.shef.ac.uk

October 2012

About the Authors

Qing-Chang Zhong received his Diploma in electrical engineering from Hunan Institute of Engineering, Xiangtan, China, in 1990, his MSc degree in electrical engineering from Hunan University, Changsha, China, in 1997, his PhD degree in control theory and engineering from Shanghai Jiao Tong University, Shanghai, China, in 1999, and his PhD degree in control and power engineering (awarded the Best Doctoral Thesis Prize) from Imperial College London, London, UK, in 2004, respectively. He holds the Chair in Control and Systems Engineering at the Department of Automatic Control and Systems Engineering, The University of Sheffield, UK. He has worked at Hunan Institute of Engineering, Xiangtan, China; Technion–Israel Institute of Technology, Haifa, Israel; Imperial College London, London, UK; University of Glamorgan, Cardiff, UK; The University of Liverpool, Liverpool, UK; and Loughborough University, Leicestershire, UK. He has been on sabbatical at the Cymer Center for Control Systems and Dynamics (CCSD), University of California, San Diego, USA; and the Center for Power Electronics Systems (CPES), Virginia Tech, Blacksburg, USA. He is the author or co-author of *Robust Control of Time-Delay Systems* (Springer-Verlag, 2006), *Control of Integral Processes with Dead Time* (Springer-Verlag, 2010) and *Control of Power Inverters in Renewable Energy and Smart Grid Integration* (Wiley-IEEE Press, 2013). His research focuses on advanced control theory and applications, including power electronics, renewable energy and smart grid integration, electric drives and electric vehicles, robust and H^∞ control, time-delay systems and process control. He is a Specialist recognised by the State Grid Corporation of China (SGCC), a Fellow of the Institution of Engineering and Technology (IET), a Senior Member of IEEE, the Vice-Chair of IFAC TC 6.3 (Power and Energy Systems) responsible for the Working Group on Power Electronics and was a Senior Research Fellow of the Royal Academy of Engineering/Leverhulme Trust, UK (2009–2010). He serves as an Associate Editor for *IEEE Transactions on Power Electronics* and the Conference Editorial Board of the IEEE Control Systems Society.

Tomas Hornik received a Diploma in Electrical Engineering in 1991 from the Technical College V Uzlabine, Prague, the BEng and PhD degree in electrical engineering and electronics from The University of Liverpool, UK, in 2007 and 2010, respectively. He was a postdoctoral researcher at the same university from 2010 to 2011. He joined Turbo Power Systems as a Control Engineer in 2011. His research interests cover power electronics, advanced control theory and DSP-based control applications. He had more than ten years working experience in industry as a system engineer responsible for commissioning and software design in power generation and distribution, control systems for central heating and building management. He is a member of the IEEE and the IET.

List of Abbreviations

abc frame	Natural Frame
$\alpha\beta$ frame	Stationary Reference Frame
dq frame	Synchronously Rotating Reference Frame
AC	Alternating Current
ADC	Analog-to-Digital Converter
APC	Active Power Compensator
APF	Active Power Filter
APFM	Amplitude Phase Frequency Model
APM	Amplitude Phase Model
C-inverters	Inverters with Capacitive Output Impedances
CHP	Combined Heat and Power
CPU	Central Processing Unit
CSI	Current Source Inverter
CSP	Concentrated Solar Power
CVCF	Constant-Voltage Constant-Frequency
DAC	Digital-to-Analog Converter
DB	Dead-Beat
DC	Direct Current
DDSRF-PLL	Decoupled Double SRF-PLL
DFIG	Doubly-Fed Induction Generator
DSC	Digital Signal Controller
DSP	Digital Signal Processor
EKF	Extended Kalman Filter
EMI	Electromagnetic Interference
EPLL	Enhanced PLL
ESR	Equivalent Series Resistance
FRF-PLL	Fixed-Reference Frame PLL
GTO	Gate-Turn Off
HAPF	Hybrid Active Power Filter

HB	Hysteresis Band
HC	Harmonics Compensator
HCS	Hill-Climb Search
IGBT	Insulated Gate Bipolar Transistor
IPM	Intelligent Power Module
KCL	Kirchhoff's Current Law
KVL	Kirchhoff's Voltage Law
L-inverters	Inverters with Inductive Output Impedances
LF	Loop Filter
LPF	Low-Pass Filter
MIMO	Multiple-Input Multiple-Output
MOSFET	Metal Oxide Semiconductor Field Effect Transistor
MPP	Maximum Power Point
MPPT	Maximum Power Point Tracking
P controller	Proportional controller
PD unit	Phase-error Detection unit
PI controller	Proportional–Integral controller
PLL	Phase-Locked Loops
PMSG	Permanent Magnet Synchronous Generators
PPF	Passive Power Filter
PR controller	Proportional–Resonant controller
PSF	Power Signal Feedback
PV	Photovoltaic
PWM	Pulse-Width Modulation
R-inverters	Inverters with Resistive Output Impedances
RMS	Root Mean Square
RPC	Railway Static Power Conditioner
SCIG	Squirrel-Cage Induction Generator
SISO	Single-Input Single-Output
SLL	Sinusoid-Locked Loops
SOA	Safe Operating Area
SOGI-PLL	Second-Order Generalised Integrator-based PLL
SOGI-QSG	SOGI-based Quadrature-Signal Generator
SPC	Static Power Conditioner
SPWM	Sinusoidal Pulse-Width Modulation
SRF-PLL	PLL in the Synchronously Rotating Reference Frame
SSM	Single-phase Synchronous Machine
STA	Sinusoidal Tracking Algorithm
SVC	Static Var Compensators

SVF	Space Vector Filter
SVPWM	Space Vector Pulse-Width Modulation
THD	Total Harmonic Distortion
TSR	Tip Speed Ratio
UPS	Uninterruptible Power Supply
VCO	Voltage Controlled Oscillator
VSI	Voltage Source Inverter
ZOH	Zero Order Hold

Conventions

A^T and A^*	transpose and complex conjugate transpose of A
A^{-1} and A^{-*}	inverse of A and shorthand for $(A^{-1})^*$
$\det(A)$	determinant of A
$G(s) = \left[\begin{array}{c\|c} A & B \\ \hline C & D \end{array}\right]$	shorthand for $G(s) = D + C(sI - A)^{-1}B$
$j\mathbb{R}$	imaginary axis
$\operatorname{Re} s$ and $\operatorname{Im} s$	real and imaginary parts of $s \in \mathbb{C}$
$\mathbb{Z}, \mathbb{R}$ and $\mathbb{C}$	fields of integral, real and complex numbers
$\in$	belong to
$\cap$	intersection
$\subset$	subset

1

Introduction

In this chapter, an overview of the book is given at first, followed by some basics about power processing and some hardware issues relevant to the design of power inverters. Moreover, wind power systems, solar power systems and smart grid integration are briefly described to set the scene for the rest of the book.

1.1 Outline of the Book

After making an introduction in this chapter and presenting preliminaries in Chapter 2, the book is divided into four Parts: Power Quality Control (Chapters 3–9), Neutral Line Provision (Chapters 10–14), Power Flow Control (Chapters 15–21) and Synchronisation (Chapters 22–23). The overall structure of the book is shown in Figure 1.1. Some chapters are related to more than one part, which is not shown in Figure 1.1 but will be mentioned below.

Part I is devoted to the power quality issues of the current fed into the grid and the output voltage of an inverter. A current controller is designed in Chapter 3 with the H^∞ repetitive control strategy so that the current injected into the grid is clean. This chapter is also directly linked to Part III under the category of current-controlled strategies. Several control strategies are presented in Chapters 4–8 to address voltage quality issues based on different mechanisms. In Chapters 4 and 5, the controllers are designed based on the H^∞ repetitive control strategy, with different sets of feedback signals and different models. Both the voltage quality and the current quality are addressed in Chapter 6 with a cascaded current–voltage controller, according to the H^∞ repetitive control strategy. In Chapters 4–6, the voltage quality issues are addressed essentially from the control point of view as a tracking problem. The voltage quality issue can also be addressed involving fundamental understanding about the degradation mechanisms of voltage quality. In Chapter 7, it is shown that the output impedance of an inverter can be changed to obtain inverters with inductive, resistive and capacitive output impedances,

---

Control of Power Inverters in Renewable Energy and Smart Grid Integration, First Edition.
Qing-Chang Zhong and Tomas Hornik.
© 2013 John Wiley & Sons, Ltd. Published 2013 by John Wiley & Sons, Ltd.

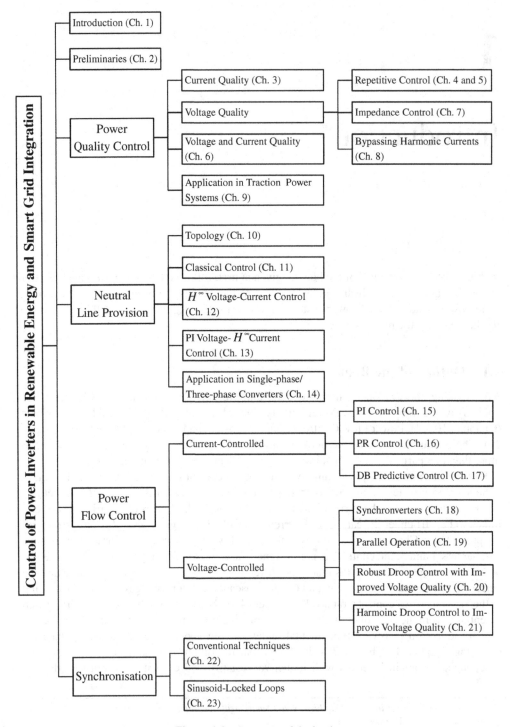

Figure 1.1 Structure of the book

which are called L-inverters, R-inverters and C-inverters, respectively. C-inverters are able to offer much better voltage quality than L-inverters and R-inverters with the same hardware. In Chapter 8, a strategy that is the same as bypassing the harmonic components in the load current is presented to improve the voltage quality. Another strategy that falls into this category is to inject the right amount of voltage harmonics into the reference voltage of an inverter so that it cancels the harmonic voltage dropped on the output impedance, which improves the quality of the output voltage. This is presented in Chapter 21, in Part III, after presenting the robust droop control in Chapter 19. As an application example, the power quality issues in traction power systems, including current harmonics, negative-sequence currents and low power factor, are addressed in Chapter 9.

Part II is devoted to the provision of an independently-controlled neutral line, which facilitates the implementation of other functions in a power electronic system. The topologies to provide a neutral line are presented in Chapter 10. In Chapter 11, a controller is designed to maintain a stable neutral point with classical control strategies, from which the parameters of the neutral leg are determined. In Chapter 12, a controller is designed with the H^∞ control strategy, taking the voltage shift of the neutral point and the current flowing into the DC-link capacitors as feedback. In Chapter 13, an H^∞ current controller is designed to minimise the current flowing into the DC-link capacitor and a PI controller is designed to bring the DC voltage shift back to the mid-point of the DC link. These two controllers are decoupled in the frequency domain and, hence, can be arranged in a parallel control structure. The provision of an independently-controlled neutral line is applied in Chapter 14, as an application example, to the generation of an independent three-phase power supply from a single-phase source.

Part III is devoted to power flow control. The control strategies can be classified into two categories: current-controlled strategies to directly control the current exchanged with the grid and voltage-controlled strategies to control the voltage of the inverter so that the power flow is indirectly controlled. Current-controlled strategies are easy to implement but the inverters equipped with current-controlled strategies cannot directly take part in the regulation of power system frequency and voltage and, hence, they may cause problems for system stability when the share of power fed into the grid is significant. The PI control, PR control and DB predictive control presented in Chapters 15–17 belong to this category. The repetitive controller presented in Chapter 3, in Part I, also belongs to this category. Voltage-controlled strategies have attracted a lot of attention from academia and industry in recent years because they are able to take part in the regulation of system frequency and voltage. In Chapter 18, a control strategy is presented to make inverters mathematically equivalent to conventional synchronous generators. Such inverters are called synchronverters. As a result, all the technologies developed for synchronous generators can be applied to inverters, which considerably facilitates the grid connection of renewable energy and smart grid integration. A highly compact controller is presented to implement the functions of frequency control, real power control, voltage control and reactive power control. In Chapter 19, the parallel operation of inverters is discussed. After presenting the conventional droop control strategies for L-, R- and C-inverters, the inherent limitations of the conventional droop control are revealed. The accuracy of power sharing greatly depends on the accuracy and consistency of the components, and the voltage regulation capability is poor. Then, robust droop control strategies for R-inverters, L-inverters and C-inverters are presented so that accurate sharing of both real power and reactive power can be achieved even if there are component mismatches, numerical errors, disturbances and noises, etc. A byproduct is that the voltage regulation capability is considerably enhanced as well. In order to improve the voltage quality of parallel-operated inverters, a strategy that combines the strategy in Chapter 8 with the robust droop control in Chapter 19 is presented in Chapter 20, and a strategy to inject the right amount of harmonic voltages into the reference voltage is presented in Chapter 21, respectively.

Part IV is devoted to the synchronisation of inverters with another source. The conventional synchronisation techniques are presented in Chapter 22, with detailed discussions about basic PLL, STA and SOGI-PLL. In Chapter 23, a synchronisation strategy based on the operation principles of synchronous generators is presented to quickly detect the amplitude, frequency and phase of the fundamental component of a periodic signal.

Most of the strategies are demonstrated with extensive experimental results and, hence, can be directly applied in practice with minimum effort.

1.2 Basics of Power Processing

Power processing is to convert a power source into a voltage or current supply that is suitable for the load, as shown in Figure 1.2. It involves the integration of power electronic devices and a controller. There are four types of power processing: AC-DC conversion, DC-DC conversion, DC-AC conversion and AC-AC conversion. These are the subject of many books on power electronics (Bose 2001; Erickson and Maksimović 2001; Fisher 1991; Mohan 2003; Rashid 1993; Thorborg 1988; Vithayathil 1995), and will be briefly described here, assuming that all the devices are ideal.

1.2.1 AC-DC Conversion

The conversion from AC to DC is often called rectification and the converter used is called a rectifier. For an ideal rectifier, it is expected that the output voltage is a pure DC signal without any ripples and the input current is in phase with the voltage and does not have harmonics. According to the power electronic devices adopted, rectifiers can be divided into uncontrolled rectifiers with diodes, phase-controlled rectifiers with thyristors and PWM-controlled rectifiers with IGBTs or MOSFETs.

1.2.1.1 Uncontrolled Rectifiers

Figure 1.3(a) shows the simplest rectifier, which consists of a diode. For the sinusoidal input voltage shown in Figure 1.3(b), the output voltage is shown in Figure 1.3(c). Only the positive

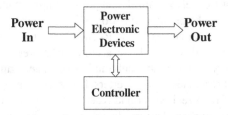

Figure 1.2 Sketch of power processing

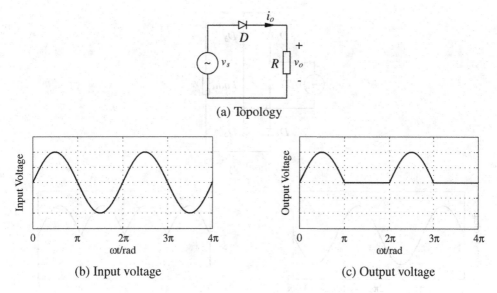

Figure 1.3 Uncontrolled rectifier with a diode

half cycle of the input voltage can pass the diode to reach the load and, hence, the output voltage is of DC but with a significant amount of ripples. The input current is not sinusoidal either so there is a significant amount of harmonic currents.

In order to reduce the ripples in the output voltage and to reduce the harmonics in the input current, several diodes are often connected to form bridge rectifiers. Figure 1.4 shows a single-phase bridge rectifier and its operation principle. Compared to the rectifier with one diode shown in Figure 1.3(a), both half cycles of the input voltage are passed to the load. As a result, the ripples in the output voltage are reduced and the harmonic components in the input current are reduced as well. In this case, the DC output voltage is

$$V_o = \frac{1}{\pi} \int_0^{\pi} \sqrt{2} V_s \sin\omega t \, d(\omega t) = \frac{2\sqrt{2}}{\pi} V_s \approx 0.9 V_s,$$

where V_s is the RMS value of the input voltage.

For three-phase applications, the bridge rectifier shown in Figure 1.5 can be adopted. The pair of diodes with the highest instantaneous line voltage conduct, in the order of $D_1 D_2 \rightarrow D_2 D_3 \rightarrow D_3 D_4 \rightarrow D_4 D_5 \rightarrow D_5 D_6 \rightarrow D_6 D_1 \rightarrow D_1 D_2$ for 120° each time. Hence, the output voltage is the envelope of the line voltages with six ripples, which further improves the performance of the DC output voltage. In this case, the DC output voltage is

$$V_o = \frac{1}{\pi/6} \int_0^{\pi/6} \sqrt{2} \times \sqrt{3} V_s \cos\omega t \, d(\omega t) = \frac{3\sqrt{6}}{\pi} V_s \approx 2.34 V_s,$$

where V_s is the RMS value of the phase input voltage.

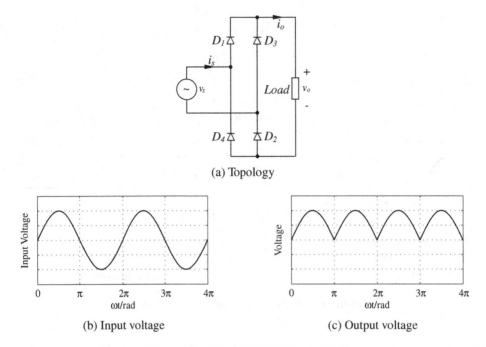

Figure 1.4 Uncontrolled bridge rectifier

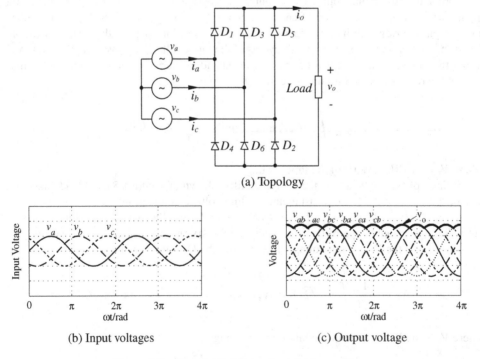

Figure 1.5 Uncontrolled three-phase bridge rectifier

In practice, capacitors are often connected to the output of a rectifier to filter out the voltage ripples and inductors are often adopted to smooth the load current.

1.2.1.2 Phase-controlled Rectifiers

Diode rectifiers can only provide fixed output voltages. In order to obtain a variable DC output voltage, thyristors, which can be turned on by applying a firing pulse when forward biased, can be adopted to form phase-controlled rectifiers. The output voltage of a phase-controlled rectifier can be changed by varying the firing angle of the thyristors. Phase-controlled rectifiers, often with an efficiency above 95%, are widely used in many industrial applications, especially in variable-speed drives.

Figure 1.6 shows a phase-controlled single-phase full-bridge rectifier and its operation. During the positive half-cycle of the input voltage, thyristors T_1 and T_2 are forward biased and the input voltage is passed to the load through T_1 and T_2 after they are fired. Because of the large inductive load, thyristors T_1 and T_2 continue conducting even when the input voltage becomes negative. Similarly, during the negative half-cycle of the input voltage, thyristors T_3 and T_4 are forward biased and the voltage is rectified and passed to the load after they are fired. Thyristors T_1 and T_2 are forced to turn off when they are backward biased and the load current is transferred from T_1 and T_2 to T_3 and T_4. It can be seen that from the firing angle α to π, the input voltage v_s and input current i_s are positive so the energy flows from the source to the load. However, during the period from π to $\pi + \alpha$, the input voltage v_s is negative and the

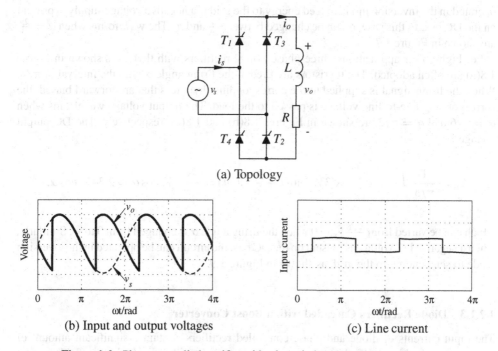

Figure 1.6 Phase-controlled rectifier with a large inductive load when $\alpha = \pi/3$

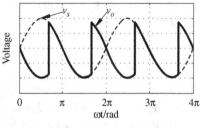
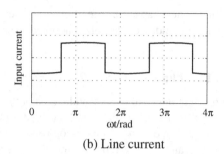

(a) Input and output voltages (b) Line current

Figure 1.7 Phase-controlled rectifier operated in the inversion mode when $\alpha = 2\pi/3$ with a negative DC bus voltage present

input current i_s is positive. Hence, the energy (stored in the large inductor) flows backwards. The DC output voltage is

$$V_o = \frac{1}{\pi}\int_\alpha^{\alpha+\pi} \sqrt{2}V_s \sin\omega t\, d(\omega t) = \frac{2\sqrt{2}}{\pi}V_s \cos\alpha,$$

which can be varied from $\frac{2\sqrt{2}}{\pi}V_s$ to 0 when the firing angle α is changed from 0 to $\frac{\pi}{2}$. If the stored energy in the inductor is not enough to maintain a continuous current, then the current becomes discontinuous and the thyristors turn off. It is worth noting that the circuit can be operated in the inversion mode to feed energy to the grid if a negative voltage supply is present on the DC bus. In this case, α can be changed between $\frac{\pi}{2}$ and π. The waveforms when $\alpha = \frac{2\pi}{3}$ are shown in Figure 1.7.

For high power applications, three-phase bridge rectifiers with thyristors shown in Figure 1.8(a) are often adopted. The thyristors are fired at the firing angle α with the interval of $\pi/3$. When the firing signal is supplied to the corresponding thyristors that are forward biased, the corresponding line-to-line voltage is passed to the load. The output voltage waveforms when $\alpha = \pi/6$ and $\alpha = \pi/2$ are shown in Figures 1.8(b) and 1.8(c), respectively. The DC output voltage is

$$V_o = \frac{1}{\pi/3}\int_{\alpha+\frac{\pi}{6}}^{\alpha+\frac{\pi}{2}} \sqrt{2}\times\sqrt{3}V_s \sin(\omega t + \frac{\pi}{6})d(\omega t) = \frac{3\sqrt{6}}{\pi}V_s \cos\alpha \approx 2.34 V_s \cos\alpha,$$

which can be varied from $\frac{3\sqrt{6}}{\pi}V_s$ to 0 when the firing angle α is changed from 0 to $\frac{\pi}{2}$. Similarly, when a negative DC voltage is present on the DC bus, the circuit can be operated in the inversion mode to send energy to the grid, as shown in Figure 1.9.

1.2.1.3 Diode Rectifiers Cascaded with a Boost Converter

The input currents of diode and phase-controlled rectifiers contain a significant amount of harmonics, which causes a low power factor as well. In order to obtain a variable output

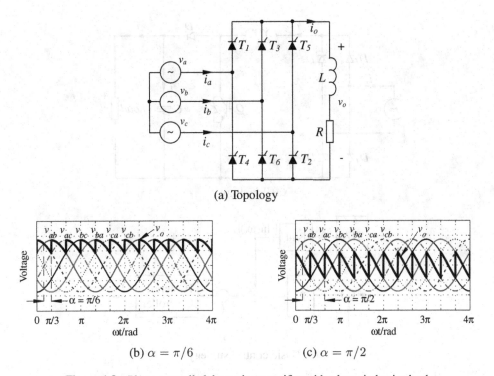

Figure 1.8 Phase-controlled three-phase rectifier with a large inductive load

voltage and a high-quality input current at the same time, many techniques based on active current control have been developed (Bollen 2000; Sankaran 2002). One option is to cascade a boost converter at the output of a diode bridge rectifier, as shown in Figure 1.10(a). The boost converter can be controlled to make the input current in phase with the input voltage while regulating the output voltage, e.g. with the basic hysteresis control strategy shown in Figure 1.10(b). Although the load current is mainly of DC, the input current is sinusoidal and the

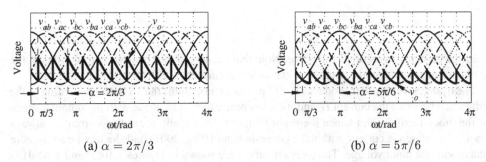

Figure 1.9 Phase-controlled three-phase rectifier operated in the inversion mode when a negative DC bus voltage is present

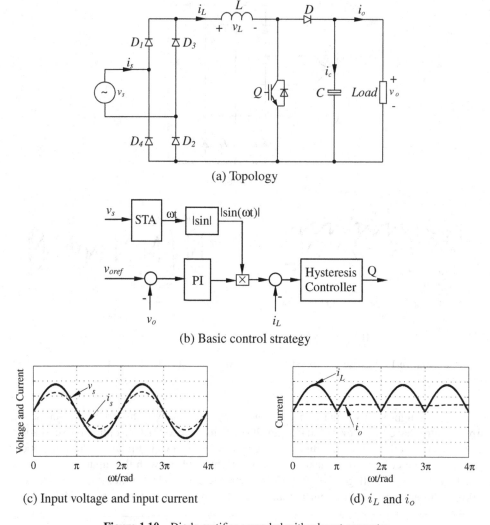

Figure 1.10 Diode rectifier cascaded with a boost converter

inductor current is a rectified full wave. Note that because of the boosted output voltage, the load current is less than the average of the inductor current.

The hysteresis controller produces PWM pulses to turn on/off the switch Q according to the difference between the current i_L and its reference, which is in phase with the output voltage of the diode rectifier (and hence the input voltage). As a result, the actual current i_L always tracks the reference current within a hysteresis band (Bose 2001) and the input current is in phase with the input voltage. The relevant curves are shown in Figures 1.10(c) and 1.10(d).

In order to obtain the phase information of the supply, an STA is adopted; see Chapter 22 for more details.

1.2.1.4 PWM-controlled Rectifiers

A diode rectifier cascaded with a boost converter is able to improve the quality of the input current but the power flow is unidirectional from the source to the load. In order to solve this problem, a bidirectional converter with fully PWM-controlled power switches can be adopted. A PWM-controlled rectifier can be operated as a rectifier or an inverter and the power can flow from the AC side to the DC side or from the DC side to the AC side, if there is energy available at the DC side.

Figure 1.11(a) shows a PWM-controlled single-phase H-bridge rectifier. It is basically operated as a boost converter so an inductor is connected to the input voltage side. A basic control strategy is shown in Figure 1.11(b), which makes the input current in phase with the

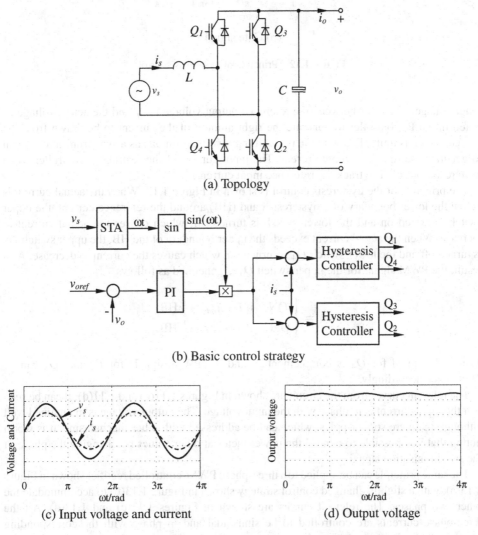

Figure 1.11   PWM-controlled single-phase full-bridge rectifier

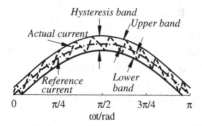

(a) Hysteresis band, actual and reference currents

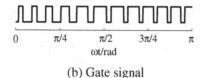

(b) Gate signal

Figure 1.12 Principle of hysteresis control

input voltage. The error between the reference output voltage v_{oref} and the actual voltage v_o is fed into a PI controller to generate the right amount of the current to be drawn from the source, which is multiplied with the per-unit input voltage $\sin(\omega t)$ as a synchronisation signal to generate the reference input current. The input current of the rectifier is controlled via a hysteresis controller to track the reference input current.

The principle of the hysteresis control is shown in Figure 1.12. When the actual current is below the lower boundary of a hysteresis band (HB) around the reference current, the upper switch is turned on and the lower switch is turned off, which causes the actual current to increase. When the actual current exceeds the upper boundary of the HB, the upper switch Q_1 is turned off and the lower switch Q_4 is turned on, which causes the current to decrease. As a result, the PWM signal for the upper switch Q_1 is generated as follows

$$Q_1 = \begin{cases} \text{ON} & \text{if } i < i_{ref} - \frac{1}{2} \text{ HB}, \\ \text{OFF} & \text{if } i > i_{ref} + \frac{1}{2} \text{ HB}. \end{cases}$$

The PWM signal for Q_4 is complementary and the PWM signals for Q_2 and Q_3 can be determined accordingly.

The relevant curves from the circuit are shown in Figures 1.11(c) and 1.11(d). It can be seen that the input current is in phase with the input voltage. The output voltage is maintained well although there are some ripples, which can be addressed with other mechanisms. It is worth noting that it is possible to control the power factor at other values by changing the phase of the synchronisation signal.

The same principle can be applied to a three-phase PWM-controlled rectifier shown in Figure 1.13(a), with a slightly changed control strategy shown in Figure 1.13(b) to accommodate the other two phases. The relevant curves are shown in Figures 1.13(c) and 1.13(d). All the three-phase currents are controlled to be sinusoidal and in phase with the corresponding phase voltages. Because the instantaneous power flowing into the converter is constant for a

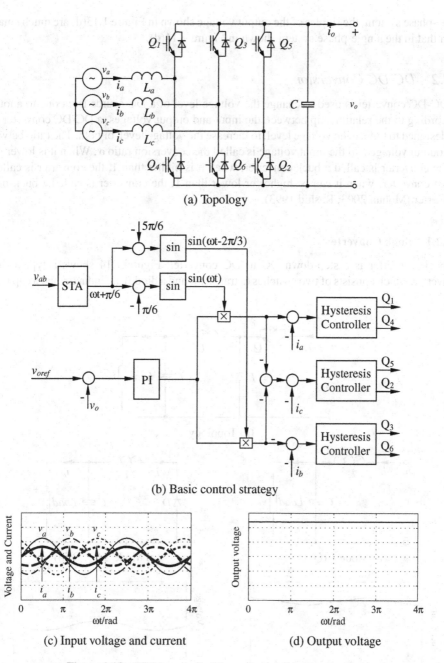

Figure 1.13 PWM-controlled three-phase full-bridge rectifier

1.2.2 DC-DC Conversion

A DC-DC converter is used to change the voltage level of a DC source from one to another. According to the relationship between the input and output voltages, a DC-DC converter can be designed to reduce the voltage level, to increase the voltage level, or both. The ratio between the output voltage and the input voltage is called the conversion ratio α. When it is lower than 1, the converter is called a buck converter; when it is higher than 1, the converter is called a boost converter; when it can be higher or lower than 1, the converter is called a buck-boost converter (Mohan 2003; Rashid 1993).

1.2.2.1 Buck Converters

A buck converter is a step-down DC to DC converter. Figure 1.14 shows a typical buck converter, which consists of two switches (a transistor and a diode), an inductor and a capacitor.

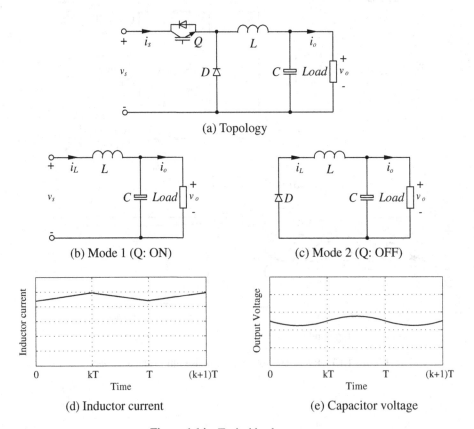

Figure 1.14 Typical buck converter

Introduction

The inductor and the capacitor act as a filter to improve the quality of the output voltage and the load current.

The switch Q is turned on and off periodically. Assume that the switching period is T and the duty cycle is k. Then, the OFF time in one period is $(1-k)T$. The circuit has two operation modes: Mode 1 when the switch Q is turned ON and Mode 2 when the switch Q is turned OFF. The equivalent circuits in both modes are shown in Figures 1.14(b) and 1.14(c), respectively. During Mode 1, the inductor current i_L increases linearly because

$$L\frac{di_L}{dt} = v_s - v_o,$$

where $v_s - v_o$ is almost constant and positive. During Mode 2, the inductor current freewheels through the diode and decreases linearly because

$$L\frac{di_L}{dt} = 0 - v_o,$$

where $-v_o$ is almost constant and negative. The corresponding inductor current waveform is shown in Figure 1.14(d). Ideally, the ripple current should flow through the capacitor and the corresponding voltage waveform is shown in Figure 1.14(e).

In the steady state, the net energy changed in the inductor should be zero during one period, which means the current increased in Mode 1 should be equal to the current decreased in Mode 2. That is,

$$\frac{kT}{L}(v_s - v_o) = \frac{(1-k)T}{L}v_o.$$

Hence, the output voltage is

$$v_o = kv_s.$$

Indeed, this is a buck converter because $\alpha = k$.

1.2.2.2 Boost Converters

Figure 1.15(a) shows a typical boost converter. A boost converter is also called a step-up converter because the output voltage is higher than the input voltage. As a result, the output current is lower than the input current because of the power balance. Similarly, there are also two operation modes when the switch Q is turned ON and OFF. The equivalent circuits in these two modes are shown in Figures 1.15(b) and 1.15(c), respectively.

During Mode 1, the inductor current increases linearly because

$$L\frac{di_L}{dt} = v_s - 0,$$

and the inductor stores energy from the power source while the capacitor discharges to supply the load. During Mode 2, both the energy stored in the inductor and from the power

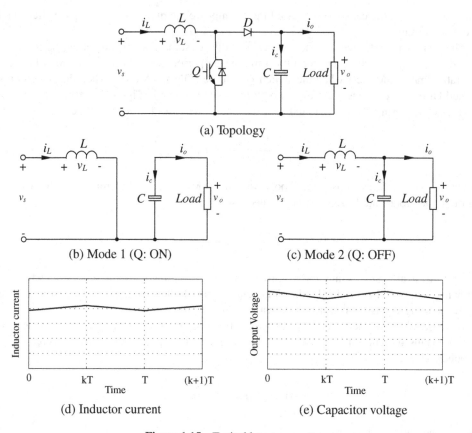

Figure 1.15 Typical boost converter

source are transferred to the load and the capacitor. The inductor current decreases linearly because

$$L\frac{di_L}{dt} = v_s - v_o.$$

Similarly, the net energy changed in the inductor should be zero during one period in the steady state, which means the current increased in Mode 1 should be equal to the current decreased in Mode 2. That is,

$$\frac{kT}{L}v_s = \frac{(1-k)T}{L}(v_o - v_s),$$

from which the output voltage can be derived as

$$v_o = \frac{1}{1-k}v_s.$$

Indeed, this is a boost converter because $\alpha = \frac{1}{1-k} > 1$ for $k \in (0, 1)$. The waveforms of the inductor current and the capacitor voltage are shown in Figures 1.15(d) and 1.15(e), respectively.

1.2.2.3 Buck-Boost Converters

A typical buck-boost converter is shown in Figure 1.16(a). Note that the polarity of the output voltage is opposite to that of the input. Similar to the buck and boost converters discussed above, this converter has two operation modes. The equivalent circuits are shown in Figures 1.16(b) and 1.16(c), and the corresponding waveforms of the inductor current and capacitor voltage are shown in Figures 1.16(d) and 1.16(e). The output voltage is

$$v_o = \frac{k}{1-k} v_s.$$

It operates in the buck mode when $k < 0.5$ and in the boost mode when $k > 0.5$.

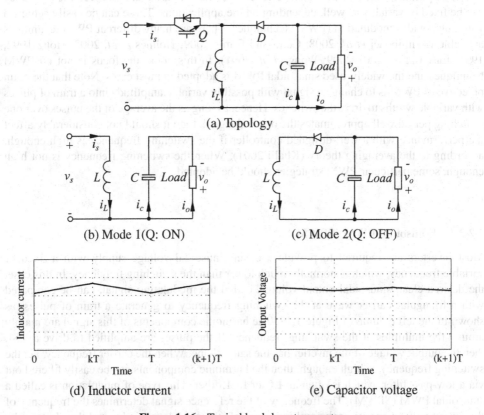

(a) Topology

(b) Mode 1(Q: ON)

(c) Mode 2(Q: OFF)

(d) Inductor current

(e) Capacitor voltage

Figure 1.16 Typical buck-boost converter

1.2.3 DC-AC Conversion

A DC-AC converter, also known as an inverter, generates an AC output from a DC source. There are different types of inverters. According to the type of the DC supply, an inverter is known as a current-source inverter (CSI) if the supply is a current source and a voltage-source inverter (VSI) if the supply is a voltage source. Typically, an inverter is a VSI if there is a large capacitor across the DC bus and is a CSI if there is a large inductor in series with the DC supply. According to the type of the inverter output, an inverter is called current-controlled if the output is controlled to be a current source and voltage-controlled if the output is controlled to be a voltage source. Hence, there are current-controlled VSIs and voltage-controlled VSIs, and there are also current-controlled CSIs and voltage-controlled CSIs. The details of CSIs can be found from textbooks about power electronics, e.g. (Bose 2001), and will not be discussed in the rest of this book. The inverters dealt with in this book are all VSIs. According to the type of commutation, inverters can be line commutated (e.g. those built with thyristors) or forced commutated (e.g. those built with IGBT and MOSFET). In the rest of this book, only forced commutated inverters are dealt with. The output voltage waveform of a voltage-controlled VSI can be a square wave, a modified square/sine wave, multi-level or a pure sine wave. In the rest of this book, voltage-controlled VSIs are expected to have a purely sinusoidal voltage output with minimal harmonic components.

The amplitude of the output of an inverter can be fixed or variable. Moreover, the frequency can be fixed or variable as well, depending on the applications. These can be easily achieved with pulse-width-modulation (PWM) techniques. There are many different PWM techniques available (Asiminoaei *et al.* 2008; Cetin and Ermis 2009; Holmes *et al.* 2003; Holtz 1992, 1994; Lascu *et al.* 2007 2009; Wong *et al.* 2001). In this book, the focus is not on PWM techniques and the widely-used sinusoidal PWM is adopted in most cases. Note that the main objective of PWM is to change a signal with possibly variable amplitude into a train of pulses with variable widths to drive the switches. Hence, as long as the average of the pulses over one switching period well approximates the original signal, then it should not considerably affect the performance with a well-designed controller if the switching frequency is high enough, according to the averaging theory (Khalil 2001). When the switching frequency is not high enough, some particular PWM strategies should be adopted.

1.2.3.1 Sinusoidal PWM (SPWM)

Most inverters are required to provide a clean sinusoidal voltage supply with a fixed or variable frequency, which is normally much lower than the switching frequency. In this case, the desired clean sinusoidal output voltage, called the modulating signal, can be compared with a triangular carrier wave at the switching frequency to generate a train of pulses, as shown in the left column of Figure 1.17. The harmonic components of this signal are mainly around the multiples of the switching frequency. If the pulses are amplified to drive a VSI, then the output voltage of the inverter has the same shape. When the carrier frequency, i.e. the switching frequency, is high enough, then the harmonic components can be easily filtered out via a low-pass filter, which is often an LC or LCL filter. This type of modulation is called a sinusoidal PWM (SPWM). The frequency of the reference signal determines the frequency of the output voltage and its peak amplitude controls the modulation index and then in turn the

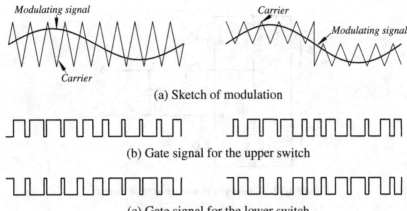

Figure 1.17 Sinusoidal PWM for a single-phase inverter: Bipolar (left column) and unipolar (right column)

RMS value of the output voltage. As a result, the amplitude and frequency of the output voltage can easily be changed by controlling the modulating signal. Because the carrier changes its sign during the positive or negative half cycle, this SPWM is bipolar. If the carrier does not change its sign during the positive or negative half cycle, then the resulting SPWM is unipolar, as shown in the right column of Figure 1.17. Note that in both cases, the upper switch and the lower switch on the same leg are operated in a complementary way.

Similarly, for three-phase applications, three modulating signals can be compared with the carrier signal to generate the gate-driving signals, as shown in Figure 1.18.

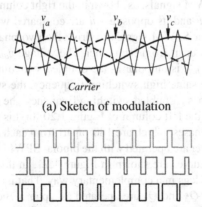

Figure 1.18 Sinusoidal PWM for a three-phase inverter

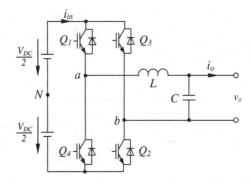

Figure 1.19 Single-phase voltage-source inverter

1.2.3.2 Operation of Single-phase Inverters

Figure 1.19 shows a single-phase inverter with a DC voltage source. The DC bus voltage is split into two halves to illustrate the operation of the inverter and the mid-point of the DC bus is the reference point for the two phase legs.

The inverter can be operated to obtain bipolar and unipolar SPWM signals for v_{ab}. When it is operated with the unipolar SPWM signal shown in the right column of Figure 1.17, the voltages v_{aN}, v_{bN}, together with v_{ab} and v_o are shown in the left column of Figure 1.20. In this case, phase-leg a is operated according to the unipolar SPWM signal and phase-leg b is operated according to the polarity of the voltage signal at its frequency. For example, when the modulating voltage u is positive, Q_1 and Q_4 are turned ON and OFF according to the unipolar SPWM signal while Q_2 is always ON and Q_3 is always OFF; when u is negative, Q_1 and Q_4 are turned ON and OFF according to the SPWM signal while Q_2 is always OFF and Q_3 is always ON. This is able to reduce switching losses because the second leg is operated at the frequency of the voltage.

It is possible to operate the inverter to obtain a unipolar SPWM for v_{ab} although the phase legs are driven by bipolar SPWM signals, as shown in the right column of Figure 1.20. In this case, the modulating voltage u and its opposite $-u$ are compared with the carrier waveform to generate two sets of bipolar SPWM signals to drive the two phase legs. As a result, the voltages v_{aN} and v_{bN} are 180° apart from each other. The voltage v_{ab}, which is the difference of the two voltages v_{aN} and v_{bN}, is unipolar at the doubled switching frequency. Since both phase legs are operated at the same high switching frequency, the switching losses are high but because the resulting v_{ab} has a doubled switching frequency, the output voltage quality is better than the case shown in the left column of Figure 1.20. In this case, the two phase legs are operated as two separate phases, which are 180° apart from each other.

When both legs of the inverter are operated with the bipolar SPWM shown in the left column of Figure 1.17, the resulting curves are shown in Figure 1.21. In this case, the same SPWM signal is sent to the two phase legs in a complementary way. That is, Q_1 and Q_2 are operated as a pair at the same time and Q_3 and Q_4 are operated as a pair at the same time. As a result, $v_{bN} = -v_{aN}$ and $v_{ab} = 2v_{aN}$.

Note that the amplitude of v_{ab} is $\pm V_{DC}$ for all three different operation modes and the maximum achievable amplitude is the same as the DC bus voltage.

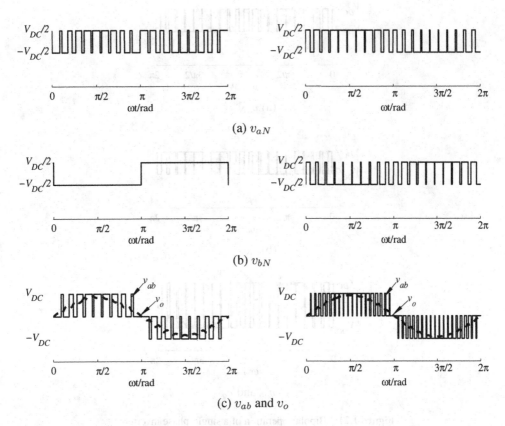

Figure 1.20 Unipolar operation of a single-phase inverter: with only one leg operated at the switching frequency (left column) and both legs operated at the switching frequency (right column)

1.2.3.3 Operation of Three-phase Inverters

For three-phase inverters shown in Figure 1.22(a), three phase voltages can be compared with the carrier waveform to generate three sets of bipolar PWM signals, as shown in Figure 1.18 to drive the three phase legs separately. The corresponding curves are shown in Figures 1.22(b) and 1.22(c). It can be seen that the maximum amplitude of the phase voltages is half of the DC-bus voltage. It is worth noting that the average voltage between the reference point N of the three phase legs and the common point N' of the capacitors over a switching period is 0 but the instantaneous voltage is not.

1.2.4 AC-AC Conversion

The AC-AC conversion can be performed indirectly via AC-DC-AC with the addition of a DC bus or directly without a DC bus. The indirect AC-AC conversion is basically the combination of AC-DC conversion and DC-AC conversion, as discussed in Sections 1.2.1

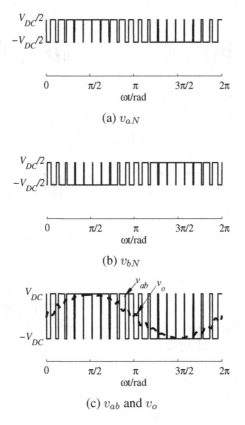

Figure 1.21 Bipolar operation of a single-phase inverter

and 1.2.3 and, hence, will not be discussed any further. The direct AC-AC conversion often involves bidirectional switches, e.g. triacs and thyristors connected in anti-parallel. One way to implement direct AC-AC conversion is to use matrix converters to generate AC outputs with arbitrary amplitude and frequency; see e.g. (Rodriguez *et al.* 2012; Wheeler *et al.* 2002). Here, the circuit shown in Figure 1.23(a) is illustrated with two major control methods: on–off control and phase control.

1.2.4.1 On–off Control

Figure 1.23(a) shows a single-phase AC-AC converter. Two thyristors are connected in anti-parallel so when they are triggered, both half cycles of the supply can be passed to the load. When it is under the on–off control, triggering pulses are provided to turn on the thyristors so that the supply is passed to the load. The thyristors are turned off when the supply is not passed to the load. Assume that the ratio of the number of ON-cycles to the number of total cycles in an operational period is k and the RMS value of the supply is V, then the RMS value of the output voltage is $\sqrt{k}V$ and the input power factor is $\sqrt{k}$. For the input voltage sketched in

Introduction

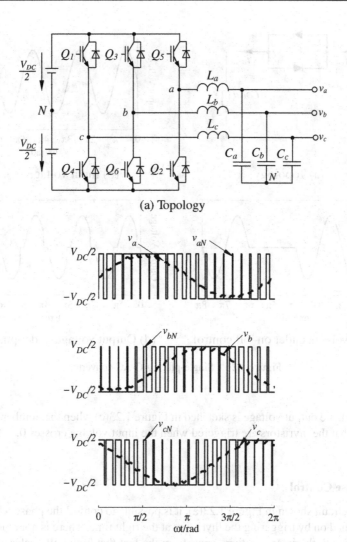

(a) Topology

(b) v_{aN}, v_{bN}, v_{cN} and the three-phase voltages v_a, v_b and v_c

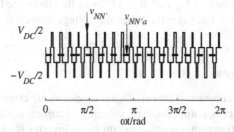

(c) $v_{NN'}$ and its average $v_{NN'a}$ over a switching period

Figure 1.22 Operation of a three-phase inverter

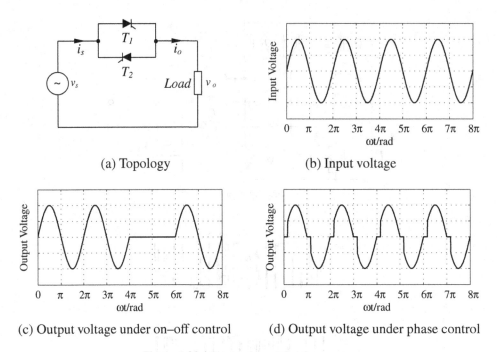

Figure 1.23 Single-phase AC-AC converter

Figure 1.23(b), the output voltage is sketched in Figure 1.23(c) when the number of off-cycles is one. Note that the thyristors are triggered when the input voltage crosses 0.

1.2.4.2 Phase Control

For the same circuit shown in Figure 1.23(a), it is possible to control the phase when both half cycles are turned on by triggering the thyristors at the right time. There is not much difference from the case with thyristor rectifiers, apart from the fact that both half cycles can be passed to the load. For the input voltage sketched in Figure 1.23(b), the output voltage is sketched in Figure 1.23(d) for a firing angle of $\frac{\pi}{6}$ rad. It can be seen that there are harmonics in the output voltage and the switches are not triggered when the voltage crosses 0.

1.3 Hardware Issues

A power inverter mainly consists of power stages, an electronic controller and the necessary auxiliary circuits for isolation, output filtering, voltage and current sensing, signal conditioning and protection, etc. The functional block diagram of an inverter is shown in Figure 1.24. In this section, some general guidelines are provided and the readers are suggested to refer to handbooks about hardware design (FUJI 2004; Kimmel and Gerke 1995; Rashid 2010; Skvarenina 2002; TI 2011).

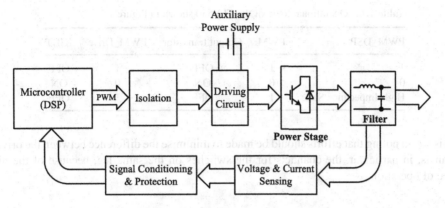

Figure 1.24 Block diagram of an inverter system

1.3.1 Isolation

In order to guarantee proper operation, the low-power-low-voltage electronic part of an inverter should be isolated from the high-power-high-voltage part. The isolation from the power part to the electronic part is often taken care of by the sensors. The isolation from the electronic part to the power part is often done to the PWM signals before entering the driving circuit, as shown in Figure 1.24. This can be easily done with optocouplers.

TLP550 is an optocoupler commonly used in the drivers for IGBT (Toshiba 2002), with a typical circuit shown in Figure 1.25. The PWM signal from a signal buffer/driver, e.g. SN74AB541, which processes the PWM signals of the Digital Signal Processor (DSP) from 3.3 V to 5 V, is connected to the cathode of the diode of the optocoupler to generate an isolated output signal PWM_Drive, which can be connected to the driving circuits. The truth table of the circuit shown in Figure 1.25, together with the logic for the operation of the IGBT, are given in Table 1.1, where PWM_DSP means the PWM signal from DSP ports. In order to avoid the damage caused by the high-impedance state of the DSP ports during reset, the ports can be pulled up with resistors. Hence, the high-impedance state of a port is equivalent to the OFF state. Note that an IGBT is turned on when the PWM signal is 0 in this case.

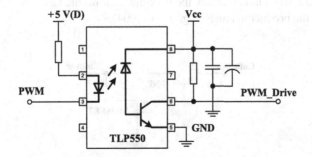

Figure 1.25 Typical circuit for TLP550. *Source:* Toshiba 2002

Table 1.1 Operational logic of the TLP550 shown in Figure 1.25

PWM_DSP	PWM	Output transistor	PWM_Drive	IGBT
1	1	OFF	1	OFF
0	0	ON	0	ON
High-Impedance State	1	OFF	1	OFF

It is worth noting that efforts should be made to minimise the difference between the driving channels, in particular, the channels for the switches on the same leg, because of the high speed of operation.

1.3.2 Power Stages

Power stages are the key part of an inverter system and are responsible for power transfer and conversion. Proper design and selection of power modules and the associated auxiliary circuits are crucial for the normal operation, reliability, lifetime and efficiency of the inverter.

1.3.2.1 Power Modules

With the development of power semiconductor technologies, power switches have experienced several stages. At present, discrete IGBT/MOSFET components and power modules built with IGBTs are commonly used in inverters. These devices are fully on–off controllable and are ideal for inverters.

Figure 1.26 shows the equivalent circuit of an ideal IGBT, which is the combination of a power transistor and a MOSFET. As a result, an IGBT combines the advantages of transistors (low conduction loss) and MOSFET (high-speed turn-on and easy drive with a low-power voltage signal). When a positive voltage is applied between the gate and the emitter, the MOSFET is turned on and hence the transistor is on. When the voltage is removed, the MOSFET is off and hence the transistor is off.

The current and voltage ratings of the IGBTs in an inverter should be selected appropriately to meet the requirements on the power and voltage levels. Moreover, discrete IGBT modules should be protected from over-currents, over-voltages, over-heating, etc. Because of the losses, heat sinks are often needed and should be designed appropriately to keep the junction temperature of the IGBT module below the maximum allowable value. Some guidelines can often be found in the product manuals, e.g. (FUJI 2004).

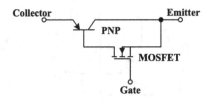

Figure 1.26 Equivalent circuit of an IGBT

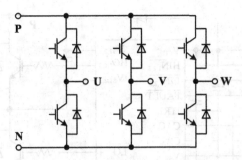

Figure 1.27 Schematic of a typical IPM

Another option is to use power modules, within which driving circuits and protection circuits are integrated with power switches. For example, intelligent power modules (IPM) include short-circuit protection and fault-detecting circuits, such as over-voltage, over-current and over-heat, in addition to driving circuits (POWEREX 2000). Due to the integrated package of the driving circuit and the power switches, the reliability and the dynamic response are considerably improved and the losses are reduced as well. The schematic of a typical power module with three phase legs is shown in Figure 1.27.

It is worth noting that the energy causing high transient voltages is $\frac{1}{2}Li^2$, which is proportional to the line inductance L and the square of the current i. Hence, the inductance of the DC buses should be designed as small and laminated bus structures are often adopted for high-current applications.

1.3.2.2 Auxiliary Power Supplies

In order to drive the switches in an inverter, isolated auxiliary power supplies are often needed. For example, for the circuit shown in Figure 1.27, the driving circuits for the three ground-connected power switches can share one power supply but the driving circuits for the three upper switches should be isolated and hence four isolated power supplies are needed. When selecting auxiliary power supplies, particular attention should be paid to their capacity to make sure that the driving current provided is enough to turn on the switches. Moreover, for each power supply to a particular driving circuit IC, a high-frequency filter capacitor and an electrolytic capacitor should be connected in parallel to the power supply of the IC, and as close as possible.

1.3.2.3 Driving Circuits

For a power semiconductor device, a driving circuit is needed to provide the right voltage level and driving current. For example, the required voltage level for IGBT is often 15 V or 20 V. Many IC companies have developed IC products to drive IGBTs at different power levels, which differ in terms of the switching frequency, current and voltage levels, etc.

The IR21xx series of drivers produced by International Rectifier (IR) are a typical set of high voltage, high speed MOSFET and IGBT drivers with independent high and low side

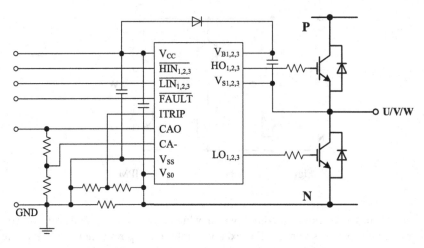

Figure 1.28 Typical connection of IR2130. *Source:* IR 2004

referenced output channels (IR 2004). IR2110 is suitable for one phase leg and IR2130 is suitable for three phase legs. Figure 1.28 shows the typical connection of IR2130 (IR 2004). Proprietary HVIC technology enables ruggedised monolithic construction. The logic inputs are compatible with CMOS or LSTTL outputs, down to 2.5 V logic. A ground-referenced operational amplifier provides analog feedback of the bridge current via an external current-sensing resistor, from which a current trip function that terminates all six outputs is also derived. An open drain $\overline{\text{FAULT}}$ signal is provided to indicate that an over-current or under-voltage shutdown has occurred. The output drivers feature a high pulse current buffer stage designed for the minimum driver cross-conduction. Propagation delays are matched to simplify use at high frequencies. The floating channels can be used to drive N-channel power MOSFETs or IGBTs in the high side configuration, which operate up to 600V.

Figure 1.29 shows a typical connection of the driving signal to the gate of an IGBT. In order to weaken the influence of the Miller capacitance and the stray inductance, turn-on and turn-off

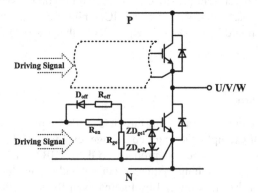

Figure 1.29 Typical connection between the gate and the emitter of an IGBT

resistors are connected between the driving signal and the gate of the IGBT. Moreover, zener didoes are connected across the gate and the emitter of the IGBT to prevent excessive driving voltage. If the driving signal does not offer an off bias with a negative voltage to speed up the turn-off process, then ZD_{ge2} is not needed. The selection of the turn-on and turn-off resistors is determined by the ratings of the IGBT. Generally, $R_{on} > R_{off}$ should be recommended to avoid a false turn-on of the IGBT caused by the Miller effect. Because the collector of the IGBT is connected to a high voltage, a resistor R_{ge} is connected across the gate and the emitter of the IGBT to avoid a false trigger caused by external high-voltage interference. This resistance should not be too small. Otherwise the gate voltage would not be high enough to trigger the IGBT and the peak voltage on the collector would also be high. Normally, this resistor is placed as close as possible to the gate and the emitter of the IGBT with $R_{ge} = 10 \text{ k}\Omega$ (FUJI 2004).

The ground of a driving circuit is connected to the emitter of the driven IGBT. Due to the line leakage inductance, there is an induced voltage between the ground-connected point and the emitter, especially in high-current applications because of the high di/dt during switching (POWEREX 2000). Therefore, the ground point of the gate signal should be connected to the emitter of the IGBT, as close as possible. Moreover, an off bias with a negative voltage V_{EE} should always be adopted. Figure 1.30(a) shows a circuit that is suitable for low-current six-pack devices, in which the power switches are integrated with minimal inductance on the negative bus and low di/dt. However, this circuit has a ground loop problem for high-current applications because the ground of the driving circuits is far from the emitters of the switches. Figure 1.30(b) shows a circuit with a common power supply for the driving circuits but with separate capacitors for each power switch, where the emitter of the switch is connected to the ground of the corresponding capacitor. This is suitable for modules rated up to 200 A. Figure 1.30(c) shows a circuit with isolated power supplies for switches with a common ground. This is recommended for IGBT modules rated 300 A or more. Because isolated power supplies are adopted, the ground loop problem is avoided.

1.3.2.4 Snubber Circuits

Power semiconductor switches are the main devices for an inverter and should be operated under safe working conditions. Snubber circuits should be placed across these devices to protect them and improve the system performance. The main functions of snubber circuits include (Severns n.d.):

1. Shaping the load line to keep it within the safe operating area (SOA).
2. Reducing or eliminating voltage or current spikes.
3. Limiting di/dt and dv/dt.
4. Reducing total losses due to switching.
5. Reducing EMI by damping voltage and current ringing.
6. Transferring power dissipation from the switch to a resistor or a useful load.

Table 1.2 shows three typical snubber circuits, together with the main features, for individual switches. Some snubber circuits can also be connected as close as possible between the collector of the upper IGBT and the emitter of the ground-connected IGBT to form lump

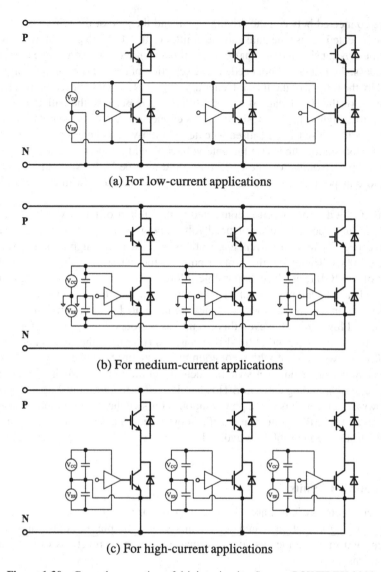

Figure 1.30 Ground connection of driving circuits. *Source:* POWEREX 2000

snubber circuits. Two such circuits, the RCD snubber circuit and the C snubber circuit, are shown in Figure 1.31. This is becoming increasingly popular due to the circuit simplification (FUJI 2004).

Note that the wiring inductance of snubber circuits is one of the main reasons for voltage spikes and, hence, the circuit should be built with the lowest possible inductance. The detailed design of snubber circuits can be found in (FUJI 2004; POWEREX 2000). Here, the design of the charge and discharge RCD snubber circuit, which is common in medium-power

Introduction

Table 1.2 Typical individual snubber circuits

Snubber circuits	Features
RC snubber circuit 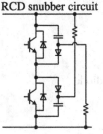	• Useful for controlling the transient voltage, parasitic oscillations and dv/dt • Not suitable for high-frequency applications because the resistor always consumes energy during the turn-ON and turn-OFF processes of the corresponding IGBT and the loss is quite high
Charge and discharge RCD snubber circuit	• The effect on the turn-off surge voltage is moderate • Not suitable for high-frequency applications • Power loss is $$P = \frac{1}{2}LI^2 f + \frac{1}{2}CV_d^2 f,$$ where L is the wiring inductance of the main circuit, I is the collector current at turn-off, C is the snubber capacitance, V_d is the DC bus voltage and f is the switching frequency. It could be high because the capacitor is charged and discharged during every switching cycle.
Discharge-suppressing RCD snubber circuit	• The effect on the turn-off surge voltage is small • Suitable for high-frequency applications • Power loss caused by the snubber circuit is $$P = \frac{1}{2}LI^2 f$$ where L is the wiring inductance of the main circuit, I is the collector current at turn-off and f is the switching frequency.

Source: FUJI 2004; POWEREX 2000.

applications is described, according to (FUJI 2004; Hossain *et al.* 1997a, b; Todd 2001). The snubber capacitance can be calculated (Todd 2001) as

$$C_s = \frac{LI^2}{\Delta V (\Delta V + 2V)}, \qquad (1.1)$$

where L is the line inductance of the power circuit and I is the current flowing through the power switch at the time it turns off. V and ΔV are the initial voltage and voltage change on

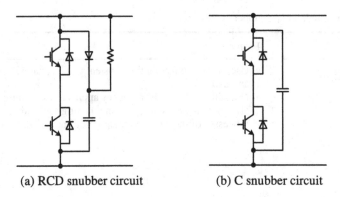

(a) RCD snubber circuit   (b) C snubber circuit

Figure 1.31 Lump snubber circuits. *Source:* FUJI 2004; POWEREX 2000

the capacitor, respectively. In practice, V is 0 and ΔV equals the acceptable overshoot of the voltage across the power switch and can be expressed as

$$\Delta V = k_m V_{CEm} - V_d,$$

where V_{CEm} is the maximum C-E withstood voltage of the power switch and V_d is the DC power supply voltage, respectively. The coefficient k_m is less than 1, often 0.7–0.8 depending on the particular IGBT, to make sure that the maximum voltage $V_d + \Delta V$ is below the rated voltage. The selection of the snubber resistance often depends on the reverse-recovery current of the freewheeling diode and the minimum snubber resistance R_{s_min} can be obtained empirically (Hossain *et al.* 1997a) as

$$R_{s\_min} = \frac{V_d}{k_s I},$$

where k_s, $0 < k_s \leq 0.2$, is the reverse-recovery factor of the freewheeling diode.

1.3.2.5 Shoot-through of Phase Legs

Power switches cannot be turned ON or OFF instantaneously, although the process is very fast. In order to avoid shoot-through between the upper and the lower switches of the same phase leg, a short period of dead time is needed between the two gate signals. It can be set in the controller, e.g. directly in a DSP or in a CPLD/FPGA chip. It can also be implemented with deadtime generator ICs. For example, IXDP630/631 are able to inject the required deadtime to convert a single-phase PWM signal into two separate logic signals required to drive the upper and lower switches in a PWM inverter. It also provides functions for output disable, and fast over-current and fault condition shutdown (IXYS 1998).

Although it is important to make sure that no shoot-through happens, it is also important to note that excessive deadtime may deteriorate the performance, e.g. increased harmonics, etc. It is also important for the deadtime to be applied symmetrically to the ON and OFF states to minimise the DC component in the output voltage.

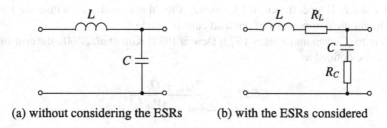

(a) without considering the ESRs   (b) with the ESRs considered

Figure 1.32 Circuit model of a passive LC filter

1.3.3 Output Filters

Since an inverter is operated with PWM signals, which contain harmonics often around the multiples of the switching frequency, it is necessary to connect a low-pass filter to the output of the inverter power switches so that the harmonics can be filtered out and the desired output voltage is recovered. This can be done with conventional passive filters (Chang et al. 2006; Das 2004; Hamadi et al. 2010), e.g. LC and LCL filters.

1.3.3.1 LC Filters

The circuit model of a passive LC filter is shown in Figure 1.32. The equivalent series resistance (ESR) of the inductor and the capacitor may not be considered during the design process because of their small values. In practice, the ESRs are able to dampen high-frequency oscillations so it is good for performance. Ideally, the smaller the inductance and the capacitance, the more cost-effective the system. However, the inductance and the capacitance should be big enough in order to filter out the switching effects, taking into account several contradictory factors (Pasterczyk et al. 2009), e.g. the cut-off frequency f_c (Hatua et al. 2012; Michels et al. 2006), size, the voltage THD, the cost function (Dewan and Ziogas 1979; Dewan 1981; Kim et al. 2000), the resonance damping, efficiency (Strom et al. 2011) and the power level, etc.

The cut-off frequency f_c of the filter is

$$f_c = \frac{1}{2\pi\sqrt{LC}}. \tag{1.2}$$

This is the most important factor to be considered. In order to filter out the switching harmonics, it should be much lower than the switching frequency while providing enough bandwidth for the controller. It is recommended to position it within $\frac{1}{3} \sim \frac{1}{2}$ of the switching frequency f_{sw} (Hatua et al. 2012), i.e.

$$\frac{f_{sw}}{3} \leq f_c \leq \frac{f_{sw}}{2}. \tag{1.3}$$

It is worth noting that this causes resonance in the output impedance of the inverter around the cut-off frequency. As will be discussed in Chapters 2 and 7, this actually amplifies the harmonic current components around the cut-off frequency and might result in high THD in the output voltage. Hence, the cut-off frequency should not be chosen within the band where the major harmonic components of the load current reside.

According to (Dewan and Ziogas 1979; Dewan 1981; Kim et al. 2000), the cost function of the filter can be defined as

$$COST = \frac{2Q_L + Q_c}{\sum_{h=1,odd}^{n} |V_{oh} I_h|}, \qquad (1.4)$$

with

$$Q_L = \sum_{h=1,odd}^{n} |I_h|^2 X_{Lh},$$

$$Q_c = \sum_{h=1,odd}^{n} \frac{|V_{Ch}|^2}{X_{Ch}}.$$

The weight of the reactive power for the inductor is taken as twice that of the capacitor, considering the real price. Intuitively, the capacitance C size should be small for high-voltage applications and the inductance L should be small for high-current applications (while keeping the same cut-off frequency). Moreover, the inductance L should be small for applications with significant amount of current harmonics and the capacitance C should be small for applications with a significant amount of voltage harmonics, e.g. when the switching frequency is low.

When the ESRs R_L and R_C are considered or intentionally increased, the LC resonance around the cut-off frequency is dampened. However, the increase of R_C and/or R_L results in excessive power losses, which might cause difficulties in the LC design, in particular, for high-power applications (Strom et al. 2011). One possible option is to adopt control strategies to add virtual resistors to achieve the same purpose (Dahono 2003; Dahono et al. 2001; Guo and Liu 2011); see Chapter 7 for more details.

Of course, the current rating of the inductor and the voltage rating of the capacitor should be chosen to meet the requirement of the current and voltage levels of the inverter.

1.3.3.2 LCL Filters

Passive LCL filters, of which the circuit model is shown in Figure 1.33, are often adopted in grid-connected inverters. Adding the grid-side inductor L_g increases the order of the filter by 1 and, hence, an LCL filter is able to attenuate the harmonics better than an LC filter.

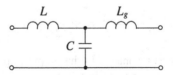

Figure 1.33 Circuit model of a passive LCL filter

More importantly, it adds a mechanism to limit the current harmonics caused by the harmonic components in the grid voltage.

For applications with a reasonably high switching frequency, the LCL filter can be designed in two steps, i.e. to design the LC filter first and then add the grid inductor. For applications with a very low switching frequency, e.g. at MW-level, extra care should be taken (Rockhill et al. 2011; Teodorescu et al. 2011). Some other guidelines about the design of LCL filters can be found in (Araujo et al. 2007; Bolsens et al. 2006; Liserre et al. 2005).

When designing the output filter, it is also important to check the reactive power of the filter capacitor. It should not considerably affect the rating of the inverter.

1.3.4 Voltage and Current Sensing

Due to the requirement on the galvanic isolation between the power part and the electronic part of an inverter, voltage and current sensors with galvanic isolation are often adopted to measure relevant voltages and currents needed by the controller. Moreover, these signals can be applied to achieve over-current and/or over-voltage protection.

It is possible to use conventional current and voltage transformers to measure currents and voltages but integrated voltage and current transducers based on the Hall effect are very popular due to the compact package, accuracy, high linearity, low temperature drift, wide frequency bandwidth, zero insertion loss, high immunity to external interference, etc. For example, LAH 25-NP from LEM is a compensated (closed-loop) multi-range current transducer (rated up to 25A) using the Hall effect (LEM n.d.a.). It can be mounted directly on a printed circuit board to measure AC, DC and pulsed currents. A sketch view of LAH 25-NP is shown in Figure 1.34(a), where R_M is the measuring resistor in the secondary circuit. The output from the sensor is a current that is in proportion to the current measured. The current flows through the measuring resistor and the resulting voltage can be processed further. The range of the measuring resistance depends on the type of the detected signal (DC or AC), the operating temperature and the supply voltage, but the range is quite wide and a suitable value can be easily chosen. The measurement range of the current can be changed via changing the connections of the input terminals, as shown in Table 1.3.

LV 25-P is a voltage transducer using the Hall effect (LEM n.d.b). Its sketch view is shown in Figure 1.34(b). In principle, this is a current transducer that works within a narrow current range around the rated primary current 10 mA (RMS). A resistor R_1 with an appropriate value

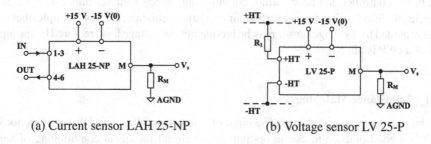

(a) Current sensor LAH 25-NP (b) Voltage sensor LV 25-P

Figure 1.34 Sketch view of voltage and current sensors

Table 1.3 Recommended PCB connections for LAH 25-NP (LEM n.d.a.)

Recommended PCB connections	Primary maximum current (A)	Primary nominal current (A)	Turns ratio	Nominal output current (mA)
(3–2–1 IN / OUT 4–5–6)	55	25	1:1000	25
(3–2, 1 IN / OUT 4–5, 6)	27	12	2:1000	24
(3, 2, 1 IN / OUT 4, 5, 6)	18	8	3:1000	24

is connected on the primary side so that the current caused by the measured voltage is around 10 mA. The output is also a current source, which can be converted into a voltage signal with the measuring resistor R_M. In order to obtain the best accuracy, the current flowing through R_1 should be designed to be around 10 mA according to the voltage measured. Note that the turns ratio is 2500:1000 and the primary coil resistance is 250 Ω.

Assume that the circuit with LV 25-P shown in Figure 1.34(b) is adopted to measure the DC bus voltage of an inverter, which is up to 800 V, and the maximum voltage for analog inputs is 3 V. If the measuring resistor is chosen as 100 Ω, then the maximum secondary current is $\frac{3V}{R_M} = 30$ mA, which corresponds to a primary current of $\frac{30 \text{ mA}}{2.5} = 12$ mA. If two 33 kΩ resistors are connected in series as the primary resistor R_1, then the maximum DC bus voltage allowed is 12 mA $\times$ 2 $\times$ 33.25 kΩ = 798 V. The maximum power dissipated by each primary resistor is $(12 \text{ mA})^2 \times 33$ kΩ = 4.752 W so the power rating can be chosen as 5W. Note that the primary current is around the rated value 10 mA.

1.3.5 Signal Conditioning

In general, signals obtained from sensors are often not compatible with the requirements of the inputs to the controller and need further conditioning, which includes impedance matching, scaling, level-shifting, filtering, converting, linearisation, isolation, etc. For example, the output voltage provided by a voltage sensor may be bipolar but the voltage level required by the analog inputs of a DSP is often $0 \sim 3$ V.

1.3.5.1 Impedance Matching

Since the input impedance of an analog input of micro-controllers and DSP is often not very high, it is a good practice to use an op-amp driver circuit for signal conditioning of analog input signals and also as a buffer. This reduces the loading effect to the sensor circuits and

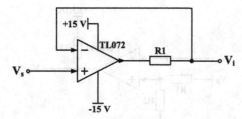
(a) Impedance matching with a voltage follower

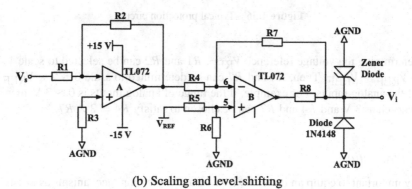
(b) Scaling and level-shifting

Figure 1.35 Typical signal conditioning circuits

offers a low output impedance to the analog inputs. The op-amp also isolates the ADC and protects the ADC inputs.

Figure 1.35(a) shows a typical voltage follower implemented with op-amp TL072. It is particularly suitable for cases where the output voltage from a sensor is already in the right range, e.g. $0 \sim 3$ V for DSP, but the impedance on both sides do not match.

1.3.5.2 Scaling and Level-shifting

If the voltage range of a signal from a sensor is not within the right range of the analog inputs, then the signal needs to be scaled and often shifted as well. In this case, the circuit shown in Figure 1.35(b) can be used. It consists of two stages: the first stage to scale (i.e. to amplify or attenuate) the signal V_s to the range of -3 V ~ 3 V and the second stage to scale and shift it to $0 \sim 3$ V. The relationship between the input V_s and the output V_i is

$$V_i = \frac{R7}{R4} \times \frac{R2}{R1} \times V_s + \frac{R4+R7}{R4} \times \frac{R6}{R6+R5} \times V_{REF}. \tag{1.5}$$

If $R4 = R5$ and $R6 = R7$, then

$$V_i = \frac{R7}{R4}(V_{REF} + \frac{R2}{R1} \times V_s).$$

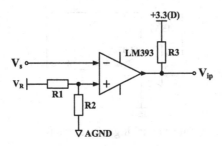

Figure 1.36 Typical protection circuit

After determining the voltage reference V_{REF}, $R1$ and $R2$ can be selected to scale V_s to the range $-V_{REF} \sim V_{REF}$. Then, $R4$ and $R7$ can be determined to scale $0 \sim 2 \times V_{REF}$ to the range of the analog inputs. For example, if the range of analog inputs is $0 \sim 3$ V, then V_{REF} can be chosen as 3 V and $R4$ and $R7$ can be chosen to satisfy $R4 = 2 \times R7$.

1.3.6 Protection

It is very important to equip an inverter with as many protection mechanisms as possible, at a reasonable cost. For the currents and voltages already measured, it is straightforward to add over-voltage and over-current protection. This can be done in the control algorithm and/or with hardware circuits.

Figure 1.36 shows a typical circuit for protection when a signal exceeds a certain value. This can be used for over-voltage and over-current protection. It is basically a comparator and the threshold can be easily adjusted with a potentiometer $R1$. The output signal V_{ip} changes from 1 to 0 when

$$V_s > \frac{R2}{R1+R2} V_R,$$

according to which appropriate actions can be taken.

1.3.7 Central Controller

Because of the complex functions of an inverter and the requirement of a high sampling frequency, e.g. to handle harmonics up to a certain order, it is often necessary to use a powerful micro-controller as the core of the electronic part of an inverter. Moreover, it is also important that many development tools and the maximum support possible are available to speed up the design process. There are many options but, in this section, the TMS320F28335 digital signal controller from Texas Instruments is described because (TI 2007)

1. it is a 32-bit single-precision floating-point processor compatible with IEEE-754 that is dedicated to demanding control applications, e.g. power and energy conversion applications;

2. it is supported with MATLAB® Embedded Coder (originally, Target Support Package) and, hence, codes for real-time execution can be generated automatically from Simulink® models, which saves a lot of time for development;
3. Texas Instruments run a worldwide university program[1] that provides excellent, and often free, support to educators, researchers and students in all phases of course curricula, senior design and research projects.

1.3.7.1 Overview of TI DSC TMS320F28335

Figure 1.37 shows the functional block diagram of TI DSC TMS320F28335 (TI 2007). It includes the same 32-bit fixed-point architecture as TI's existing C28x DSCs, but also include a single-precision (32-bit) IEEE 754 floating-point unit (FPU). It is a very efficient C/C++ engine, enabling users to develop their system control software in a high-level language. It also enables math algorithms to be developed using C/C++. The device is as efficient at DSP math tasks as it is at system control tasks that typically are handled by micro-controller devices. This efficiency removes the need for a second processor in many systems. The 32×32-bit MAC 64-bit processing capabilities enable the controller to handle high numerical resolution problems efficiently. Add to this the fast interrupt response with automatic context save of critical registers, resulting in a device that is capable of servicing many asynchronous events with minimal latency. The device has an 8-level-deep protected pipeline with pipelined memory accesses. This pipelining enables it to execute at high speeds without resorting to expensive high-speed memories. Special branch-look-ahead hardware minimises the latency for conditional discontinuities. Special store conditional operations further improve performance. TMS320F28335 comes with 256 K×16 Flash, 34 K×16 SARAM on-chip memory. There are two lower versions: F28334 with 128 K×16 Flash, 34 K×16 SARAM on-chip memory and F28332 with 64 K×16 Flash, 26 K×16 SARAM on-chip memory. The equivalent versions that do not include the FPU are F2823x.

The 2833x/2823x devices implement the standard IEEE 1149.1 JTAG interface. Additionally, the devices support real-time mode of operation whereby the contents of memory, peripheral and register locations can be modified while the processor is running and executing code and servicing interrupts. The user can also single step through non-time critical code while enabling time-critical interrupts to be serviced without interference. The device implements the real-time mode in hardware within the CPU. This is a feature unique to the 2833x device, requiring no software monitor. Additionally, special analysis hardware is provided that sets the hardware breakpoint or data/address watch-points and generates various user-selectable break events when a match occurs.

The 2833x/2823x devices provide options to boot normally or to download new software from an external connection or to select boot software that is programmed in the internal Flash/ROM. The Boot ROM also contains standard tables, such as SIN/COS waveforms, for use in math-related algorithms. The 2833x devices support high levels of security to protect the user firmware from being reverse engineered.

---

[1] http://e2e.ti.com/group/universityprogram/default.aspx?DCMP=univ&HQS=university

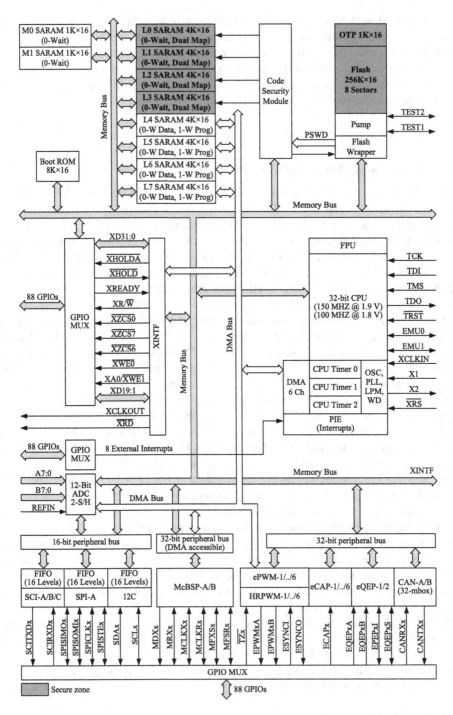

Figure 1.37 Functional block diagram of TI DSC TMS320F28335. *Source:* TI 2007

The 2833x/2823x devices support eight external interrupts and up to 96 peripheral interrupts through the Peripheral Interrupt Expansion (PIE) Block. The devices contain a watchdog timer and support three low-power modes: IDLE, STANDBY and HALT.

The 2833x/2823x devices support a range of peripherals for embedded control, including:

ePWM (up to 6-channel): The enhanced PWM peripheral supports independent/complementary PWM generation, adjustable dead-band generation for leading/trailing edges, latched/cycle-by-cycle trip mechanism. Three of the six PWM pins support HRPWM features. The ePWM registers are supported by the DMA to reduce the overhead for servicing this peripheral.

eCAP (up to six modules): The enhanced capture peripheral uses a 32-bit time base and registers up to four programmable events in continuous/one-shot capture modes. This peripheral can also be configured to generate an auxiliary PWM signal.

eQEP (up to 2 modules): The enhanced QEP peripheral uses a 32-bit position counter, supports low-speed measurement using capture unit and high-speed measurement using a 32-bit unit timer. This peripheral has a watchdog timer to detect motor stall and input error detection logic to identify simultaneous edge transition in QEP signals.

ADC (one module with 16 channels): The ADC block is a 12-bit converter with 16 single-ended (0-3V) channels multiplexed through two built-in S/H. Conversion rate can be up to 80 ns at 25-MHz ADC clock, 12.5 MSPS. It contains two sample-and-hold units for simultaneous sampling. The ADC registers are supported by the DMA to reduce the overhead for servicing this peripheral.

The 2833x/2823x devices also support a range of peripherals for communication, including:

eCAN: The enhanced CAN peripheral supports 32 mailboxes, time stamping of messages, and is CAN 2.0B-compliant.

McBSP: The multichannel buffered serial port (McBSP) connects to E1/T1 lines, phone-quality codecs for modem applications or high-quality stereo audio DAC devices. The McBSP receive and transmit registers are supported by the DMA to significantly reduce the overhead for servicing this peripheral. Each McBSP module can be configured as an SPI as required.

SPI: The SPI is a high-speed, synchronous serial I/O port that allows a serial bit stream of programmed length (one to sixteen bits) to be shifted into and out of the device at a programmable bit-transfer rate. Normally, the SPI is used for communications between the DSC and external peripherals or another processor. Typical applications include external I/O or peripheral expansion through devices such as shift registers, display drivers, and ADCs. Multi-device communications are supported by the master/slave operation of the SPI. On the 2833x/2823x, the SPI contains a 16-level receive and transmit FIFO for reducing interrupt servicing overhead.

SCI: The serial communications interface is a two-wire asynchronous serial port, commonly known as UART. The SCI contains a 16-level receive and transmit FIFO for reducing interrupt servicing overhead.

I2C: The inter-integrated circuit (I2C) module provides an interface between a DSC and other devices compliant with Philips Semiconductors Inter-IC bus (I2C-bus) specification version 2.1 and connected by way of an I2C-bus. External components attached to this 2-wire serial bus can transmit/receive up to 8-bit data to/from the DSC through the I2C module. On the 2833x/2823x, the I2C contains a 16-level receive and transmit FIFO for reducing interrupt servicing overhead.

More details about TMS320F28335 can be found in (TI 2007).

1.3.7.2 TMS320F28335 ControlCARD

Texas Instruments offer a controlCARD[2] that is equipped with a TMS320F28335 to be used for initial software development and short run builds for system prototypes, test stands, and many other projects that require easy access to high-performance controllers (TI 2002). The controlCARDs are complete board-level modules that utilise an industry-standard DIMM form factor to provide a low-profile single-board controller solution. All of the C2000 controlCARDs use the same 100-pin connector footprint to provide the analog and digital I/Os on-board controller and are completely interchangeable. Each controlCARD provides an isolated RS-232 interface for communications. The host system needs to provide only 5 V power to the controlCARD.

1.3.7.3 TMS320F28335 Experimenter Kit

Texas Instruments also offer an experimenter kit[3] that is equipped with a TMS320F28335 DSC. It is an ideal product to be used for initial device exploration and testing. The kit has a docking station with access to all controlCARD signals, breadboard areas, RS-232, JTAG connector, and on board USB JTAG emulation. Each kit contains a 28335 controlCARD and is complete with Code Composer Studio™ IDE v3.3 C28x™ Free 32 K Byte Version. C2000 applications software with example codes and full hardware details are also available. Note that no separate JTAG emulator is required, as the docking station features on-board USB JTAG emulation.

1.3.8 Test Equipment

In order to test the working condition and performance of inverters, many instruments are needed. In addition to conventional multi-meters and oscilloscopes, etc, power analysers/meters are often needed.

---

[2] http://www.ti.com/tool/tmdscncd28335
[3] http://www.ti.com/tool/tmdsdock28335

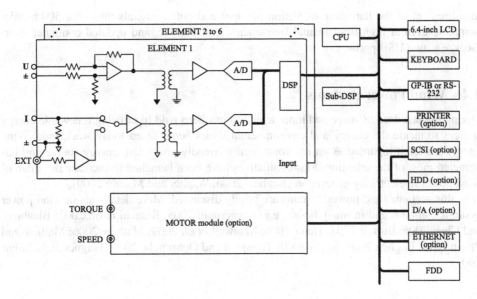

Figure 1.38 Functional block diagram of WT1600. *Source:* Iwase *et al.* 2003

Yokogawa WT1600 is a high-precision, wide-bandwidth digital power meter (Iwase *et al.* 2003) that is able to measure DC and 0.5 Hz to 1 MHz AC signals with a basic power accuracy of 0.1%. With the maximum of six input elements installed, a single WT1600 is able to measure the efficiency of a three-phase inverter, in addition to the measurement of voltages, currents, power, waveform quality, and waveform display. There are two different input elements: 5 A and 50 A. Both elements are standard-equipped with direct voltage and current inputs, as well as a compatible current sensor input covering from a shunt resistor to various current probes. Both elements can be installed together so both extremely small currents and large currents can be measured. The measuring ranges are 1.5 V to 1000 V for voltage (DC and 0.5 Hz to 1 MHz AC), and 10 mA to 5 A on the 5 A input elements (DC and 0.5 Hz to 1 MHz AC signals) and 1 A to 50 A on the 50 A input elements (0.5 Hz to 1 MHz AC signals) for current.

Figure 1.38 shows the basic configuration of the WT1600. It consists of four main components: input elements, CPU, Sub-DSP and display. All the input signals are first processed by the input block. After being electrically insulated and digitally converted, signals are passed to a DSP to determine measured values. The CPU block receives the measured values from the DSP and prepares them for display, communication, D/A output and other purposes. The Sub-DSP block performs calculations that require phase information between the elements, including harmonic measurement, and delta-Y conversion for three-phase, three-wire systems (Iwase *et al.* 2003). It is also possible to synchronise the measurement with two WT1600 units for 12 voltages and 12 currents.

An upgraded model WT1800 is now available,[4] with 5 MHz bandwidth for voltage and current, capable of measuring low frequency AC signals down to 0.1 Hz, simultaneous

---

[4] http://tmi.yokogawa.com/products/digital-power-analyzers/

measurement of the harmonic distortion of input and output signals up to the 500th-order harmonic even at high fundamental frequencies, e.g. 400 Hz, and updated computer interfaces, e.g. two USB ports.

1.4 Wind Power Systems

During the last decade, more and more attention has been paid to utilising renewable energy sources to tackle the energy and environmental issues being faced today worldwide. Wind energy has been regarded as an environmentally friendly alternative energy source and has attracted most of the attention. Many initiatives have been launched to increase the share of wind power in electricity generation (Mathew 2006; Wagner and Mathur 2009).

In this section, wind power systems are briefly discussed. More details about wind power systems can be found in many books, e.g. (Ackerman 2005; Bianchi et al. 2007; Blaabjerg and Chen 2006; Burton 2001; Heier 2006; Manwell et al. 2009; Mathew 2006; Mathew and Philip 2011; Ragheb 2009; Spera 2009; Thongam and Ouhrouche 2011; Wagner and Mathur 2009).

1.4.1 Basics of Wind Power Generation

Assume that the wind speed is v_w m/s and the area swept by a wind turbine is A m². Then the volume of the air swept through in unit time is Av_w. If the air density is ρ kg/m³, then the mass m of the air passing through the area in unit time is $\rho A v_w$ kg. The kinetic energy of this mass of the air moving at velocity v_w in unit time is

$$\frac{1}{2}mv_w^2 = \frac{1}{2}\rho A v_w^3.$$

This is actually the same as the power carried by the wind motion. For a wind turbine with rotor blades of R m long, the area swept is $A = \pi R^2$ and hence the wind power available is

$$P_w = \frac{1}{2}\rho \pi R^2 v_w^3.$$

In reality, it is impossible to convert all the energy into electricity. The actual power produced by a wind turbine can be calculated as

$$P_m = \frac{1}{2}\rho \pi R^2 v_w^3 C_p(\lambda, \beta), \qquad (1.6)$$

where $C_p(\lambda, \beta)$ is the power coefficient that is dependent on the turbine design, the pitch angle β and the tip-speed ratio λ defined as

$$\lambda = \omega_r R / v_w, \qquad (1.7)$$

where ω_r is the angular speed of the wind turbine. The tip-speed ratio plays a vital role in extracting power from wind. If the rotor turns too slowly, most of the wind passes through

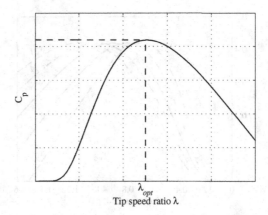

Figure 1.39 Power coefficient C_p as a function of the tip-speed ratio λ

the gap between the rotor blades without doing any work; if the rotor turns too quickly, the blurring blades block the wind like a solid wall. Hence, wind turbines are designed to operate at optimal tip-speed ratios so that as much power as possible can be extracted.

The power coefficient C_p is a highly non-linear function of λ and β. For wind turbines with a fixed pitch angle β, the relationship between C_p and the tip speed ratio λ often has the shape shown in Figure 1.39. The power coefficient reaches its maximum point at the optimum tip-speed ratio λ_{opt}, which depends on the number of blades in the wind turbine rotor. The fewer the number of blades, the faster the wind turbine rotor needs to turn to extract the maximum power from the wind. A two-bladed rotor has an optimum tip-speed ratio of around 6, a three-bladed rotor around 5, and a four-bladed rotor around 3 (Burton 2001). According to (1.7), the curves of C_p against different wind speeds have similar shapes, as shown in Figure 1.40(a), where six C_p curves are shown for six different wind speeds with $v_{w1} > v_{w2} > v_{w3} > v_{w4} > v_{w5} > v_{w6}$. The corresponding power P_m is shown in Figure 1.40(b). Therefore, the operational points of a wind turbine at different wind speeds are different, which are determined by the optimal tip-speed ratio λ_{opt} and the wind speed. See (Heier 2006) for more details. It is worth noting that the power coefficient C_p of a wind turbine is limited by $\frac{16}{27} \approx 0.593$, according to the Betz law.[5]

1.4.2 Wind Turbines

A wind turbine is a device that captures the kinetic energy of wind. Historically, a wind turbine was frequently used as a mechanical device with a number of blades to drive machinery. Nowadays, it is often used to drive a generator so that the kinetic energy is converted to electricity. The main types of wind turbines are shown in Figure 1.41 (Heier 2006). Most modern wind turbines use a horizontal axis configuration with two or three blades, operating either downwind or upwind (Manwell *et al.* 2009). The typical structure of a horizontal axis

---

[5] http://en.wikipedia.org/wiki/Betz%27_law

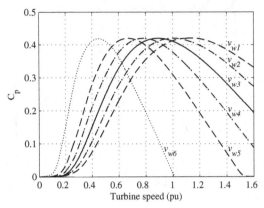

(a) C_p as a function of the turbine speed normalised to the rated speed

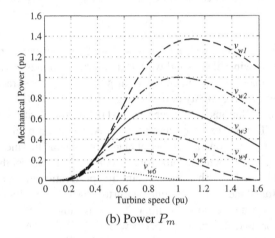

(b) Power P_m

Figure 1.40 Power coefficient C_p and mechanical power P_m at different wind speeds

wind turbine is shown in Figure 1.42, with major components including blades, a rotor hub, drivetrain (bearing and gears, etc.), a generator and the associated control system.

Wind turbines can be used for stand-alone applications, connected to a utility power grid or even combined with photovoltaic systems, batteries and diesel generators, etc. to form hybrid systems. Small-scale wind turbines are often used in stand-alone applications, e.g. for water pumping, communication stations, and supply of electricity to farms and light towers, etc. that are far from the utility grid. For utility-scale applications of wind power, a large number of turbines are usually built together to form wind farms to fully utilise the available wind power and to reduce the investment cost on infrastructure.

A wind turbine can be designed for fixed-speed or variable-speed operation. Variable-speed wind turbines can produce more energy than fixed-speed ones but power electronic converters are needed to provide a voltage at a fixed frequency and a fixed amplitude.

Most turbine manufacturers have opted for a direct drive configuration to remove the gears between the low speed turbine rotor and the high speed three-phase generator. This

Introduction

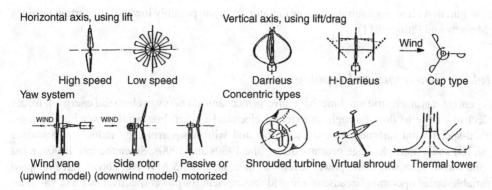

Figure 1.41 Main turbine types. *Source:* Heier 2006, Grid Integration of Wind Energy Conversion Systems: Second Edition

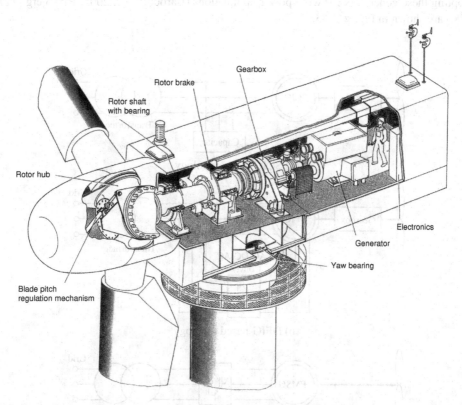

Figure 1.42 Structure of a typical wind turbine. *Source:* Heier 2006, Grid Integration of Wind Energy Conversion Systems: Second Edition

configuration offers high reliability, low maintenance, and possibly low cost for certain turbines (Mathew and Philip 2011).

1.4.3 Generators and Topologies

A generator is an electric machine that converts mechanical energy to electrical energy. It forces electric charge to flow through an external electrical circuit. For wind power applications, fixed-speed wind turbines were mostly operated with a squirrel-cage induction generator (SCIG) and a multiple-stage gearbox during the 1980s and 1990s. Since the late 1990s, most wind turbines, in which the power level was increased to 1.5 MW and above, have adopted variable-speed operation because of the grid requirement for power quality. For these variable-speed applications, doubly-fed induction generators (DFIG) are commonly used together with a multi-stage gearbox and power electronic converters. Permanent magnet synchronous generators (PMSG) are becoming increasingly popular because of their ability to reduce failures in the gearbox and lower maintenance problems (Spera 2009). The common topologies adopting these generators for wind power applications (Baroudi *et al.* 2005; Blaabjerg *et al.* 2006) are shown in Figure 1.43.

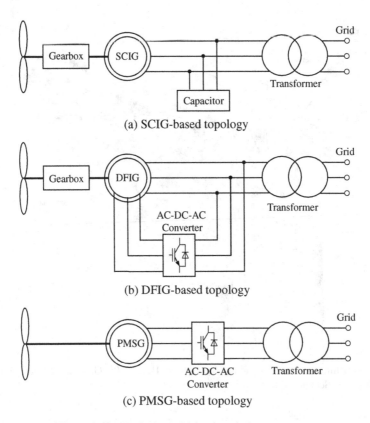

Figure 1.43 Typical topologies for wind power systems

1.4.3.1 Squirrel-Cage Induction Generators (SCIG)

A squirrel-cage induction machine is often operated as a motor but it can be operated as a generator when driven by a prime mover to a speed exceeding the synchronous speed. Induction machines are widely applied as generators in wind power applications due to the reduced unit cost and size, ruggedness, lack of brushes, absence of a separate DC source, ease of maintenance, self-protection against severe overloads and short circuits, etc. (Bansal 2005).

An induction generator produces real power but it needs reactive power to establish the excitation (the magnetic field). This leads to a low power factor, which is often penalised by utility companies. The reactive power needed for excitation can be provided by a capacitor bank, the grid or a solid-state power electronic converter. The connection of an SCIG, in particular a big one, to the grid often causes a large inrush current that is $7 \sim 8$ times of the rated current and a soft-starter is often needed. The pole pair number of SCIG used in commercial fixed-speed wind turbines is often equal to 2 or 3, which corresponds to a synchronous speed of 1500 rpm or 1000 rpm for a 50 Hz system. As a result, a three-stage gearbox is often required in the drive train.

SCIGs are often applied in fixed-speed wind turbine systems directly connected to the grid through a transformer, as shown in Figure 1.43(a). With this topology, the rotor blades are directly fixed to the hub and adjusted only once when the turbine is erected. The power limitation over the rated wind speed is achieved by stalling the rotor blades. Wind turbines with this topology are completely passive and, hence, this topology is called passive stall control or shortly stall control. In most cases, capacitors are connected in parallel to provide the reactive power needed for excitation.

There are obvious advantages of using SCIGs. However, there are also disadvantages. The speed of operation is not controllable and it can be varied only within a very narrow range because the rotor circuit is not accessible, which makes it difficult to extract the maximum available wind power. The need for a three-stage gearbox in the drive train considerably increases the weight of the nacelle, and the investment and maintenance costs. Moreover, it is necessary to obtain the excitation current from the grid, which makes impossible to support the grid voltage.

1.4.3.2 Doubly-fed Induction Generators (DFIG)

The fact that the rotor circuit of an SCIG is not accessible can be changed if the rotor circuit is wound and made accessible via slip rings, which offers the possibility of controlling the rotor circuit so that the operational speed range of the generator can be increased in a controlled manner. The rotor circuit is often connected to back-to-back power electronic converters, which consists of a rotor-side converter and a grid-side converter sharing the same DC bus, so that the difference between the mechanical speed of the rotor and the electrical speed of the grid can be compensated via injecting a current with a variable frequency into the rotor circuit. Hence, the operation during both normal and faulty conditions can be regulated by controlling the converters.

A DFIG can be excited via the rotor windings and does not have to be excited via the stator windings. If needed, the reactive power needed for the excitation from the stator windings can be generated by the grid-side converter. As a result, a wind power plant equipped with DFIGs can easily take part in the regulation of grid voltage. The stator always feeds real power to the grid but the real power in the rotor circuit can flow bidirectionally, from the grid to the rotor or from the rotor to the grid, depending on the operational condition. Ignoring the losses, the power handled by the rotor circuit is (Tazil et al. 2010)

$$P_{\text{rotor}} = -s \cdot P_{\text{stator}},$$

where s is the slip, and the power sent to the grid is

$$P_{\text{grid}} = P_{\text{rotor}} + P_{\text{stator}} = (1-s)P_{\text{stator}}.$$

Since most of the power flows through the stator circuit, the power processed by the rotor circuit can be reduced to roughly 30%. This means the great advantage of a sufficient range of operational speed can be achieved at a reasonably low cost.

DFIGs are often applied in variable speed wind turbine systems with a multi-stage gearbox, as shown in Figure 1.43(b). Its basic operating principle is the same as an SCIG-based system but the rotor active power is controlled by the power electronic converters so that a speed range of $\pm 30\%$ around the synchronous speed can be obtained. The choice of the rated power for the rotor converter is a trade-off between cost and the desired speed range. Moreover, the converter compensates the reactive power and smooths the grid connection.

Although a DFIG offers a sufficient range of operational speed and many other merits, it is very sensitive to voltage disturbances, especially voltage sags. Abrupt voltage drops at the terminals often cause large voltage disturbances on the rotor, which may exceed the voltage rating of the rotor-side converter (RSC), make the rotor current uncontrollable, and even damage the RSC. Many strategies are available to improve the low-voltage ride-through capability of DFIGs; see (Guo et al. 2012).

1.4.3.3 Permanent Magnet Synchronous Generator (PMSG)

A PMSG adopts a permanent magnet to generate the magnetic field needed for electricity generation. Hence, there is no need to provide an external power supply for excitation and there is no need to have a rotor circuit. This simplifies the structure and reduces the maintenance cost. PMSGs are more efficient than induction generators and the power factor can be made unity or even leading. Moreover, PMSGs have very high power density and are becoming cost-effective because the price of rare-earth magnets has reduced by more than an order of magnitude in the last 10 years. As a result, PMSGs are becoming increasingly popular for wind power applications. A PMSG runs at the synchronous speed and the frequency of electricity generated is directly in proportion to the mechanical speed and hence the slip is zero. This could be used to eliminate the need for a mechanical sensor to measure the speed of the turbine.

PMSGs are often applied in variable speed wind turbine systems, which are direct-driven or with a single stage gearbox, as shown in Figure 1.43(c). Full-scale back-to-back converters are often used for the AC-DC-AC conversion. It is also possible to use an uncontrolled rectifier cascaded with a DC-DC converter and an inverter.

Because permanent magnets are used, care should always be taken to avoid possible demagnetisation caused by too high currents and/or too high temperature.

1.4.4 Control of Wind Power Systems

1.4.4.1 Rotor Power Control at High Wind Speeds

At high wind speeds, the power transferred to the rotor should be controlled to protect the rotor, the generator and the power electronic converters, if any, from overloading and to protect the rotor from damage when the generator loses its electrical load. Hence, rotor power control plays a very important role in wind turbine systems.

The following strategies are often adopted (Bianchi *et al.* 2007; Ragheb 2009) for rotor power control:

1. Yaw control: Wind turbines are oriented perpendicular to the wind stream during normal operation using wind orientation mechanism or yaw control. During high wind speeds, the rotor axis can be turned out of the wind direction to protect the wind turbine. This is often used in small-scale wind turbines.
2. Pitch control: Instead of turning the whole rotor out of the wind direction during high wind speeds, rotor blades can be individually rotated around their longitudinal axis to furl, that is, to reduce the angle of attack. A fully furled turbine blade, when stopped, has the edge of the blade facing into the wind. With the pitch control mechanism in place, it can also be used during normal operation to maximise the power extracted.
3. Stall control: The blades, which are attached to the hub at a fixed angle, are aerodynamically designed to take advantage of the stall effect that occurs at high wind speeds. When this happens, turbulence occurs at the back side of the blade. As a result, the lift drops and the drag increases, which leads to reduced driving torque and power production. Stall-controlled wind turbines do not have moving parts introduced into the rotor but often have additional aerodynamic brakes.
4. Active stall control: The blades are pitched to increase the angle of attack during high wind speeds so that the stall effect is created. This is opposite to the pitch control, where the angle of attack is reduced during high wind speeds.

1.4.4.2 Rotor Speed Control during Normal Operation

It was shown at the beginning of this section that the operational condition of a wind turbine should be changed according to the wind speed in order to extract the maximum power available, which requires the rotor blades to run at such a speed that the tip-speed ratio is kept at the optimum value. This is often called maximum power point tracking (MPPT).

The maximum achievable power of a wind turbine system can be written as

$$P_{max} = K_{opt}\omega_{ropt}^3,$$

where

$$K_{opt} = \frac{0.5\pi\rho C_{pmax} R^5}{\lambda_{opt}^3}$$

is a constant related to the optimal tip-speed ratio λ_{opt} and the maximum power coefficient C_{pmax}, and

$$\omega_{ropt} = \frac{\lambda_{opt} v_w}{R}$$

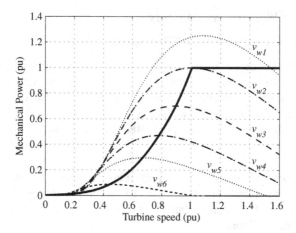

Figure 1.44 Desired operational power curve

is the optimal turbine speed corresponding to the wind speed v_w. In order to extract the maximum power from the wind, the turbine should always be operated with λ_{opt} via controlling the turbine speed.

A typical desired operational power curve corresponding to different wind speeds is shown by the thick line in Figure 1.44: it increases in proportion to ω_{ropt}^3 when the rotor speed is below the rated speed and is maintained at the rated power to protect the wind turbine system when the rotor speed exceeds the rated speed. If the wind speed reaches the cut-off speed, then the turbine should be shut down. This is not shown in Figure 1.44. An MPPT technique should be applied to extract the maximum power from the wind turbine when the rotor speed is below the rated speed and a rotor power control mechanism should be activated when the rotor speed exceeds the rated power.

MPPT controllers can be classified into three types (Thongam and Ouhrouche 2011): tip-speed ratio (TSR) control, power signal feedback (PSF) control and hill-climb search (HCS) control. For TSR control, both the wind speed and the turbine speed are measured or estimated to maintain the TSR at λ_{opt} so that the maximum possible power is extracted. For PSF control, the power is measured and controlled to reach the maximum power, according to the power curve of the wind turbine obtained through either simulation or off-line experiments. The HCS method is different from the TSR and PSF control methods, in which the maximum power point is continuously looked for, according to the current operating power point and the relationship between the changes in power and speed.

1.4.4.3 Grid Integration

Once the power is extracted, then it is important to feed it to the grid. Apart from the SCIG-based topology, power electronic converters are involved in the grid integration of wind power. How to control power electronic converters so that the integration of wind power into the grid can be done in a grid-friendly manner and imposes minimum impact on the grid is the main subject of this book and will be discussed in detail.

1.5 Solar Power Systems

1.5.1 Introduction to Solar Power

Solar energy, radiant light and heat from the sun, has been utilised since ancient times using a range of ever-evolving technologies. The total solar energy absorbed by the Earth's atmosphere, oceans and land masses is approximately 3,850,000 exajoules (EJ) per year. In 2002, this was more energy in one hour than the world used in one year.[6]

When matter (metals and non-metallic solids, liquids or gases) absorbs energy from electromagnetic radiation of very short wavelength, such as visible or ultraviolet radiation, electrons are emitted (such electrons are often referred to as photoelectrons). This effect is called the photoelectric effect. Based on this, solar cells or photovoltaic (PV) cells, which consist of one or two layers of semi-conducting material, can be made to directly convert the radiation from the sun into electricity. When light shines on the cell it creates an electric field across the layers, which causes electricity to flow. The greater the intensity of the light, the greater the flow of electricity is (Carrasco et al. 2006b; EPIA 2010).

Due to the growing demand for renewable energy sources, the manufacturing of solar cells and photovoltaic arrays has advanced considerably in recent years. Over the past decade, the photovoltaic market has experienced unprecedented growth. In 2010, the capacity grew from 7.2 GW installed in 2009 to 16.6 GW. The total installed global capacity in 2011 amounts to around 40 GW, producing some 50 TWh of electrical power every year (EPIA 2011).

The performance of a solar cell is measured in terms of its efficiency at turning sunlight into electricity. A typical commercial solar cell has an efficiency of about 15%. The amount of power produced by a solar power installation depends on the location of the sun in the sky and on the amount of cloud cover. The variations and predictability in cloud cover are similar to that of wind speed. The location of the sun in the sky shows a predictable daily and seasonal variation caused by the rotation of the earth on its axis and around the sun. This makes solar power more predictable than wind power but there is limited experience with prediction for solar power to verify this (Bollen and Hassan 2011; EPIA 2010).

The energy from the sun can be focused (concentrated) using large arrays of mirrors in concentrated solar power (CSP) plants to boil water that in turn powers a turbine as in a conventional thermal power station. One advantage of the CSP plants is that the energy can be more easily stored in large amounts of thermal energy with minimal losses and, thus, they can provide energy on demand during day and night. As a result, CSP plants can contribute to stabilising electricity grids by compensating fluctuations of renewable energy sources if they are part of the same network (Jacobson 2009; Moreno 2011).

PV systems can provide clean power for small or large applications. They are already installed and generating energy around the world in individual homes, housing developments, offices and public buildings. Although PV systems can operate as stand-alone systems, where it is difficult to connect to the grid or where there is no energy infrastructure, they are mostly connected to the grid for homes and businesses in developed areas. In this case, any excessive power produced can be fed into the electricity grid. Electricity can be imported from the network when there is no sunlight. Such small installations are also easy to set up and connect to the grid. The rules about grid connection vary from country to country, but almost in all countries it is compulsory to contact the local network operator before connection. Small

---

[6] http://en.wikipedia.org/wiki/Solar_energy

rooftop installations are close to the consumption of electricity and they, therefore, reduce the power transport through both the transmission and the distribution networks.

1.5.2 Processing of Solar Power

Similar to wind power, the source of solar power is not controllable and hence there is a need to maximise the power generated from the sunlight. This can be done with MPPT strategies implemented at an appropriate stage of power processing.

PV systems can be categorised according to the number of power processing stages, the location of power decoupling capacitors, with or without transformers, and types of grid interfaces etc (Carrasco *et al.* 2006b; Kjaer *et al.* 2005; Li and Wolfs 2008). Some typical topologies for PV systems are shown in Figure 1.45. Power electronic inverters are essential for converting the DC power produced by PV cells into AC power that is compatible with the electricity distribution network and the majority of common electrical appliances. For a grid-connected PV system, the inverter often plays two major roles: (1) to ensure that the PV system captures the maximum power from the sunlight with a maximum-power-point-tracking (MPPT) algorithm; and (2) to feed the energy into the grid, nowadays often as a clean current,

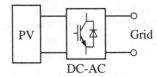

(a) Single-stage processing

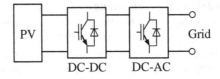

(b) Two-stage processing

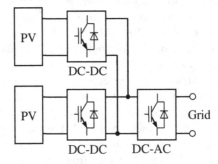

(c) Two-stage processing with a shared DC bus

Figure 1.45 Some typical topologies for PV systems

according to utility regulations, e.g. on power quality control, reactive power control, fault ride-through etc. (Kjaer *et al.* 2005). If a DC/DC converter is introduced, then the MPPT function is often embedded into the controller of the DC/DC converter.

1.6 Smart Grid Integration

1.6.1 Operation Paradigms of Power Systems

1.6.1.1 Centralised Generation

Electrical power systems have been in existence all over the world for more than 100 years. A typical power system, as shown in Figure 1.46, consists of facilities to generate, transmit and distribute electrical power to consumers or loads (Karady and Holbert 2004) because power plants, which generate electricity from energy sources such as fossil fuels, hydro, nuclear, etc, are far from consumers. Although there are often interconnections at the transmission level to form a strong grid, to which massive power plants are connected, electricity normally flows uni-directionally from generation to loads. In particular, the electricity flows through the distribution network is unidirectional.

1.6.1.2 Distributed Generation

For economic, technical and environmental reasons, there is today a trend towards the use of small power generating units connected to the low-voltage distribution systems in addition to the traditional large generators connected to the high-voltage transmission systems (Jenkins

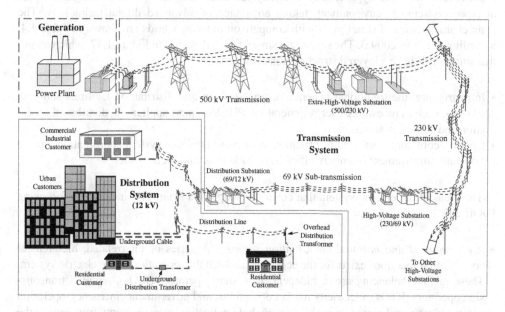

Figure 1.46 Overview of a current electrical power system. *Source:* Karady G and Holbert K, 2004. *Electrical Energy Conversion and Transport: An Interactive Computer-Based Approach,* © John Wiley & Sons

et al. 2000), following the large-scale utilisation of renewable energy sources, energy storage systems, electrical vehicles, etc. Not only is there a change of scale but also a change of technology. Large generators are almost exclusively 50/60 Hz synchronous machines. Distributed power generators include variable speed (variable frequency) sources, high speed (high frequency) sources and direct energy conversion sources that produce DC. For example, wind turbines are most effective if free to generate at variable frequency and so they require conversion from AC (variable frequency) to DC to AC (50/60 Hz) (Chen and Spooner 2001); small gas turbines with direct drive generators operate at high frequencies and also require AC to DC to AC conversion (Etezadi-Amoli and Choma 2001), and photovoltaic arrays require DC-AC conversion (Enslin *et al.* 1997).

There are several operating regimes possible for distributed generation. One such is for distributed generators to form microgrids before being connected to the public grid. As a result, local consumers are largely supplied by the local distributed generation with shortfalls or surpluses exchanged through a connection to the public electricity supply system (Lasseter 2002; Venkataramanan and Illindala 2002). The use of a microgrid opens up the possibility of making the distributed generator responsible for local power quality in a way that is not possible with conventional generators (Green and Prodanović 2003). Another option is to connect distributed generation and storage systems directly to the grid.

1.6.2 Introduction to Smart Grids

The change of the operation paradigms of power systems does not stop at distributed generation. A more advanced concept, the smart grid, has been introduced to power systems to further improve reliability, quality, operating efficiency, resilience to threats while reducing the impact of power systems to environment, taking advantage of advanced digital technology. The main characteristics of smart grids with comparison to today's grids are shown in Table 1.4, according to (DOE 2009a). The scope of a smart grid is depicted in Figure 1.47, which shows that smart grids have a layered structure consisting of:

- Infrastructure: the traditional generation, transmission and distribution facilities and new add-ons, such as renewable energy generators, PHEVs, smart appliances, distributed generation and storage systems, etc.
- Control, communication and information systems to facilitate system coordination, operation, and improvement of energy efficiency, marketing, and security etc.

The areas of the electric system that cover the scope of a smart grid include the following (DOE 2009b):

- *Area, regional and national coordination regimes:* A series of interrelated, hierarchical coordination functions exists for the economic and reliable operation of the electric system. These include balancing areas, independent system operators (ISOs), regional transmission operators (RTOs), electricity market operations, and government emergency-operation centres. Smart-grid elements in this area include collecting measurements from across the system to determine system state and health, and coordinating actions to enhance economic efficiency, reliability, environmental compliance, or response to disturbances.

Table 1.4 Comparison of today's grid with smart grid

Characteristic	Today's Grid	Smart Grid
Enables active participation by consumers	Consumers are uninformed and non-participative with power system	Informed, involved, and active consumers, demand response and distributed energy resources
Accommodates all generation and storage options	Dominated by central generation, many obstacles for distributed energy resources interconnection	Many distributed energy resources with plug-and-play convenience, focus on renewables
Enables new products, services and markets	Limited wholesale markets, not well integrated, limited opportunities for consumers	Mature, well-integrated wholesale markets, growth of new electricity markets for consumers
Provides power quality for the digital economy	Focus on outages, slow response to power quality issues	Power quality is a priority with a variety of quality/price options, rapid resolution of issues
Optimises assets and operates efficiently	Little integration of operational data with asset management, business process silos	Greatly expanded data acquisition of grid parameters, focus on prevention, minimising impact to consumers
Anticipates and responds to system disturbances (self-heals)	Responds to prevent further damage, focus on protecting assets following faults	Automatically detects and responds to problems, focus on prevention, minimising impact to consumers
Operates resiliently against attack and natural disaster	Vulnerable to malicious acts of terror and natural disasters	Resilient to attack and natural disasters with rapid restoration capabilities

Source: DOE 2009a, The Smart Grid: An Introduction

- *Distributed-energy resource technology:* This area includes the integration of distributed-generation, storage, and demand-side resources for participation in electric-system operation. Consumer products such as smart appliances and electric vehicles are expected to become important components of this area as are renewable-generation components such as those derived from solar and wind sources. Aggregation mechanisms of distributed-energy resources are also considered.
- *Transmission and distribution (T&D) infrastructure:* T&D represents the delivery part of the electric system. Smart grid items at the transmission level include substation automation, dynamic limits, relay coordination, and the associated sensing, communication, and coordinated action. Distribution-level items include distribution automation (such as feeder-load balancing, capacitor switching, and restoration) and advanced metering (such as meter reading, remote-service enabling and disabling, and demand-response gateways).
- *Central generation:* Generation plants already contain sophisticated plant automation systems because the production-cost benefits provide clear signals for investment. While

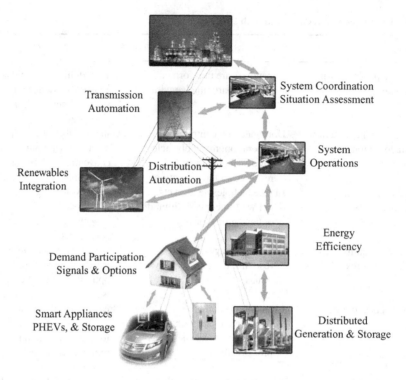

Figure 1.47 Scope of smart-grid concerns. *Source:* DOE 2009b Smart Grid System Report

technological progress is related to the smart grid, change is expected to be incremental rather than transformational.
- Information networks and finance: Information technology and pervasive communications are the cornerstones of a smart grid. Though the requirements (capabilities and performance) on the information networks will be different in different areas, their attributes tend to transcend application areas. Examples include interoperability and the ease of integration of automation components as well as cyber-security concerns. Information technology-related standards, methodologies, and tools also fall into this area. In addition, the economic and investment environment for procuring smart grid-related technology is an important part of the discussion concerning implementation progress.

Arguably, the integration of renewable and distributed energy sources, energy storage and demand-side resources into smart grids is the largest "new frontier" for smart grid advancements (DOE 2009b). Control and power electronics are two key enabling technologies for this. Power electronics is a part of the grid, and control is where "smart" is found (Ekanayake *et al*. 2012; Hopkins and Safiuddin 2010). Together with the power systems infrastructure, they form the backbone of smart grids.

This book is devoted to smart grid integration. For other aspects of smart grids, see e.g. (DOE 2009a 2009b; Ekanayake *et al*. 2012; Farhangi 2010; Momoh 2012).

1.6.3 Requirements for Smart Grid Integration

As mentioned above, the integration of renewable and distributed energy sources, energy storage and demand-side resources into smart grids is arguably the largest "new frontier" for smart grid advancements (DOE 2009b), in particular, when these sources are connected to smart grids via inverters. Several challenging technical problems should be addressed in order to fully maximise the benefits of smart grids.

1.6.3.1 Synchronisation

One of the most important problems in renewable energy and smart grid integration is how to synchronise the inverters with the grid (Blaabjerg *et al.* 2006; Rodriguez *et al.* 2007b; Shinnaka 2008; Wildi 2005). There are two different scenarios: one is before connecting an inverter to the grid and the other is during the operation. If an inverter is not synchronised with the grid or another power source, to which it is to be connected, then large transient currents may appear at the time of connection, which may cause damage. During normal operation, the inverter needs to be synchronised with the source it is connected to so that the system can work properly. In both scenarios, the grid information is needed accurately and in a timely manner so that the inverter is able to synchronise with the grid voltage. Depending on the control strategies adopted, the information needed can be any combination of the phase, the frequency and the voltage amplitude of the grid.

1.6.3.2 Power Flow Control

A simple reason for integrating renewable energy, distributed generation and storage systems, etc. into a grid is to inject power to the grid. This should be done in a controlled manner.

Naturally, this is done via directly controlling the current injected into the grid. Another option is to control the voltage difference between the inverter output voltage and the grid voltage. As a result, there are current-controlled strategies and voltage-controlled strategies. Current-controlled strategies are easy to implement but the inverters equipped with current-controlled strategies do not take part in the regulation of power system frequency and voltage and, hence, they may cause problems for the system stability when the share of power fed into the grid is significant. It is more difficult to control voltage than current but voltage-controlled inverters can easily take part in the regulation of system frequency and voltage, which is very important when the penetration level of renewable energy, distributed generation and storage systems, etc. reaches a certain level. The closer to conventional synchronous generators these sources behave, the smoother the operation of the grid is.

1.6.3.3 Power Quality Control

Power quality is a set of electrical properties that may affect the proper function of electrical systems. It is used to describe the electric power that drives an electrical load. Without proper power quality, an electrical device (or load) may malfunction, fail prematurely or not operate at all. Poor power quality can be described in different ways, e.g. the continuity of power, variations in magnitude and frequency, transient changes, harmonic contents in the waveform, low power factor, imbalance of phases, etc.

The integration of renewable energy, distributed generation and storage systems, etc. into smart grids via inverters may cause serious power quality issues. A major power quality issue in these applications is the harmonics in the voltage provided by the inverters and the current injected into the grid, which can be caused by the PWM switching effect, and the load current.

1.6.3.4 Neutral Line Provision

For applications in renewable energy, distributed generation and smart grids, there is often a need to have a neutral line to work with inverters so that a current path is provided for unbalanced loads. The provision of a neutral line also facilitates the independent operation of the three phases so that the coupling effect among the phases is minimised.

1.6.3.5 Fault Ride-through

When the penetration level of renewable energy, distributed generation and storage systems, etc. to the grid reaches a certain level, it is required to be able to successfully negotiate the short faults that have occurred in the grid, e.g. voltage sags, voltage dips, phase jumps, frequency variations, etc. (Rodriguez et al. 2007a; Song and Huang 2010; Timbus et al. 2006b). They can only disconnect from the grid when the faults are serious. As an example, Figure 1.48 illustrates the boundaries for connecting/disconnecting wind turbines in the Danish power system under different voltage disturbances with different time scales.

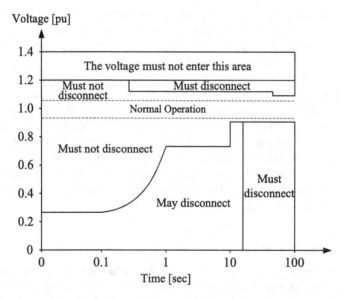

Figure 1.48 Operational boundaries for wind farms in the Danish power system under different voltage disturbances with different time scales. *Source:* Timbus A, Teodorescu T, Blaabjerg F, Liserre M and Rodriguez P 2006 PLL algorithm for power generation systems robust to grid voltage faults in *Proceedings of the 37th IEEE Power Electronics Specialists Conference (PESC)*, pp. 1–7

Table 1.5 Some islanding detection methods

Passive methods	Active methods	Utility-based methods
Under-/over-voltage	Impedance measurement	Manual disconnection
Under-/over-frequency	Impedance measurement at a specific frequency	Automated disconnection
Phase jump detection	Slip mode frequency shift	Transfer-trip method
Harmonics detection	Frequency bias	Impedance insertion

1.6.3.6 Anti-islanding

Here, islanding refers to unexpected situations when renewable energy, distributed generation or storage systems continue feeding power to a grid that has lost power. Islanding can be dangerous for the utility workers, who may not realise that a circuit is still powered, and it may prevent automatic re-connection of devices.[7] Hence, islanding must be detected and the renewable energy, distributed generation and storage systems involved must be disconnected from the grid. This is referred to as anti-islanding. However, intentional islanding is often used by backup power systems to power the local circuit when it is disconnected from the grid.

As mentioned before, synchronisation is one of the most important requirements for smart grid integration. A good synchronisation method could be utilised for fault ride-through (Rodriguez *et al.* 2006b; Timbus *et al.* 2006b) and anti-islanding. Islanding detection is the subject of considerable research and there are many methods available, which can be classified into passive methods, active methods and utility-based methods.[8] Passive methods attempt to detect transient changes on the grid caused by grid failures while active methods attempt to detect grid failures by injecting small signals into the grid and then detecting whether or not the signal changes. The utility can also apply a variety of methods to force systems offline in the event of a failure. Some of these methods are listed in Table 1.5.

The first four problems are addressed in this book in detail and the last two problems are not discussed any further.

---

[7] http://en.wikipedia.org/wiki/Islanding
[8] http://en.wikipedia.org/wiki/Islanding

2

Preliminaries

In this chapter, some preliminaries that are common to more than one chapter are discussed. These include power quality issues, repetitive control and reference frames.

2.1 Power Quality Issues

2.1.1 Introduction

Power quality is a set of electrical properties that may affect the proper function of electrical systems. It is used to describe the electric power that drives an electrical load. Without proper power quality, an electrical device (or load) may malfunction, fail prematurely or not operate at all. Although the term "power quality" is commonly used, it mainly refers to the quality of voltages because in conventional power systems the supplies are voltage sources instead of current sources and the quality of currents is determined by the loads.

Poor power quality can be described in different ways., e.g., the continuity of power, variations in magnitude and frequency, transient changes, harmonic contents in the waveform, low power factor, imbalance of phases etc. Some commonly used terms are shown in Table 2.1. Poor power quality can be the consequence due to different reasons, e.g. the result of shared infrastructure in power systems. A fault at one site may cause poor power quality to loads at other sites. Nowadays, more and more distributed generation and renewable energy sources, e.g. wind, solar and tidal power, are developed. They often form microgrids via power inverters (Guerrero et al. 2009; Iyer et al. 2010; Lasseter, 2002; Weiss et al. 2004; Zhong and Weiss 2011), which may or may not be connected to the grid. A major power quality problem in these applications is the harmonics in the voltage provided by the inverters. This is the main power quality issue to be addressed in this book.

There are two sources of harmonics: one is from the inverters (e.g. because of the pulse-width-modulation and the switching) and the other is from the loads or the grid. The majority

---

Control of Power Inverters in Renewable Energy and Smart Grid Integration, First Edition.
Qing-Chang Zhong and Tomas Hornik.
© 2013 John Wiley & Sons, Ltd. Published 2013 by John Wiley & Sons, Ltd.

Table 2.1 Common terms to describe power quality

Term	Meaning
swell	The RMS voltage exceeds the nominal value by 10% to 80% for 0.5 cycle to 1 minute.
dip or sag	The RMS voltage is below the nominal value by 10% to 90% for 0.5 cycle to 1 minute.
flicker	Random or repetitive variations in the RMS voltage between 90% and 110% of the nominal value, which cause rapid visible changes of light level in lighting equipment.
spikes, impulses, or surges	Abrupt, very brief increases in voltage, which are generally caused by large inductive loads being turned off.
under-voltage	The voltage drops below 90% of the nominal value for more than 1 minute.
over-voltage	The voltage rises above 110% of the nominal value for more than 1 minute.
brownout	An apt description for voltage drops somewhere between full power (bright lights) and a blackout (no power – no light).
harmonics	The wave shape is distorted.

of loads today are non-linear and generate harmonic currents when a purely sinusoidal voltage supply is provided. These harmonic currents then cause harmonic components in the voltage because of the impedances in the distribution network and, also, inside the voltage sources. Of course, harmonic voltages then cause harmonic currents as well. The odd multiples of the 3rd harmonic (3rd, 9th, 15th, 21st ...), i.e. the $6n - 3$ harmonics, are called triplen harmonics. These currents on a three-phase system are zero-sequence harmonics, which are additive in the neutral line and cause particular concern. The $6n - 1$ harmonics are negative-sequence harmonics and can cause problems to electrical machines because these harmonics create a negative torque and attempt to drive the machines in reverse. Harmonics are not desirable because they cause overheating, increased losses, decreased power capacity, neutral line overloading, distorted voltage and current waveforms, etc. It has become a very serious issue in modern power systems. Hence, stringent regulations have been put into place (Hornik and Zhong, 2011; Yousefpoor et al. 2012). The total harmonic distortion (THD) of voltages and currents needs to be maintained low, often below 5%. Table 2.2 shows the maximum THD allowed in the currents fed to the grid (Blooming and Carnovale, 2007). Since harmonics tend to be cumulative in power systems (Irwin, 1996), the controllers used should have very good capability in harmonic rejection, in order to meet the operator's requirements.

A natural solution is to introduce some controllers and, as a result, several feedback control schemes are available to reduce the voltage THD of inverters. In principle, this is a tracking problem with a sinusoidal reference while rejecting other harmonic components. Deadbeat

Table 2.2 Maximum THD allowed in currents fed to the grid

Odd harmonics	$< 11^{th}$	$11^{th}-15^{th}$	$17^{th}-21^{th}$	$23^{rd}-33^{rd}$	$> 33^{rd}$
Maximum current THD	< 4%	< 2%	< 1.5%	< 0.6%	< 0.3%

or hysteresis controllers (Blaabjerg *et al.* 2006) are some examples. Repetitive control theory (Hara *et al.* 1988; Nakano *et al.* 1989), which is regarded as a simple learning control method, provides an alternative to eliminate periodic errors in dynamic systems, based on the internal model principle (Francis and Wonham, 1975). Such a closed-loop system can deal with a very large number of harmonics simultaneously, as it has high gains at the fundamental and all harmonic frequencies of interest. It has been successfully applied to constant-voltage constant-frequency (CVCF) PWM inverters (Chen *et al.* 2008; Tzou *et al.* 1999; Wang *et al.* 2007; Ye *et al.* 2007, 2006; Zhang *et al.* 2008), grid-connected inverters (Weiss *et al.* 2004; Hornik and Zhong, 2011, 2010) and active filters (Costa-Castello *et al.* 2007; Garcia-Cerrada *et al.* 2007) to obtain very low THD. Strategies have been developed to maintain low THD in the current exchanged with the grid (Hornik and Zhong, 2011), in the microgrid voltage (Hornik and Zhong, 2010b; Weiss *et al.* 2004) and in both the current exchanged with the grid and the microgrid voltage at the same time (Hornik and Zhong, 2010a; Zhong and Hornik, 2012). A cooperative harmonic filtering strategy was proposed in (Lee and Cheng, 2007) to share the harmonic Var (Watanabe *et al.* 1993) among distributed inverters, which has the potential to improve the voltage THD. The drawback is that extensive experiments are needed to determine the reference value for the harmonic Var.

Another way is to analyse the degradation mechanisms of the voltage quality, so that some strategies can be developed to address the power quality problems. For example, the output impedance of an inverter plays a critical role in reducing the THD of the output voltage (Zhong *et al.* 2011) and the voltage THD could be improved by reducing the output impedance. This will be discussed further in the next subsection.

As mentioned before, the main power quality issue addressed in this book is the voltage THD. It is worth mentioning that, for the harmonics in load currents, active power filters can be adopted. See, for example, (Grino *et al.* 2007; Lascu *et al.* 2007; Routimo *et al.* 2007).

2.1.2 Degradation Mechanisms of Voltage Quality

It has been widely recognised that the output voltage of an inverter as modelled in Figure 2.1 can be written as

$$v_o = v_r - Z_o(s) \cdot i,$$

where v_r is the voltage reference, v_o is the output voltage across the capacitor and i is the output current flowing through the inductor; see e.g. (Zhong 2012b). In general, there are

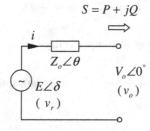

Figure 2.1 Model of an inverter

harmonics in the current i because of non-linear loads and/or the pulse width modulation built in the inverter, which cause harmonic voltage drops on the output impedance Z_o. If there are no corresponding harmonic voltage components provided by the reference v_r, then the harmonic voltage components appear in the output voltage, which degrades the voltage quality and causes a high THD. Another source of high THD of the output voltage is from the voltage reference v_r if it contains significant harmonic components.

In order to obtain low THD for v_o, there are three options:

1. to keep v_r clean and maintain a small output impedance Z_o over the frequency range of the major harmonic current components;
2. to bypass the harmonic current components so that the current i is clean; and
3. to make sure that the reference voltage v_r is able to provide the right amount of harmonic voltages to compensate the harmonic voltage dropped on the output impedance.

The first option will be discussed in detail in the next subsection, with control strategies presented in Chapter 7. The second option will be discussed in Chapter 8 while the third option will be discussed in Chapter 21.

It is believed that the repetitive control strategies discussed in (Chen et al. 2008; Costa-Castello et al. 2004; Escobar et al. 2008; Garcia-Cerrada et al. 2007; Hornik and Zhong, 2010b,2011; Tzou et al. 1999; Weiss et al. 2004; Zhou and Wang, 2003; Zhou et al. 2009) and those in Chapters 3–6 should lead to small output impedances but this fact has not been well documented. Note that the harmonics injection discussed in (Borup et al. 2001) belongs to the third option as well.

2.1.3 Role of Inverter Output Impedance

Assume that the output current is

$$i = \sqrt{2}\Sigma_{h=1}^{\infty} I_h \sin(h\omega t + \phi_h),$$

where ω is the system frequency. Then the amplitude of the h-th harmonic voltage dropped on the output impedance is $\sqrt{2}I_h |Z_o(jh\omega)|$. Moreover, assume that the voltage reference v_r is clean and sinusoidal and is described as

$$v_r = \sqrt{2}E \sin(\omega t + \delta).$$

Then the fundamental component of the output voltage is[1]

$$v_1 = \sqrt{2}E \sin(\omega t + \delta) - \sqrt{2}I_1 |Z_o(j\omega)| \sin(\omega t + \phi_1 + \theta)$$
$$= \sqrt{2}V_1 \sin(\omega t + \beta),$$

---

[1] See the trigonometric identities at http://en.wikipedia.org/wiki/List_of_trigonometric_identities.

where

$$V_1 = \sqrt{E^2 + I_1^2 |Z_o(j\omega)|^2 - 2EI_1 |Z_o(j\omega)| \cos(\phi_1 + \theta - \delta)},$$

$$\beta = \arctan\left(\frac{I_1 |Z_o(j\omega)| \sin(\phi_1 + \theta - \delta)}{I_1 |Z_o(j\omega)| \cos(\phi_1 + \theta - \delta) - E}\right).$$

The sum of all harmonic components in the output voltage is

$$v_H = \sqrt{2} \Sigma_{h=2}^{\infty} I_h |Z_o(jh\omega)| \sin(h\omega t + \phi_h + \angle Z_o(jh\omega)).$$

It is clear that v_1 and v_H do not affect each other. v_1 is determined by the clean reference voltage, the fundamental current and the output impedance at the fundamental frequency. v_H is determined by the harmonic current components and the output impedance at the harmonic frequencies. This is an important feature and enables the design of the impedance to meet different requirements.

According to the definition of THD, the THD of the output voltage is

$$\text{THD} = \frac{\sqrt{\Sigma_{h=2}^{\infty} I_h^2 |Z_o(jh\omega)|^2}}{V_1} \times 100\%. \qquad (2.1)$$

Hence, the THD is mainly affected by the output impedance at the harmonic frequencies. As a result, it is feasible to optimise the design of the output impedance at high frequencies to minimise the THD of the output voltage, without affecting the impedance at the fundamental frequency. In other words, the design of the output impedance can be decoupled to meet two different requirements in the frequency domain. For example, the output impedance at the fundamental frequency can be designed to meet the requirements of the droop controller for proportional load sharing, as shown in Chapter 19, and the output impedance at harmonic frequencies can be designed to reduce the THD of the voltage, as shown in Chapter 7. Hence, one parameter can be used to meet two requirements at the same time (one qualitatively and the other quantitatively).

2.2 Repetitive Control

2.2.1 Basic Principles

One of the main objectives of a control system is to asymptotically track a reference signal while rejecting disturbances. According to the well-known internal model principle (Francis and Wonham, 1975), there is a need to include a model of the reference signal or the disturbance signal in the closed loop in order to achieve perfect tracking or disturbance rejection. For example, when the reference signal and/or the disturbances are step signals, it is necessary to include an integrator $\frac{1}{s}$ in the closed loop to achieve zero static error. Similarly, for periodic signals (such as those in the case of controlling inverters), there is a need to include a pair of conjugate imaginary poles at the frequency ω of the signal in the closed loop. A pair of conjugate imaginary poles can be provided by $\frac{1}{s^2+\omega^2}$, as used in the proportional-resonant

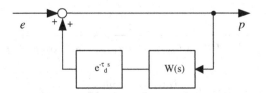

Figure 2.2 Internal model for repetitive control

control scheme discussed in Chapter 16. The output voltage of an inverter is full of harmonics and it is important to reduce the level of major harmonics in order to obtain a low THD. One possible solution is to include a bundle of models in the form of $\frac{1}{s^2 + \omega^2}$ at different harmonic frequencies.

Another solution is the so-called repetitive control (Chen et al. 2008; Hara et al. 1988; Nakano et al. 1989; Weiss, 1997; Weiss and Hafele, 1999), which adopts an infinite-dimensional internal model $M(s)$ shown in Figure 2.2 to provide a series of conjugate poles at all harmonic frequencies. The internal model consists of a local positive feedback via a delay line cascaded with a low-pass filter $W(s)$, which is added to improve the stability of a repetitive control system because the system without it is a neutral-type delay system (Logemann and Pandolfi, 1994; Logemann and Townley, 1996; Partington and Bonnet, 2004). As shown in the next subsection, this internal model has a series of poles very close to the harmonic frequencies and hence is able to improve THD.

2.2.2 Poles of the Internal Model $M(s)$

The transfer function of the internal model is

$$M(s) = \frac{1}{1 - W(s)e^{-\tau_d s}}, \qquad (2.2)$$

where $W(s)$ is often chosen as a low-pass filter

$$W(s) = \frac{\omega_c}{s + \omega_c}$$

with the cut-off frequency ω_c. The poles of the internal model $M(s)$ from (2.2) are the solutions of the transcendental equation

$$\frac{s + \omega_c}{\omega_c} = e^{-\tau_d s},$$

which has infinitely many roots $s_k, k \in \mathbb{Z}$. Substitute $s_k = \tilde{s}_k - \omega_c$, then $\frac{1}{\omega_c}\tilde{s}_k = e^{-\tau_d \tilde{s}_k} e^{\tau_d \omega_c}$, i.e.,

$$\tau_d \tilde{s}_k e^{\tau_d \tilde{s}_k} = a, \qquad \text{with} \qquad a = \tau_d \omega_c e^{\tau_d \omega_c}. \qquad (2.3)$$

The Lambert W function, whose history and properties are beautifully presented in (Corless et al. 1996), is needed to solve this equation. The Lambert W function is a multi-valued analytic function, with infinitely many branches denoted W_k, $k \in \mathbb{Z}$. The function W_k may be defined for $z \in \mathbb{C}\setminus(-\infty, 0]$ as the (unique) solution of

$$W_k(z) = \ln z + j2k\pi - \ln W_k(z), \tag{2.4}$$

where ln is the principal branch of the logarithm. Note that (2.4) implies that if $w = W_k(z)$ then $we^w = z$, which is the basic equation satisfied by all the branches W_k. For $z > 0$, $W_0(z)$ is the only positive solution of $we^w = z$. The equation (2.4) can be thought of as a fixed-point equation for the function

$$T_{k,z}(w) = \ln z + j2k\pi - \ln w.$$

For $k \neq 0$ and large $|z|$, the iterations defined by $w_{n+1} = T_{k,z}(w_n)$ converge fast to $W_k(z)$. The approximation formula (4.11) given in (Corless et al. 1996) can be written as $W_k(z) \approx T_{k,z}(T_{k,z}(1))$. Since, for large real z, $W_0(z)$ is a better initial approximation of $W_k(z)$ than 1, the following approximation is used for $W_k(z)$:

$$W_k(z) \approx T_{k,z}(T_{k,z}(W_0(z))) = \ln z + j2k\pi - \ln(\ln z + j2k\pi - \ln W_0(z)). \tag{2.5}$$

From (2.3) and the definition of the Lambert W function, there is

$$\tau_d \tilde{s}_k = W_k(a) \quad (k \in \mathbb{Z}), \quad \tau_d \omega_c = W_0(a),$$

and

$$s_k = \frac{W_k(a) - \tau_d \omega_c}{\tau_d}. \tag{2.6}$$

Since a is normally very large, $W_k(a)$ can be well approximated by (2.5), i.e.,

$$W_k(a) \approx \ln(\tau_d \omega_c) + \tau_d \omega_c + j2k\pi - \ln(\tau_d \omega_c + j2k\pi).$$

Now from (2.6),

$$\tau_d s_k \approx \ln(\tau_d \omega_c) - \ln\sqrt{(\tau_d \omega_c)^2 + (2k\pi)^2} + j\left(2k\pi - \tan^{-1}\frac{2k\pi}{\tau_d \omega_c}\right).$$

Hence, the real part and the imaginary part of the poles s_k are

$$\operatorname{Re} s_k \approx \frac{1}{\tau_d} \ln \frac{\tau_d \omega_c}{\sqrt{(\tau_d \omega_c)^2 + (2k\pi)^2}} = -\frac{1}{2\tau_d} \ln\left(1 + \left(\frac{2k\pi}{\tau_d \omega_c}\right)^2\right), \tag{2.7}$$

$$\operatorname{Im} s_k \approx \frac{2k\pi - \tan^{-1}\frac{2k\pi}{\tau_d \omega_c}}{\tau_d} \approx \frac{2k\pi}{\tau_d}\left(1 - \frac{1}{\tau_d \omega_c}\right). \tag{2.8}$$

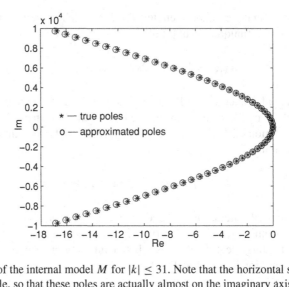

Figure 2.3 Poles of the internal model M for $|k| \leq 31$. Note that the horizontal scale is much smaller than the vertical scale, so that these poles are actually almost on the imaginary axis

The last approximation holds because if $\left|\frac{2k\pi}{\tau_d \omega_c}\right| \ll 1$, then $\tan^{-1}$ can be approximated by the identity function.

The approximated poles and the true poles are shown in Fig. 2.3 for $|k| \leq 31$. Actually, the approximation is very good even for $|k| \leq 1000$. Ideally, it is expected that $s_k = j\frac{2\pi}{\tau}k$. At least, this is approximately true for small $|k|$.

2.2.3 Selection of the Delay in the Internal Model

In order to make $\text{Im } s_k \approx \frac{2\pi}{\tau}k$, according to (2.8), τ_d needs to satisfy

$$\tau_d^2 = \tau_d \tau - \frac{1}{\omega_c}\tau. \tag{2.9}$$

Solving this equation,

$$\tau_d = \frac{1 \pm \sqrt{1 - \frac{4}{\omega_c \tau}}}{2}\tau. \tag{2.10}$$

The solution with a minus sign is not reasonable, since it would lead to $\tau_d \omega_c \approx 1$ and then many of the approximations used earlier would break down. The reasonable solution corresponds to the plus sign in (2.10), which leads to a good recommendation for τ_d as

$$\tau_d \approx \frac{1 + 1 - \frac{2}{\omega_c \tau}}{2}\tau = \tau - \frac{1}{\omega_c}.$$

Preliminaries

This coincides with the recommendation in (Weiss and Hafele, 1999, Section 2). This recommendation is good when $\omega_c \tau$ is very large, which is normally the case. When $\omega_c \tau$ is not so large, τ_d should be chosen according to (2.10) with the plus sign.

2.3 Reference Frames

For a three-phase system, the voltages and currents can be described in different reference frames (also called coordinate systems). As a result, the controller can be designed in different coordinates. Here, voltages are taken as an example to describe the different reference frames but the analysis can be applied to currents as well, even to some other quantities; see (Bose, 2001, 2009). The case with the phase sequence of abc will be discussed first in detail and the case with the phase sequence of acb will be discussed briefly.

2.3.1 Natural (abc) Frame

The voltages of a balanced three-phase system in the natural frame, also called the abc frame, can be represented as

$$\begin{bmatrix} v_a \\ v_b \\ v_c \end{bmatrix} = \begin{bmatrix} V_m \cos(\theta) \\ V_m \cos\left(\theta - \frac{2\pi}{3}\right) \\ V_m \cos\left(\theta + \frac{2\pi}{3}\right) \end{bmatrix}, \tag{2.11}$$

where V_m is the peak value of the voltage. θ is the phase of Phase a voltage and it changes with time t. Hence, the voltages are functions of time t (and frequency). Note that the phase sequence here is abc.

The three phases (coordinates) a, b and c in the natural frame can be regarded as spatially distributed away from each other by $\frac{2\pi}{3}$ rad, as shown in Figure 2.4(a). It is drawn to be consistent with the phase sequence abc. The three-phase voltages v_a, v_b and v_c at a particular time t or phase θ can be expressed as vectors according to their values spatially distributed on the corresponding coordinates, as shown in Figure 2.4(a). That is, v_a is on the horizontal line; v_b is in line with the vector $\alpha = e^{-j\frac{2\pi}{3}}$ and v_c is in line with the vector $\alpha^2 = e^{-j\frac{4\pi}{3}}$. The resulting diagrams are referred to as spatial diagrams in the sequel. Note that this is different from the widely used phasor diagrams, where the length of a vector is the amplitude of the voltage and the angle of the vector is the phase of the voltage. Here, the length of a vector on a spatial diagram is the instantaneous value of the voltage and the angle of the vector is fixed as the angle of the coordinate, i.e. 0 for Phase a, $-\frac{2\pi}{3}$ radians for Phase b and $\frac{2\pi}{3}$ radians for Phase c in Figure 2.4(a). Moreover, a phasor diagram is independent of time but the spatial diagram is dependent on time (and frequency). When θ changes with time t, the vectors v_a, v_b and v_c on the spatial diagram shown in Figure 2.4(a) change their lengths (and direction) but do *not* rotate.

It is worth noting that

$$v_a + v_b + v_c = 0$$

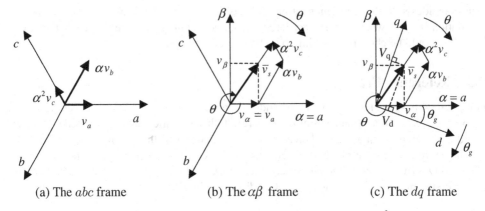

(a) The abc frame (b) The $\alpha\beta$ frame (c) The dq frame

Figure 2.4 Reference frames for three-phase abc systems: with $\alpha = e^{-j\frac{2\pi}{3}}$ (preferred)

for a balanced system, which means that there are only two independent variables and that the three-phase balanced voltages can be expressed on a two-dimensional space.

Another possible way to spatially distribute the three phases a, b and c is shown in Figure 2.5(a), with $\alpha = e^{j\frac{2\pi}{3}}$. It is worth noting that the spatial order of the phases should not be confused with the phase sequence. They may or may not be the same.

2.3.2 Stationary Reference ($\alpha\beta$) Frame

Taking into account the spatial feature of the voltages on a spatial diagram, the vectors having the instantaneous voltage values in the natural frame (2.11) can be adopted to form a space vector

$$\bar{v}_s = k(v_a + \alpha v_b + \alpha^2 v_c), \tag{2.12}$$

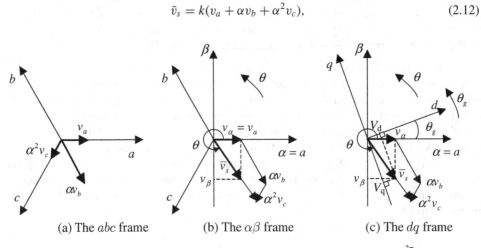

(a) The abc frame (b) The $\alpha\beta$ frame (c) The dq frame

Figure 2.5 Reference frames for three-phase abc systems: with $\alpha = e^{j\frac{2\pi}{3}}$

Preliminaries

which can be decomposed into a real part v_α and an imaginary part v_β, as shown in Figure 2.4(b). That is,

$$\bar{v}_s = v_\alpha + jv_\beta.$$

Note that the instantaneous values of v_a, v_b and v_c, not their phasors, are used to form the space vector $\bar{v}_s$. As a result, both v_α and v_β are all functions of time (and frequency) as well.

For $\alpha = e^{-j\frac{2\pi}{3}}$ and $k = \frac{2}{3}$, the voltage space vector (2.12) is

$$\bar{v}_s = k(v_a + \alpha v_b + \alpha^2 v_c)$$
$$= kv_a + k\left(\cos\frac{2\pi}{3} - j\sin\frac{2\pi}{3}\right)v_b + k\left(\cos\frac{4\pi}{3} - j\sin\frac{4\pi}{3}\right)v_c$$
$$= k\left(v_a - \frac{1}{2}v_b - \frac{1}{2}v_c\right) + jk\left(-\frac{\sqrt{3}}{2}v_b + \frac{\sqrt{3}}{2}v_c\right).$$

Hence,

$$\begin{bmatrix} v_\alpha \\ v_\beta \end{bmatrix} = T_{\alpha\beta} \begin{bmatrix} v_a \\ v_b \\ v_c \end{bmatrix} \tag{2.13}$$

with the $abc \to \alpha\beta$ transformation matrix

$$T_{\alpha\beta} = \frac{2}{3}\begin{bmatrix} 1 & -\frac{1}{2} & -\frac{1}{2} \\ 0 & -\frac{\sqrt{3}}{2} & \frac{\sqrt{3}}{2} \end{bmatrix}. \tag{2.14}$$

The reference frame with v_α and v_β as coordinates is often referred to as the stationary reference frame or the $\alpha\beta$ frame. The transformation (2.13) is called the Clarke transform or the $abc \to \alpha\beta$ transformation. It transforms the voltages in the natural frame into the stationary frame or the $\alpha\beta$ frame.

For balanced systems, $v_a + v_b + v_c = 0$. Hence,

$$\begin{bmatrix} v_\alpha \\ v_\beta \end{bmatrix} = \begin{bmatrix} v_a \\ \frac{\sqrt{3}}{3}(-v_b + v_c) \end{bmatrix}.$$

The α-component in the $\alpha\beta$ frame is always the same as v_a in the abc frame. In other words, v_a remains unchanged and stationary in the $\alpha\beta$ frame. Moreover, according to (2.11),

$$v_\beta = \frac{\sqrt{3}V_m}{3}\left(\cos\left(\theta + \frac{2\pi}{3}\right) - \cos\left(\theta - \frac{2\pi}{3}\right)\right)$$
$$= -V_m \sin\theta.$$

Hence, the space vector (2.12) is

$$\bar{v}_s = V_m \cos\theta - jV_m \sin\theta = V_m e^{-j\theta}.$$

When θ changes with time t, the length of the space vector $\bar{v}_s$ does not change but the angle changes. As a result, the vector $\bar{v}_s$ *rotates* clockwise when θ increases with time t. The direction of the rotation is the same as the spatial order abc shown in Figure 2.4(a). Note that the $\alpha\beta$ reference frame itself is stationary and does not rotate, hence the name stationary reference frame. Here, the introduction of the space vector has converted translational movements on a spatial diagram into rotational movements on the $\alpha\beta$ reference frame.

For the case shown in Figure 2.5(a) with the spatial operator $\alpha = e^{j\frac{2\pi}{3}}$, the corresponding transformation matrix is

$$\tilde{T}_{\alpha\beta} = \frac{2}{3}\begin{bmatrix} 1 & -\frac{1}{2} & -\frac{1}{2} \\ 0 & \frac{\sqrt{3}}{2} & -\frac{\sqrt{3}}{2} \end{bmatrix} = \begin{bmatrix} 1 & 0 \\ 0 & -1 \end{bmatrix} T_{\alpha\beta}.$$

In this case, the real part v_α remains unchanged but the imaginary part v_β changes its sign. As a result, the space vector is $\bar{v}_s = V_m e^{j\theta}$ and it rotates counterclockwise when θ increases with time t, as shown in Figure 2.5(b). Again, the direction of the rotation is the same as the spatial order abc shown in Figure 2.5(a).

2.3.3 Synchronously Rotating Reference (dq) Frame

The voltages v_α and v_β in the stationary frame are still functions of time although the space vector $\bar{v}_s$ rotates with time t at the speed of ω. If a reference frame that synchronously rotates at the same speed of the space vector is introduced, then the space vector $\bar{v}_s$ on this reference frame is no longer a function of time and does not rotate. Such a reference frame is often called the dq reference frame, or the synchronously rotating reference frame (Bose, 2001). Assume that the new set of coordinates dq rotate in the same direction as the space vector with a phase angle θ_g. Then, as shown in Figure 2.4(c), the angle of the space vector $\bar{v}_s$ on the dq frame is $-\theta - (-\theta_g) = \theta_g - \theta$ and the space vector is

$$\bar{v}_s = V_d + jV_q = (v_\alpha + jv_\beta)e^{j\theta_g} = V_m e^{j(\theta_g - \theta)}. \tag{2.15}$$

Hence,

$$\begin{bmatrix} V_d \\ V_q \end{bmatrix} = T_{dq}\begin{bmatrix} v_\alpha \\ v_\beta \end{bmatrix} \tag{2.16}$$

with the rotating matrix

$$T_{dq} = \begin{bmatrix} \cos\theta_g & -\sin\theta_g \\ \sin\theta_g & \cos\theta_g \end{bmatrix}. \tag{2.17}$$

Substitute (2.13) into (2.16), then

$$\begin{bmatrix} V_d \\ V_q \end{bmatrix} = \frac{2}{3} \begin{bmatrix} \cos\theta_g & -\sin\theta_g \\ \sin\theta_g & \cos\theta_g \end{bmatrix} \begin{bmatrix} 1 & -\frac{1}{2} & -\frac{1}{2} \\ 0 & -\frac{\sqrt{3}}{2} & \frac{\sqrt{3}}{2} \end{bmatrix} \begin{bmatrix} v_a \\ v_b \\ v_c \end{bmatrix}$$

$$= \frac{2V_m}{3} \begin{bmatrix} \cos\theta_g & -\sin\theta_g \\ \sin\theta_g & \cos\theta_g \end{bmatrix} \begin{bmatrix} 1 & -\frac{1}{2} & -\frac{1}{2} \\ 0 & -\frac{\sqrt{3}}{2} & \frac{\sqrt{3}}{2} \end{bmatrix} \begin{bmatrix} \cos(\theta) \\ \cos\left(\theta - \frac{2\pi}{3}\right) \\ \cos\left(\theta + \frac{2\pi}{3}\right) \end{bmatrix}$$

$$= \frac{2V_m}{3} \begin{bmatrix} \cos\theta_g & -\sin\theta_g \\ \sin\theta_g & \cos\theta_g \end{bmatrix} \begin{bmatrix} \cos\theta - \dfrac{\cos\left(\theta - \frac{2\pi}{3}\right) + \cos\left(\theta + \frac{2\pi}{3}\right)}{2} \\ \sqrt{3}\dfrac{\cos\left(\theta + \frac{2\pi}{3}\right) - \cos\left(\theta - \frac{2\pi}{3}\right)}{2} \end{bmatrix}$$

$$= \frac{2V_m}{3} \begin{bmatrix} \cos\theta_g & -\sin\theta_g \\ \sin\theta_g & \cos\theta_g \end{bmatrix} \begin{bmatrix} \frac{3}{2}\cos\theta \\ -\frac{3}{2}\sin\theta \end{bmatrix}$$

$$= V_m \begin{bmatrix} \cos\theta_g \cos\theta + \sin\theta_g \sin\theta \\ \sin\theta_g \cos\theta - \cos\theta_g \sin\theta \end{bmatrix}$$

$$= \begin{bmatrix} V_m \cos(\theta_g - \theta) \\ V_m \sin(\theta_g - \theta) \end{bmatrix}.$$

This is consistent with (2.15). Since θ_g and θ synchronously change at the same speed, $\theta_g - \theta$ is a constant and so are V_d and V_q. They are no longer dependent on time, which facilitates controller design and analysis. Actually, V_d and V_q can be regarded as the real and reactive power components of the voltage, respectively. For inverter control, θ_g is often chosen as the phase of the grid voltage. The introduction of a synchronously rotating reference frame has removed the rotational movement on the $\alpha\beta$ frame and made it stationary.

For the case shown in Figure 2.5(a) with the spatial operator $\alpha = e^{j\frac{2\pi}{3}}$, the space vector is $\bar{v}_s = V_m e^{j\theta}$ on the $\alpha\beta$ frame and it rotates counterclockwise when θ increases with time t, as shown in Figure 2.5(b). On the corresponding dq frame shown in Figure 2.5(c), the space vector is

$$\bar{v}_s = V_d + jV_q = (v_\alpha + jv_\beta)e^{-j\theta_g} = V_m e^{j(\theta - \theta_g)}.$$

Hence, the rotation matrix is

$$\tilde{T}_{dq} = \begin{bmatrix} \cos\theta_g & \sin\theta_g \\ -\sin\theta_g & \cos\theta_g \end{bmatrix} \quad (2.18)$$

and

$$\begin{bmatrix} V_d \\ V_q \end{bmatrix} = \frac{2}{3} \begin{bmatrix} \cos\theta_g & \sin\theta_g \\ -\sin\theta_g & \cos\theta_g \end{bmatrix} \begin{bmatrix} 1 & -\frac{1}{2} & -\frac{1}{2} \\ 0 & \frac{\sqrt{3}}{2} & -\frac{\sqrt{3}}{2} \end{bmatrix} \begin{bmatrix} v_a \\ v_b \\ v_c \end{bmatrix}$$

$$= \frac{2V_m}{3} \begin{bmatrix} \cos\theta_g & \sin\theta_g \\ -\sin\theta_g & \cos\theta_g \end{bmatrix} \begin{bmatrix} 1 & -\frac{1}{2} & -\frac{1}{2} \\ 0 & \frac{\sqrt{3}}{2} & -\frac{\sqrt{3}}{2} \end{bmatrix} \begin{bmatrix} \cos(\theta) \\ \cos\left(\theta - \frac{2\pi}{3}\right) \\ \cos\left(\theta + \frac{2\pi}{3}\right) \end{bmatrix}$$

$$= \frac{2V_m}{3} \begin{bmatrix} \cos\theta_g & \sin\theta_g \\ -\sin\theta_g & \cos\theta_g \end{bmatrix} \begin{bmatrix} \cos\theta - \dfrac{\cos\left(\theta - \frac{2\pi}{3}\right) + \cos\left(\theta + \frac{2\pi}{3}\right)}{2} \\ \sqrt{3} \dfrac{\cos\left(\theta - \frac{2\pi}{3}\right) - \cos\left(\theta + \frac{2\pi}{3}\right)}{2} \end{bmatrix}$$

$$= \frac{2V_m}{3} \begin{bmatrix} \cos\theta_g & \sin\theta_g \\ -\sin\theta_g & \cos\theta_g \end{bmatrix} \begin{bmatrix} \frac{3}{2}\cos\theta \\ \frac{3}{2}\sin\theta \end{bmatrix}$$

$$= V_m \begin{bmatrix} \cos\theta_g \cos\theta + \sin\theta_g \sin\theta \\ -\sin\theta_g \cos\theta + \cos\theta_g \sin\theta \end{bmatrix}$$

$$= \begin{bmatrix} V_m \cos(\theta_g - \theta) \\ -V_m \sin(\theta_g - \theta) \end{bmatrix}.$$

The V_d component remains unchanged but the V_q component has changed its sign. In order to minimise confusion, it is preferred to use the case shown in Figure 2.4 with the spatial operator $\alpha = e^{-j\frac{2\pi}{3}}$ for a system with the phase sequence abc. In this case, there is no negative sign in V_q.

2.3.4 The Case with Phase Sequence acb

For three-phase voltages with the phase sequence of acb described as

$$\begin{bmatrix} v_a \\ v_b \\ v_c \end{bmatrix} = \begin{bmatrix} V_m \cos(\theta) \\ V_m \cos\left(\theta + \frac{2\pi}{3}\right) \\ V_m \cos\left(\theta - \frac{2\pi}{3}\right) \end{bmatrix},$$

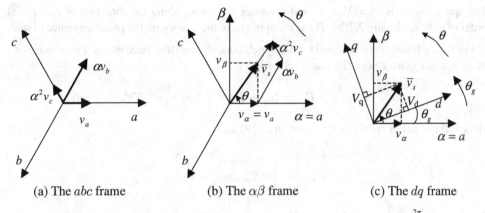

Figure 2.6 Reference frames for three-phase *acb* systems: with $\alpha = e^{-j\frac{2\pi}{3}}$

what is discussed in the natural frame still holds. There are two different spatial diagrams corresponding to $\alpha = e^{-j\frac{2\pi}{3}}$ and $\alpha = e^{j\frac{2\pi}{3}}$, respectively, which are the same as the case with the *abc* sequence.

For the $\alpha\beta$ frame, v_α is still the same as v_a. The transition matrices $T_{\alpha\beta}$ for $\alpha = e^{-j\frac{2\pi}{3}}$ and $\tilde{T}_{\alpha\beta}$ for $\alpha = e^{j\frac{2\pi}{3}}$ are not affected, either. However, the actual β-component v_β is affected, due to the change of the phase sequence. For $\alpha = e^{-j\frac{2\pi}{3}}$, it is

$$v_\beta = \frac{\sqrt{3}}{3}(-v_b + v_c)$$

$$= \frac{\sqrt{3} V_m}{3}\left(\cos\left(\theta - \frac{2\pi}{3}\right) - \cos\left(\theta + \frac{2\pi}{3}\right)\right) = V_m \sin\theta.$$

Hence, the space vector is $\bar{v}_s = V_m e^{j\theta}$ and it rotates counterclockwise along the direction of the spatial order of *acb*; see Figure 2.6(b). For $\alpha = e^{j\frac{2\pi}{3}}$, the β-component v_β is $v_\beta = -V_m \sin\theta$.

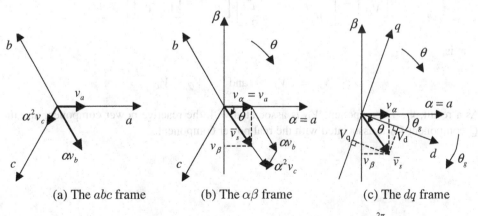

Figure 2.7 Reference frames for three-phase *acb* systems: with $\alpha = e^{j\frac{2\pi}{3}}$ (preferred)

The space vector is $\bar{v}_s = V_m e^{-j\theta}$ and it rotates clockwise along the direction of the spatial order of acb; see Figure 2.7(b). These are opposite to the cases with the phase sequence of abc.

For the dq frame, the frame still rotates synchronously with the space vector. For $\alpha = e^{-j\frac{2\pi}{3}}$, as shown in Figure 2.6(c), the space vector is

$$\bar{v}_s = V_d + jV_q = (v_\alpha + jv_\beta)e^{-j\theta_g} = V_m e^{j(\theta-\theta_g)}.$$

Hence, the rotation matrix is $\tilde{T}_{dq}$ given in (2.18) and

$$\begin{bmatrix} V_d \\ V_q \end{bmatrix} = \begin{bmatrix} V_m \cos(\theta_g - \theta) \\ -V_m \sin(\theta_g - \theta) \end{bmatrix}.$$

For $\alpha = e^{j\frac{2\pi}{3}}$, as shown in Figure 2.7(c), the space vector is

$$\bar{v}_s = V_d + jV_q = (v_\alpha + jv_\beta)e^{j\theta_g} = V_m e^{j(\theta_g - \theta)}.$$

The rotation matrix is T_{dq} given in (2.17) and

$$\begin{bmatrix} V_d \\ V_q \end{bmatrix} = \begin{bmatrix} V_m \cos(\theta_g - \theta) \\ V_m \sin(\theta_g - \theta) \end{bmatrix}.$$

In summary, when the phase sequence is consistent with the spatial order of the phases, as shown in Figures 2.4 and 2.7, the space vector always rotates clockwise and $V_q = V_m \sin(\theta_g - \theta)$. These are preferred. When the phase sequence is opposite to the spatial order of the phases, the space vector always rotates counterclockwise and $V_q = -V_m \sin(\theta_g - \theta)$. In all cases, $V_d = V_m \cos(\theta_g - \theta)$.

It is worth noting that, in the literature, the d-axis is often the reversed q-axis here and the q-axis is the d-axis here. This is equivalent to rotating the dq frame here clockwise by $\frac{\pi}{2}$ rad. Denote the dq frame in the literature as the DQ frame, then

$$\begin{bmatrix} V_D \\ V_Q \end{bmatrix} = \begin{bmatrix} \cos\frac{\pi}{2} & -\sin\frac{\pi}{2} \\ \sin\frac{\pi}{2} & \cos\frac{\pi}{2} \end{bmatrix} \begin{bmatrix} V_d \\ V_q \end{bmatrix} = \begin{bmatrix} 0 & -1 \\ 1 & 0 \end{bmatrix} \begin{bmatrix} V_d \\ V_q \end{bmatrix}.$$

That is,

$$V_D = -V_q \quad \text{and} \quad V_Q = V_d.$$

As a result, the D-component V_D is associated with the reactive power component and the Q-component V_Q is associated with the real power component.

Part I

Power Quality Control

3

Current H^∞ Repetitive Control

In this chapter, a current controller is designed for grid-connected inverters based on the H^∞ and repetitive control techniques. As a result, the inverter is able to inject clean currents into the grid. Moreover, a simple and effective mechanism for the inverter to quickly synchronise with the grid is also discussed. Experimental results show that the controller is able to achieve very low current THD.

3.1 System Description

The control system, as shown in Figure 3.1, adopts an individual controller for each phase in the *abc* frame and the system is equipped with a neutral point controller discussed in (Zhong et al. 2006) and Part II. It has a current loop including a repetitive controller so that the current injected into the grid could track the reference current i_{ref}, which is generated from the dq-current references I_d^* and I_q^* using the $dq \rightarrow abc$ transformation. A PLL is used to provide the phase information of the grid voltage needed to generate i_{ref}. The real power and reactive power exchanged with the grid are determined by I_d^* and I_q^*. The inverter is powered by a constant DC power source so no controller is needed to regulate the DC-link voltage. Otherwise, a controller can be introduced to regulate the DC-link voltage and to generate I_d^* accordingly.

When the references I_d^* and I_q^* are all equal to 0, the generated voltage should be equal to the grid voltage, i.e., the inverter should be synchronised with the grid and the circuit breaker could be closed at any time if needed. In order to achieve this, the grid voltages (u_{ga}, u_{gb} and u_{gc}) are feed-forwarded and added to the output of the repetitive current controller via a phase-lead low-pass filter

$$F(s) = \frac{33(0.05s + 1)}{(s + 300)(0.002s + 1)}, \tag{3.1}$$

Control of Power Inverters in Renewable Energy and Smart Grid Integration, First Edition.
Qing-Chang Zhong and Tomas Hornik.
© 2013 John Wiley & Sons, Ltd. Published 2013 by John Wiley & Sons, Ltd.

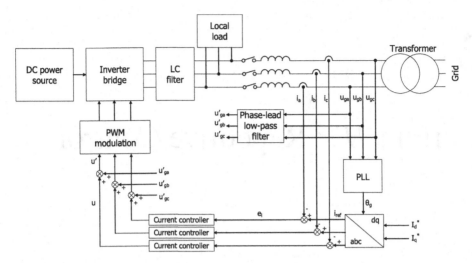

Figure 3.1 Block diagram of a current-controlled VSI with the H^∞ repetitive current controller in the natural frame

which has a gain slightly higher than 1 and a phase lead at the fundamental frequency. It is introduced to compensate the phase shift and gain attenuation caused by the computational delay, PWM modulation, the inverter bridge and the LC filter. It also attenuates the harmonics in the feed-forwarded grid voltages and improves the dynamics during grid voltage fluctuations (Timbus et al. 2009). Such a structure is capable of coping with unbalanced grid voltages and voltage sags within the range given by the nature, and waveform quality requirements. This filter could be designed analytically but here it is chosen by trial-and-error according to the principles just mentioned. Moreover, it does not affect the independence of each phase. Once the circuit breaker is closed, there should be no current exchanged with the grid until the current references are changed to be non-zero. When the inverter generates power, the repetitive current controller makes appropriate contributions on top of the feed-forwarded grid voltages. Although the controller is a current controller, the output of the inverter is still a voltage around the grid voltage. Hence, local loads can be connected. When the grid is down, the grid voltages (u_{ga}, u_{gb} and u_{gc}) fed to the filter $F(s)$ can be replaced with rated three-phase voltages and the current references I_d^* and I_q^* can be set as 0. This allows the local loads to work properly, although in an open-loop manner. This is different from the case without a feed-forwarded voltage added to the output of the current controller. It is possible to introduce a voltage loop to improve the local load voltage; see Chapter 6.

3.2 Controller Design

In this section, the current controller is designed based on the H^∞ and repetitive control techniques. The main objective of the H^∞ repetitive current controller is to inject a clean and balanced current to the grid in the presence of grid voltage distortion. The block diagram of the H^∞ repetitive current control scheme is shown in Figure 3.2, where P is the transfer function of the control plant, C is the stabilising compensator to be designed and M is the transfer

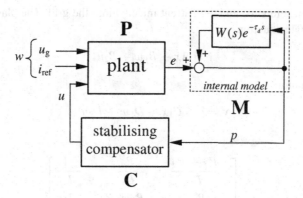

Figure 3.2 Block diagram of the H^∞ repetitive current control scheme

function of the internal model. The internal model M is implemented with the frequency adaptive mechanism discussed in Chapter 5. The external signal w contains the grid voltage u_g and the current reference i_{ref}. Both are assumed to be periodic with a fundamental frequency of 50 Hz.

3.2.1 State-space Model of the Control Plant P

The control plant consists of the inverter bridge, an LC filter (L_f and C_f) and a grid interface inductor L_g; see the single-phase diagram of the system shown in Figure 3.3. The PWM block together with the inverter bridge are modelled by using an average voltage approach with the limits of the available DC-link voltage (Weiss et al. 2004) so that the fundamental component of u_f is equal to $u' = u + u'_g$. The feed-forwarded grid voltage u'_g actually provides a base output voltage and the same voltage appears on both sides of the LCL filter. It does not affect the controller design and can be ignored during the design process. The only contribution that needs to be considered during the design process is the output u of the repetitive current controller; see Figures 3.1 and 3.2.

The currents of the two inductors and the voltage of the capacitor are chosen as state variables $x = \begin{bmatrix} i_1 & i_2 & u_c \end{bmatrix}^T$. The external input is $w = \begin{bmatrix} u_g & i_{ref} \end{bmatrix}^T$ and the control input is u. The output signal from the plant P is the tracking error $e = i_{ref} - i_2$, i.e., the difference

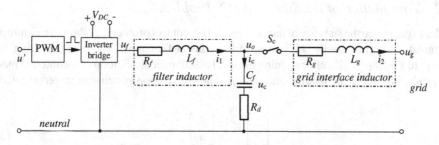

Figure 3.3 Single-phase representation of the plant P

between the current reference and the current injected into the grid. The plant P can then be described by the state equation

$$\dot{x} = Ax + B_1 w + B_2 u \qquad (3.2)$$

and the output equation

$$y = e = C_1 x + D_1 w + D_2 u \qquad (3.3)$$

with

$$A = \begin{bmatrix} -\dfrac{R_f + R_d}{L_f} & \dfrac{R_d}{L_f} & -\dfrac{1}{L_f} \\ \dfrac{R_d}{L_g} & -\dfrac{R_g + R_d}{L_g} & \dfrac{1}{L_g} \\ \dfrac{1}{C_f} & -\dfrac{1}{C_f} & 0 \end{bmatrix},$$

$$B_1 = \begin{bmatrix} \dfrac{1}{L_f} & 0 \\ -\dfrac{1}{L_g} & 0 \\ 0 & 0 \end{bmatrix}, \quad B_2 = \begin{bmatrix} \dfrac{1}{L_f} \\ 0 \\ 0 \end{bmatrix},$$

$$C_1 = \begin{bmatrix} 0 & -1 & 0 \end{bmatrix},$$
$$D_1 = \begin{bmatrix} 0 & 1 \end{bmatrix}, \quad D_2 = 0.$$

The corresponding plant transfer function is then

$$P = \begin{bmatrix} D_1 & D_2 \end{bmatrix} + C_1 (sI - A)^{-1} \begin{bmatrix} B_1 & B_2 \end{bmatrix}. \qquad (3.4)$$

In the sequel, the following notation is used for transfer functions:

$$P = \left[\begin{array}{c|cc} A & B_1 & B_2 \\ \hline C_1 & D_1 & D_2 \end{array} \right]. \qquad (3.5)$$

Detailed manipulation of state-space realisations of systems can be found in (Zhong 2006).

3.2.2 Formulation of the Standard H^∞ Problem

In order to guarantee the stability of the system, an H^∞ control problem, as shown in Figure 3.4, is formulated to minimise the H^∞ norm of the transfer function $T_{\tilde{z}\tilde{w}} = \mathcal{F}_l(\tilde{P}, C)$ from $\tilde{w} = [v \ w]^T$ to $\tilde{z} = [z_1 \ z_2]^T$, after opening the local positive feedback loop of the internal model and introducing weighting parameters ξ and μ. The closed-loop system can be represented as

$$\begin{bmatrix} \tilde{z} \\ \tilde{y} \end{bmatrix} = \tilde{P} \begin{bmatrix} \tilde{w} \\ u \end{bmatrix},$$
$$u = C\tilde{y}, \qquad (3.6)$$

Current H^∞ Repetitive Control

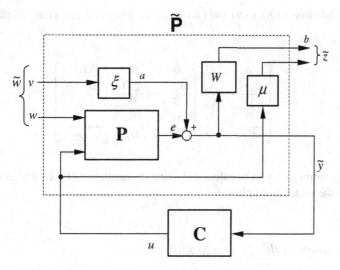

Figure 3.4 Formulation of the H^∞ control problem

where $\tilde{P}$ is the generalised plant and C is the controller to be designed. The generalised plant $\tilde{P}$ consists of the original plant P together with the low-pass filter W and weighting parameters ξ and μ. The additional parameters ξ and μ are added to provide more freedom in design and to minimise $\frac{\gamma_0}{1-\gamma}$, where $\gamma_0 = \|T_{ew}\|_\infty$, $\gamma = \|T_{ba}\|_\infty$, while keeping $\gamma < 1$.

Assume that W is realised as

$$W = \left[\begin{array}{c|c} A_w & B_w \\ \hline C_w & 0 \end{array}\right] = \left[\begin{array}{c|c} -\omega_c & \omega_c \\ \hline 1 & 0 \end{array}\right]. \tag{3.7}$$

From Figure 3.4, the following equations can be obtained:

$$\tilde{y} = e + \xi v = \xi v + \left[\begin{array}{c|cc} A & B_1 & B_2 \\ \hline C_1 & D_1 & D_2 \end{array}\right]\begin{bmatrix} w \\ u \end{bmatrix}$$

$$= \left[\begin{array}{c|ccc} A & 0 & B_1 & B_2 \\ \hline C_1 & \xi & D_1 & D_2 \end{array}\right]\begin{bmatrix} v \\ w \\ u \end{bmatrix}, \tag{3.8}$$

$$z_1 = W(e + \xi v)$$

$$= \left[\begin{array}{c|c} A_w & B_w \\ \hline C_w & 0 \end{array}\right]\left[\begin{array}{c|ccc} A & 0 & B_1 & B_2 \\ \hline C_1 & \xi & D_1 & D_2 \end{array}\right]\begin{bmatrix} v \\ w \\ u \end{bmatrix} \tag{3.9}$$

$$= \left[\begin{array}{cc|ccc} A & 0 & 0 & B_1 & B_2 \\ B_w C_1 & A_w & B_w \xi & B_w D_1 & B_w D_2 \\ \hline 0 & C_w & 0 & 0 & 0 \end{array}\right]\begin{bmatrix} v \\ w \\ u \end{bmatrix},$$

$$z_2 = \mu u. \tag{3.10}$$

Combining equations (3.8), (3.9) and (3.10), the realisation of the generalised plant is then obtained as

$$\tilde{P} = \left[\begin{array}{cc|cc|c} A & 0 & 0 & B_1 & B_2 \\ B_w C_1 & A_w & B_w \xi & B_w D_1 & B_w D_2 \\ 0 & C_w & 0 & 0 & 0 \\ 0 & 0 & 0 & 0 & \mu \\ \hline C_1 & 0 & \xi & D_1 & D_2 \end{array}\right]. \qquad (3.11)$$

The stabilising controller C can be calculated using the results on H^∞ controller design (Zhou et al. 1996) for the generalised plant $\tilde{P}$.

3.2.3 Evaluation of the System Stability

According to (Weiss and Hafele 1999; Weiss et al. 2004), the closed-loop system in Figure 3.2 is exponentially stable if the closed-loop system from Figure 3.4 is stable and its transfer function from a to b, denoted T_{ba}, satisfies $\|T_{ba}\|_\infty < 1$. Assume that the state-space realisation of the controller is

$$C = \left[\begin{array}{c|c} A_c & B_c \\ \hline C_c & D_c \end{array}\right]. \qquad (3.12)$$

Note that the controller obtained from the H^∞ design is always strictly proper. However, after controller reduction, the reduced controller may not be strictly proper. The realisation of the transfer function from a to b, assuming that $w = 0$ and noting that $D_2 = 0$, can be found as follows:

$$\begin{aligned} T_{ba} &= \left(1 - \left[\begin{array}{c|c} A & B_2 \\ \hline C_1 & D_2 \end{array}\right] C\right)^{-1} W \\ &= \left(1 - \left[\begin{array}{c|c} A & B_2 \\ \hline C_1 & 0 \end{array}\right] \left[\begin{array}{c|c} A_c & B_c \\ \hline C_c & D_c \end{array}\right]\right)^{-1} \left[\begin{array}{c|c} A_w & B_w \\ \hline C_w & 0 \end{array}\right] \\ &= \left[\begin{array}{ccc|c} A + B_2 D_c C_1 & B_2 C_c & B_2 D_c C_w & 0 \\ B_c C_1 & A_c & B_c C_w & 0 \\ 0 & 0 & A_w & B_w \\ \hline C_1 & 0 & C_w & 0 \end{array}\right]. \end{aligned}$$

Once the controller C is obtained, the stability of the system can be verified by checking $\|T_{ba}\|_\infty$. The realisation of the transfer function from w to e, assuming that $v = 0$, can be found as follows:

$$T_{ew} = (1 - P_2C)^{-1} P_1$$

$$= \left(1 - \left[\begin{array}{c|c} A & B_2 \\ \hline C_1 & 0 \end{array}\right]\left[\begin{array}{c|c} A_c & B_c \\ \hline C_c & D_c \end{array}\right]\right)^{-1} \left[\begin{array}{c|c} A & B_1 \\ \hline C_1 & D_1 \end{array}\right]$$

$$= \left[\begin{array}{cc|c} A + B_2D_cC_1 & B_2C_c & B_1 + B_2D_cD_1 \\ B_cC_1 & A_c & B_cD_1 \\ \hline C_1 & 0 & D_1 \end{array}\right].$$

3.3 Design Example

The parameters of an inverter are given in Table 3.1. The low-pass filter W is chosen as $W = \begin{bmatrix} -2550 & 2550 \\ \hline 1 & 0 \end{bmatrix}$ for $f = 50$ Hz. The weighting parameters are chosen to be $\xi = 44$ and $\mu = 0.26$. Using the MATLAB® $hinfsyn$ algorithm, the H^∞ controller C which nearly minimises the H^∞ norm of the transfer matrix from $\tilde{w}$ to $\tilde{z}$ is obtained as

$$C(s) = \frac{604785473.5899(s + 300)(s^2 + 9189s + 4.04 \times 10^8)}{(s + 3.41 \times 10^8)(s + 2550)(s^2 + 1.236 \times 10^4 s + 3.998 \times 10^8)}. \quad (3.13)$$

The factor $s + 3.41 \times 10^8$ in the denominator can be approximated by the constant 3.41×10^8 without causing any noticeable performance change (Weiss and Hafele 1999). The poles and zeros that are close to each other can be cancelled as well. The resulting reduced controller is

$$C(s) = \frac{1.7736(s + 300)}{s + 2550}. \quad (3.14)$$

The Bode plots of the original and reduced controllers in the continuous time domain are shown in Figure 3.5, which shows little difference at low frequencies. The Bode plots in the discrete time domain are almost identical at the sampling frequency of 5 kHz that was used for implementation.

Table 3.1 Parameters of an inverter

Parameter	Value	Parameter	Value
L_f	150 μH	L_g	450 μH
R_f	0.045 Ω	R_g	0.135 Ω
C_f	22 μF	R_d	1 Ω
f_s	5 kHz	f_{sw}	12 kHz

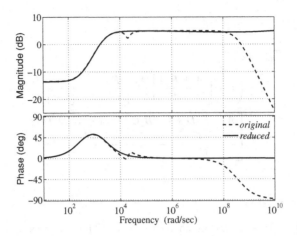

Figure 3.5 Bode plots of the original and reduced controllers

The resulting $\|T_{ba}\|_\infty$ is 0.4634 for the reduced controller and $\|T_{ba}\|_\infty = 0.2265$ for the original controller. Hence, the closed-loop system is stable. The corresponding norm of the transfer function from w to e is $\|T_{ew}\|_\infty = 1.4434$ and $\|T_{ew}\|_\infty = 1.0014$, respectively. Using the MATLAB® c2d (ZOH) algorithm, the discretised controller can be obtained as

$$C(z) = \frac{1.7736(z - 0.953)}{z - 0.6005}.$$

3.4 Experimental Results

Various experiments were carried out in the grid-connected mode under different scenarios.

3.4.1 Synchronisation Process

As explained before, the grid voltages (u_{ga}, u_{gb} and u_{gc}) are feed-forwarded through a phase-lead low-pass filter and added to the control signal for the inverter to synchronise with the grid. The grid voltage u_{ga}, the inverter output voltage u_A and the voltage e_u dropped on the circuit breaker and the grid interface inductor are shown in Figure 3.6 before and after the circuit breaker was turned on. The inverter was started at $t = 0$ second and, immediately, it was synchronised and connected to the grid.

3.4.2 Steady-state Performance

3.4.2.1 Without a Load

In the steady state, the current reference I_d^* was set at 3 A. The reactive power was set at 0 Var ($I_q^* = 0$) so that the power factor is unity. Since there was no local load, all the generated

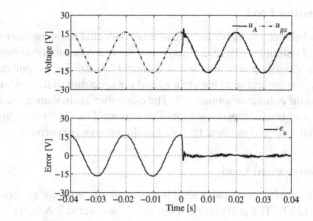

Figure 3.6 Start-up and synchronisation process: inverter output voltage u_A and grid voltage u_{ga} (upper) and the voltage e_u dropped on the circuit breaker and the grid interface inductor (lower)

active power was injected into the grid via a step-up transformer. The output current i_a, the reference current i_{ref}, the current tracking error e_i and the spectrum of the grid output current i_a are shown in the left column of Figure 3.7. The controller demonstrated very good harmonics rejection and tracking performance. It is worth mentioning that the quality of the current injected into the grid was better than that of the grid voltage.

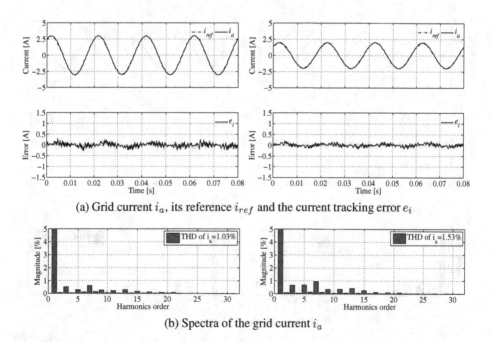

(a) Grid current i_a, its reference i_{ref} and the current tracking error e_i

(b) Spectra of the grid current i_a

Figure 3.7 Experimental results in the grid-connected mode: without a local load (left column) and with a resistive load (right column)

3.4.2.2 With a Resistive Load

In this experiment, a balanced resistive local load was connected to the inverter ($R_A = R_B = R_C = 12\,\Omega$). The grid output current reference I_d^* was set at 2 A (after connecting the inverter to the grid) and the reactive power was set at 0 Var ($I_q^* = 0$). The output current i_a, the reference current i_{ref}, the current tracking error e_i and the spectrum of the grid output current i_a are shown in the right column of Figure 3.7. The controller again demonstrated very good harmonics rejection and tracking performance. The current THD is slightly higher because the current injected into the grid is lower than that in the previous experiment.

3.4.2.3 With an Unbalanced Load

In this experiment, an unbalanced resistive local load was connected to the system ($R_A = 12\,\Omega$, $R_B = \infty$ and $R_C = 12\,\Omega$). The grid current reference I_d^* was set at 2 A (after connecting the inverter to the grid) and the reactive power was set at 0 Var ($I_q^* = 0$). The output current i_a, the reference current i_{ref}, the current tracking error e_i and the spectrum of the grid output current i_a are shown in the left column of Figure 3.8. Because of the available neutral line, the imbalance of the load did not affect the performance of the system.

3.4.2.4 With a Non-linear Load

In this experiment, a non-linear load (a three-phase uncontrolled rectifier loaded with an LC filter $L = 150\,\mu H$, $C = 1000\,\mu F$ and a resistor $R = 20\,\Omega$) was connected to the inverter. The grid

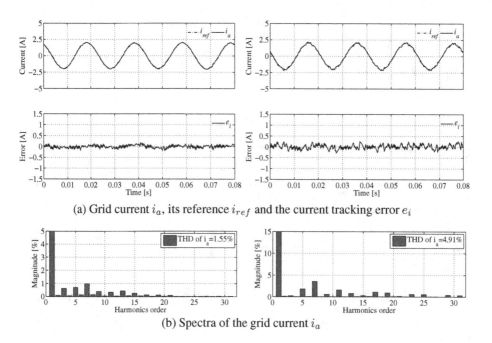

(a) Grid current i_a, its reference i_{ref} and the current tracking error e_i

(b) Spectra of the grid current i_a

Figure 3.8 Experimental results in the grid-connected mode: with unbalanced local loads (left column) and with a non-linear local load (right column)

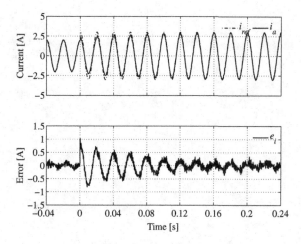

Figure 3.9 Transient response of the current H^∞ repetitive controller in the grid-connected mode without a local load

output current reference I_d^* was set at 2 A (after connecting the inverter to the grid) and the reactive power was set at 0 Var ($I_q^* = 0$). The output current i_a, the reference current i_{ref}, the current tracking error e_i and the spectrum of the grid output current i_a are shown in the right column of Figure 3.8. The tracking error e_i remained almost unchanged. Although the load is heavily non-linear, the current fed to the grid is quite clean, which means that the harmonic currents of the non-linear loads were contained inside the microgrid and was not injected into the grid.

3.4.3 Transient Response (without a Load)

In order to evaluate the transient performance of the controller, a step change in the current reference I_d^* from 2 A to 3 A was applied (while keeping $I_q^* = 0$). Since there was no local load connected to the system in this experiment, all the generated active power was injected into the grid via a step-up transformer. The output current i_a, reference current i_{ref} and the corresponding current tracking error e_i are shown in Figure 3.9. It took about 5 cycles to settle down. This reflects the inherent property of repetitive control, i.e. the compromise between the response speed and low THD.

3.5 Summary

Based on (Hornik and Zhong 2009, 2011), a current H^∞ repetitive controller is designed, implemented and evaluated with experiments in this chapter. The experimental results show that the controller offers significant improvement in terms of THD, even in the presence of non-linear local loads. Because of the feed-forwarded grid voltage to the inverter controller, the synchronisation process is quite straightforward. The main drawback is the compromise between slow dynamics and low THD.

Note that the processing delay W_d taken into account in Chapter 5 is not considered here but no visible degradation of performance is shown.

4

Voltage and Current H^∞ Repetitive Control

In the previous chapter and Chapters 15–17, several control strategies are presented to inject currents into the grid. These inverters are current-controlled VSIs and their output voltages are maintained by the grid. As a result, these strategies are not the best for stand-alone operation without a grid because the voltage is determined by the load. In this chapter through to Chapter 6, several control strategies are presented to maintain clean and stable output voltages. Such VSIs are voltage controlled and can be operated in both the stand-alone mode and the grid-connected mode. Moreover, the operation mode can be changed between stand-alone and grid-connected without changing the controller.

In this chapter, the repetitive control technique is applied to design a controller for inverters based on the H^∞ control strategy. Both the output voltage and the load current are adopted for feedback. The controller contains an infinite-dimensional internal model discussed in Chapter 2 and is able to reject all periodic disturbances having the same period as the grid voltage. This leads to a very low harmonic distortion for the output voltage, even in the presence of non-linear loads and/or grid distortions.

4.1 System Description

In this chapter, a three-phase system is considered. It consists of local loads, a grid interface inductor, an (external) power grid and a 4-wire-3-phase inverter that consists of IGBT bridges, an LC filter to attenuate the switching frequency voltage components and the controller. This forms a microgrid, of which the load contains both linear and distorting elements, lumped together into one linear load and a current sink which generates the harmonic components of the load current. Figure 4.1 shows the single-phase representation of the inverter system to be controlled. The microgrid can be supplied solely by the local generator, or solely by the grid, or by both in combination. The local generator can supply the microgrid and export power to the grid. Two isolators, S_c and S_g are provided to facilitate this and a grid interface inductor is provided to allow separation of the (sinusoidal) microgrid voltage and the (possibly distorted)

Control of Power Inverters in Renewable Energy and Smart Grid Integration, First Edition.
Qing-Chang Zhong and Tomas Hornik.
© 2013 John Wiley & Sons, Ltd. Published 2013 by John Wiley & Sons, Ltd.

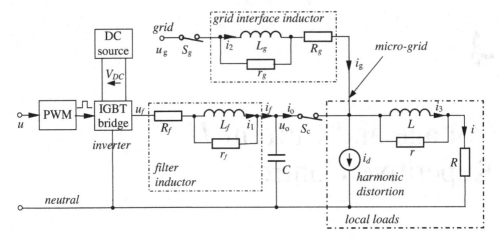

Figure 4.1 Single-phase representation of a three-phase inverter system

grid voltage and also to facilitate the control of the real and reactive power exchanged between the microgrid and the grid.

In this chapter the attention is paid to the microgrid voltage control, using the H^∞ repetitive control theory developed in (Weiss and Hafele 1999). It is assumed that there is an outer control loop to regulate the power exchanged between the microgrid and the grid by developing appropriate reference voltages in terms of magnitude and phase shift with respect to the grid. See Part III for more details. These reference voltages for the three phases of the microgrid voltage are sinusoidal. It is then the task of the voltage controller to track these reference voltages accurately so that the resulting THD is small. This controller will be subject to disturbances which include non-sinusoidal currents, changes in the load current, changes and distortions in the grid voltage, and changes in the DC-link voltage.

4.2 Modelling of an Inverter

It is assumed that a balanced DC link (see Part II) is present and the system can be regarded as three independent single-phase systems shown in Figure 4.1. The filter inductor and other inductors in the system are modelled to include two parasitic resistances: a series resistor to model the winding resistance and a parallel resistor to model the core losses.

The local loads are represented by a single linear load in parallel with a current source i_d with harmonics. The pulse-width-modulation (PWM) block is designed such that for $|u(t)| < \frac{V_{DC}}{2}$, the average of the bridge output voltage u_f over a switching period equals u. This makes it possible to model the PWM block and the inverter with an average voltage approach. The model for the PWM and inverter is thus a simple saturated unity gain, where the saturation models the limit of the available DC-link voltage with respect to the neutral line ($\pm\frac{V_{DC}}{2}$).

The control objective is to maintain the microgrid voltage u_o as close as possible to the sinusoidal reference voltage u_{ref} so that the THD of u_o is small. The two circuit breakers S_c and S_g appearing in Figure 4.1 are needed in the start-up and shut-down processes of the inverter. They are assumed to be closed for controller design.

The state variables are taken as the currents of the three inductors and the voltage of the capacitor (u_o, since S_c is closed). The external input variables (disturbances and references) are i_d, u_g and u_{ref} and the control input is u. Thus,

$$x = \begin{bmatrix} i_1 \\ i_2 \\ i_3 \\ u_o \end{bmatrix}, \quad \begin{bmatrix} w \\ u \end{bmatrix} = \begin{bmatrix} i_d \\ u_g \\ u_{ref} \\ \hline u \end{bmatrix}.$$

The state equations of the plant are

$$\dot{x} = Ax + [B_1 \quad B_2] \begin{bmatrix} w \\ u \end{bmatrix} \tag{4.1}$$

with

$$A = \begin{bmatrix} -\dfrac{R_f r_f}{(R_f+r_f)L_f} & 0 & 0 & -\dfrac{r_f}{(R_f+r_f)L_f} \\ 0 & -\dfrac{R_g r_g}{(R_g+r_g)L_g} & 0 & -\dfrac{r_g}{(R_g+r_g)L_g} \\ 0 & 0 & -\dfrac{Rr}{(R+r)L} & \dfrac{r}{(R+r)L} \\ \dfrac{r_f}{(R_f+r_f)C} & \dfrac{r_g}{(R_g+r_g)C} & -\dfrac{r}{(R+r)C} & -\left(\dfrac{1}{R+r}+\dfrac{1}{R_f+r_f}+\dfrac{1}{R_g+r_g}\right)\dfrac{1}{C} \end{bmatrix},$$

$$B = [B_1 \quad B_2] = \begin{bmatrix} 0 & 0 & 0 & \dfrac{r_f}{(R_f+r_f)L_f} \\ 0 & \dfrac{r_g}{(R_g+r_g)L_g} & 0 & 0 \\ 0 & 0 & 0 & 0 \\ -\dfrac{1}{C} & \dfrac{1}{(R_g+r_g)C} & 0 & \dfrac{1}{(R_f+r_f)C} \end{bmatrix}.$$

The output signals from the plant are the tracking error $e = u_{ref} - u_o$ and the current i_o, so that $y = [e \quad i_o]^T$. The output equations are

$$y = \begin{bmatrix} C_1 \\ C_2 \end{bmatrix} x + \begin{bmatrix} D_{11} & D_{12} \\ D_{21} & D_{22} \end{bmatrix} \begin{bmatrix} w \\ u \end{bmatrix} \tag{4.2}$$

with

$$C = \begin{bmatrix} C_1 \\ C_2 \end{bmatrix} = \begin{bmatrix} 0 & 0 & 0 & -1 \\ 0 & -\dfrac{r_g}{R_g+r_g} & \dfrac{r}{R+r} & \dfrac{1}{R+r}+\dfrac{1}{R_g+r_g} \end{bmatrix},$$

$$D = \begin{bmatrix} D_{11} & D_{12} \\ D_{21} & D_{22} \end{bmatrix} = \begin{bmatrix} 0 & 0 & 1 & 0 \\ 1 & -\dfrac{1}{R_g+r_g} & 0 & 0 \end{bmatrix}.$$

The corresponding plant transfer function is

$$\mathbf{P} = C(sI - A)^{-1}B + D = \left[\begin{array}{c|cc} A & B_1 & B_2 \\ \hline C_1 & D_{11} & D_{12} \\ C_2 & D_{21} & D_{22} \end{array}\right] = \begin{bmatrix} \mathbf{P}_{yw} & \mathbf{P}_{yu} \end{bmatrix}. \quad (4.3)$$

4.3 Controller Design

4.3.1 Formulation of the H^∞ Control Problem

The H^∞ control-based design procedure for repetitive controllers proposed in (Weiss and Hafele 1999) can be applied to design the controller. It uses additional measurement information from the plant. The block diagram of the control system is shown in Figure 4.2, where the controller consists of an internal model M and a stabilising compensator $\mathbf{C}$. The stabilising compensator assures the exponential stability of the entire system, which implies that the error e converges to a small steady-state error, according to (Weiss and Hafele 1999). The three external signals (the components of w) are assumed to be periodic, with a fundamental frequency of 50 Hz.

According to Chapter 2, the internal model M is obtained from a low-pass filter $W(s) = \dfrac{\omega_c}{s + \omega_c}$ with $\omega_c = 10000$ rad/sec, cascaded with a delay element $e^{-\tau_d s}$, where τ_d is slightly less than the fundamental period $\tau = 20$ ms given as

$$\tau_d = \tau - \frac{1}{\omega_c} = 19.9 \text{ ms}.$$

A realisation of W is $W = \left[\begin{array}{c|c} A_w & B_w \\ \hline C_w & 0 \end{array}\right] = \left[\begin{array}{c|c} -\omega_c & \omega_c \\ \hline 1 & 0 \end{array}\right]$. The choice of ω_c is based on a compromise: if ω_c is too low, only a few poles of the internal model are close to the imaginary axis, leading to poor tracking and disturbance rejection at higher frequencies; if ω_c is too high, the system is

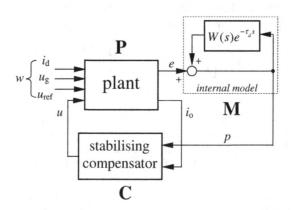

Figure 4.2 Repetitive control for voltage tracking

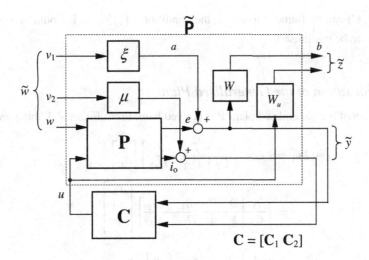

Figure 4.3 Formulation of the H^∞ control problem

difficult to stabilise (a stabilising compensator may not exist, or it may need an unreasonably high bandwidth).

In order to find a controller **C**, a standard H^∞ problem is formulated as shown in Figure 4.3, with $\tilde{w} = [v_1 \quad v_2 \quad w]^T$ and

$$\begin{bmatrix} \tilde{z} \\ \tilde{y} \end{bmatrix} = \tilde{\mathbf{P}} \begin{bmatrix} \tilde{w} \\ u \end{bmatrix}, \qquad u = \mathbf{C}\,\tilde{y}.$$

The nonzero parameter ξ is introduced to offer more freedom in the design process. The small positive parameter μ is introduced to satisfy a rank condition needed to make the H^∞ problem solvable and $W_u = \begin{bmatrix} A_u & B_u \\ C_u & D_u \end{bmatrix}$ is a weighting function whose value at infinity, $D_u = W_u(\infty) \neq 0$, is also needed to meet a rank condition. The problem formulated here is a slight improvement over the one in (Weiss and Hafele 1999), where W_u was a constant. The fact that W_u is a frequency-dependent high-pass filter has the effect of reducing the controller gains at high frequencies.

According to (Weiss and Hafele 1999), the closed-loop system in Figure 4.2 is exponentially stable if the finite-dimensional closed-loop system shown in Figure 4.3 is stable and its transfer function from a to b, denoted $\mathbf{T}_{ba}$, satisfies $\|\mathbf{T}_{ba}\|_\infty < 1$. The intuitive explanation for this is that, in the control system of Figure 4.2, a delay line is connected from the output b to the input a appearing in Figure 4.3. To make this interconnected system stable, it is sufficient to make the gain from a to b less than 1 at all frequencies according to the small gain theorem (Zhou et al. 1996). Thus, the controller **C** needs to be designed so that the above two conditions are satisfied. Moreover, in order to obtain a small steady-state error, it is necessary to minimise $\frac{\gamma_0}{1-\gamma}$, where $\gamma_0 = \|\mathbf{T}_{ew}\|_\infty$, $\gamma = \|\mathbf{T}_{ba}\|_\infty < 1$, according to (Weiss and Hafele 1999).

Using the μ-analysis and synthesis toolbox from MATLAB® (the routine *hinfsyn*), a controller **C** which nearly minimises the H^∞-norm of the transfer matrix from $\tilde{w}$ to $\tilde{z}$,

$\mathbf{T}_{\tilde{z}\tilde{w}} = \mathcal{F}_l(\tilde{\mathbf{P}}, \mathbf{C})$ can be found. Moreover, the condition $\|\mathbf{T}_{ba}\|_\infty < 1$ should be satisfied and $\dfrac{\gamma_0}{1-\gamma}$ needs to be minimised.

4.3.2 Realisation of the Generalised Plant

The realisation of the generalised plant $\tilde{\mathbf{P}}$ is derived here. From Figure 4.3, there are

$$\tilde{y}_1 = e + \xi v_1 = \xi v_1 + \left[\begin{array}{c|cc} A & B_1 & B_2 \\ \hline C_1 & D_{11} & D_{12} \end{array}\right]\begin{bmatrix} w \\ u \end{bmatrix}$$

$$= \left[\begin{array}{c|ccc c} A & 0 & 0 & B_1 & B_2 \\ \hline C_1 & \xi & 0 & D_{11} & D_{12} \end{array}\right]\begin{bmatrix} v_1 \\ v_2 \\ w \\ u \end{bmatrix},$$

$$\tilde{y}_2 = i_o + \mu v_2 = \left[\begin{array}{c|cc} A & B_1 & B_2 \\ \hline C_2 & D_{21} & D_{22} \end{array}\right]\begin{bmatrix} w \\ u \end{bmatrix} + \mu v_2$$

$$= \left[\begin{array}{c|cccc} A & 0 & 0 & B_1 & B_2 \\ \hline C_2 & 0 & \mu & D_{21} & D_{22} \end{array}\right]\begin{bmatrix} v_1 \\ v_2 \\ w \\ u \end{bmatrix},$$

$$\tilde{z}_1 = W(\xi v_1 + e) = \left[\begin{array}{c|c} A_w & B_w \\ \hline C_w & 0 \end{array}\right]\left[\begin{array}{c|cccc} A & 0 & 0 & B_1 & B_2 \\ \hline C_1 & \xi & 0 & D_{11} & D_{12} \end{array}\right]\begin{bmatrix} v_1 \\ v_2 \\ w \\ u \end{bmatrix}$$

$$= \left[\begin{array}{cc|cccc} A & 0 & 0 & 0 & B_1 & B_2 \\ B_w C_1 & A_w & B_w \xi & 0 & B_w D_{11} & B_w D_{12} \\ \hline 0 & C_w & 0 & 0 & 0 & 0 \end{array}\right]\begin{bmatrix} v_1 \\ v_2 \\ w \\ u \end{bmatrix},$$

$$\tilde{z}_2 = W_u \cdot u = \left[\begin{array}{c|cccc} A_u & 0 & 0 & 0 & B_u \\ \hline C_u & 0 & 0 & 0 & D_u \end{array}\right]\begin{bmatrix} v_1 \\ v_2 \\ w \\ u \end{bmatrix}.$$

Combining the above equations, the generalised plant $\tilde{\mathbf{P}}$ can be represented as

$$\tilde{\mathbf{P}} = \left[\begin{array}{ccc|cccc} A & 0 & 0 & 0 & 0 & B_1 & B_2 \\ B_w C_1 & A_w & 0 & B_w \xi & 0 & B_w D_{11} & B_w D_{12} \\ 0 & 0 & A_u & 0 & 0 & 0 & B_u \\ \hline 0 & C_w & 0 & 0 & 0 & 0 & 0 \\ 0 & 0 & C_u & 0 & 0 & 0 & D_u \\ \hline C_1 & 0 & 0 & \xi & 0 & D_{11} & D_{12} \\ C_2 & 0 & 0 & 0 & \mu & D_{21} & D_{22} \end{array}\right]. \quad (4.4)$$

4.3.3 State-space Realisation of T_{ew}

Denote the central sub-optimal controller for a given H^∞-norm of $T_{\tilde{z}\tilde{w}}$ by

$$\mathbf{C} = \left[\begin{array}{c|cc} A_c & B_{c1} & B_{c2} \\ \hline C_c & 0 & 0 \end{array}\right] = [\,C_1 \quad C_2\,].$$

Note that the feed-through matrix is equal to 0. Assume in Figure 4.3 that $v_1 = 0$ and $v_2 = 0$, then $u = C_c x_c$, where x_c satisfies

$$\dot{x}_c = A_c x_c + B_{c1} e + B_{c2} i_o.$$

Substitute $u = C_c x_c$ into (4.1), then

$$\dot{x} = Ax + B_2 C_c x_c + B_1 w$$

and, from (4.2),

$$e = C_1 x + D_{12} C_c x_c + D_{11} w, \quad i_o = C_2 x + D_{22} C_c x_c + D_{21} w.$$

Furthermore,

$$\dot{x}_c = (A_c + B_{c1} D_{12} C_c + B_{c2} D_{22} C_c) x_c + (B_{c1} C_1 + B_{c2} C_2) x + (B_{c1} D_{11} + B_{c2} D_{21}) w.$$

Hence, the transfer matrix from w to e is

$$\mathbf{T}_{ew} = \left[\begin{array}{cc|c} A & B_2 C_c & B_1 \\ B_{c1} C_1 + B_{c2} C_2 A_c + (B_{c1} D_{12} + B_{c2} D_{22}) C_c & B_{c1} D_{11} + B_{c2} D_{21} \\ \hline C_1 & D_{12} C_c & D_{11} \end{array}\right]. \quad (4.5)$$

4.3.4 State-space Realisation of T_{ba}

Assume $w = 0$ and $v_2 = 0$ in Figure 4.3, then

$$\dot{x} = Ax + B_2 u, \quad e = C_1 x + D_{12} u, \quad i_o = C_2 x + D_{22} u$$

and

$$u = C_c x_c.$$

Hence

$$e = C_1 x + D_{12} C_c x_c$$

and

$$\dot{x}_c = A_c x_c + B_{c1}(e + a) + B_{c2} i_o$$
$$= A_c x_c + (B_{c1} C_1 + B_{c2} C_2) x + (B_{c1} D_{12} + B_{c2} D_{22}) C_c x_c + B_{c1} a.$$

The transfer matrix from a to b can then be written as

$$\mathbf{T}_{ba} = W \begin{bmatrix} A & B_2 C_c & 0 \\ B_{c1} C_1 + B_{c2} C_2 & A_c + (B_{c1} D_{12} + B_{c2} D_{22}) C_c & B_{c1} \\ \hline C_1 & D_{12} C_c & 1 \end{bmatrix}$$

$$= \begin{bmatrix} A & B_2 C_c & 0 & 0 \\ B_{c1} C_1 + B_{c2} C_2 & A_c + (B_{c1} D_{12} + B_{c2} D_{22}) C_c & 0 & B_{c1} \\ B_w C_1 & B_w D_{12} C_c & A_w & B_w \\ \hline 0 & 0 & C_w & 0 \end{bmatrix}. \quad (4.6)$$

It is worth noting that the formulae (4.4), (4.5) and (4.6) are valid for the general case, regardless of the dimension of the measurement vector, which is the scalar i_o here, and for any W and W_u.

4.4 Design Example

The parameters of a system are shown in Table 4.1. The switching frequency for the IGBT bridge is 10 kHz. A nearly optimal controller, for which the Bode plots are shown in Figure 4.4, was obtained for $W_u = \begin{bmatrix} -10000 & 1 \\ -500 & 0.05 \end{bmatrix}$, $W = \begin{bmatrix} -10000 & 10000 \\ 1 & 0 \end{bmatrix}$, $\xi = 37$ and $\mu = 0.5$. The Bode plots show that this controller is not realistic, because it has a very large bandwidth. Normally, the high-frequency poles can be reduced using various controller/model reduction techniques, which is a topic of wide interest. Here, a different approach from that used in (Naim *et al.* 1997; Zhong *et al.* 2004, 2006) is adopted to decrease the controller bandwidth: instead of minimising the H^∞-norm of $\mathcal{F}_l(\tilde{\mathbf{P}}, \mathbf{C})$ but finding a suboptimal controller $\mathbf{C}$ such that $\|\mathcal{F}_l(\tilde{\mathbf{P}}, \mathbf{C})\|_\infty \leq 25$ (which is larger than the minimal value 17.37). Such a controller that was obtained using the *hinfsyn* routine in MATLAB® is:

$$\mathbf{C}(s) = \begin{bmatrix} \dfrac{156751.405(s^2 + 2172s + 1.185 \times 10^6)(s^2 + 5274s + 3.767 \times 10^7)}{(s+9533)(s+2331)(s+989.2)(s^2 + 2.76 \times 10^4 s + 3.07 \times 10^8)} \\ \dfrac{266474.3105(s+10^4)(s+2958)(s+989.3)(s+20.43)}{(s+9533)(s+2331)(s+989.2)(s^2 + 2.76 \times 10^4 s + 3.07 \times 10^8)} \end{bmatrix}^T . \quad (4.7)$$

Table 4.1 Parameters of the inverter for simulation

Parameter	Value	Parameter	Value
R_f	0.053 Ω	R_g	0.1 Ω
L_f	1.3 mH	L_g	0.3 mH
r_f	30.5 Ω	r_g	7 Ω
R	5 Ω	C	50 μF
L	5 mH	u_g	230 V, 50 Hz
r	500 Ω	V_{DC}	850 V

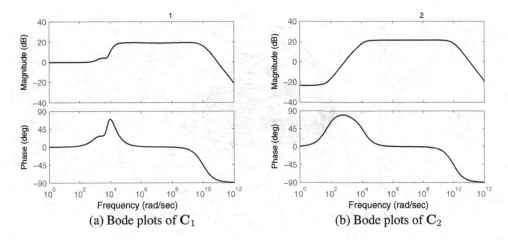

Figure 4.4 Bode plots of a nearly optimal controller **C**

It is easy to check that $\|T_{ba}\|_\infty < 1$ is satisfied for this **C**. The pole p with the highest frequency creates a peak on the controller Bode plots at $|p| = \sqrt{3.07 \times 10^8} \approx 1.752 \times 10^4$ rad/s, as can be seen in Figure 4.5. This is well below half of the switching frequency $10 \times 10^3 \times 2\pi \approx 6.28 \times 10^4$ rad/sec, which is the same as the sampling frequency to be used to implement the controller, and hence the controller is implementable. This **C** is used as the stabilising compensator in Figure 4.2 for the simulations presented in the next section.

The loop gain of the control system shown in Figure 4.2 is the transfer function **L** from u back to u. It is clear that

$$\mathbf{L} = \mathbf{C} \begin{bmatrix} M & 0 \\ 0 & 1 \end{bmatrix} \mathbf{P}_{yu},$$

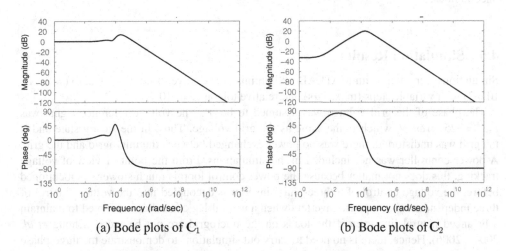

Figure 4.5 Bode plots of a more realistic suboptimal controller **C** from (4.7)

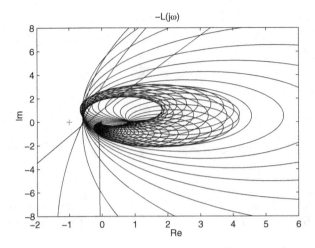

Figure 4.6 Nyquist plot of the loop gain $-\mathbf{L}(j\omega)$ from u to u in Figure 4.2, shown for $\omega \geq 0$

where $\mathbf{P}_{yu}$ is the plant transfer function from u to y given in (4.3) and M is the internal model from (2.2). Hence,

$$\mathbf{L} = \left[\begin{array}{c|c} A_c & B_{c1} \\ \hline C_c & 0 \end{array}\right] \left[\begin{array}{c|c} A & B_2 \\ \hline C_1 & D_{12} \end{array}\right] M + \left[\begin{array}{c|c} A_c & B_{c2} \\ \hline C_c & 0 \end{array}\right] \left[\begin{array}{c|c} A & B_2 \\ \hline C_2 & D_{22} \end{array}\right].$$

The Nyquist plot of $-\mathbf{L}(j\omega)$ for $\omega \geq 0$ is shown in Figure 4.6. This is a complicated curve which does not encircle the critical point -1. The phase margin is about $40°$ and the gain margin is about 4.49 dB. Hence, the system has very good robustness w.r.t. to parameter uncertainties.

4.5 Simulation Results

Simulations were done with MATLAB®/Simulink®. The solver used was ode23tb (stiff/TR-BDF2) with variable steps (max. $1\mu s$) and relative tolerance of 10^{-6}.

The phase of the grid voltage was assumed to be $0°$. The voltage reference signal was $u_{ref} = 325 \sin \omega t$ V, which is the same as the grid voltage. Thus, in the steady state and if the grid was undistorted, there was no power exchanged between the microgrid and the grid. A power controller was not included in the simulations. From the point of view of voltage tracking, this does not matter because the power control loop is much slower. As mentioned before, the voltage control of a three-phase inverter is decoupled into the voltage control of three independent single-phase inverters when a neutral-leg controller is adopted to maintain a balanced neutral line even if the loads on the microgrid are not balanced (Zhong et al. 2004a, 2006). Hence, there is no need to carry out simulations to demonstrate the three-phase behaviour.

4.5.1 Nominal Responses

Two simulations were conducted to assess the steady-state tracking performance of the system with the nominal load, with no disturbance current and an undistorted grid. The nominal load is shown in Figure 4.1, with the values of the components given in Table 4.1.

The first simulation was conducted with the PWM block and the inverter modelled as a simple saturated unity gain with saturation levels of $\pm V_{DC}/2 = \pm 425$ V, as described in Section 4.2. The output voltage and the tracking error are shown in Figure 4.7(a), which demonstrates that the tracking error was reduced to very small after approximately 5 mains cycles. The steady-state error, which is about 0.15 V(peak), is shown in Figure 4.7(c). The second simulation used a detailed inverter model including a PWM block with a switching frequency of 10 kHz. The response is shown in Figure 4.7(b). The results are similar but there are switching noises present, which increases the steady-state tracking error to approximately 7 V (peak), as shown in Figure 4.7(c). The controller was not able to suppress the switching noises, because it could only update the input to the pulse-width modulator once

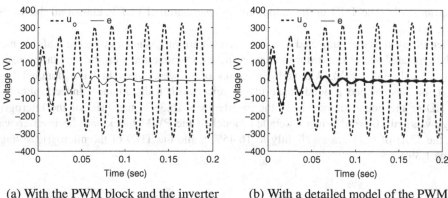

(a) With the PWM block and the inverter modelled as a simple saturation

(b) With a detailed model of the PWM block and the inverter ($f_s = 10$ kHz)

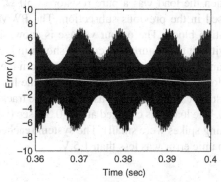

(c) The steady-state tracking error in Case (b) above. The white line shows the steady-state tracking error in Case (a) above.

Figure 4.7 Output voltage u_o and tracking error e with the nominal load when $i_d = 0$

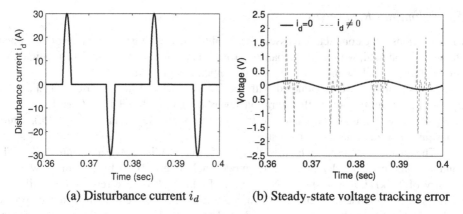

(a) Disturbance current i_d (b) Steady-state voltage tracking error

Figure 4.8 Disturbance current and steady-state voltage tracking error with the nominal load when the PWM block and the inverter were modelled as a unity gain with saturation

per carrier cycle. The THD of the output voltage u_o is around 1.37%, almost all due to the switching effect.

When the system was subjected to a disturbance current i_d shown in Figure 4.8(a), which has the typical shape of the distortion caused by a capacitive rectifier, the simulation result is shown in Figure 4.8(b). Although the peaks of i_d were about half the nominal load current, the system had a good capability to reject such a disturbance. Indeed, although the THD of the output current i_o of the inverter was 16.38%, the THD of the microgrid voltage was only 0.16%. When a detailed model of the PWM block and the inverter was used, the THD of i_o increased slightly to 16.45% while the THD of the microgrid voltage became 1.39%.

4.5.2 Response to Load Changes

Simulations were done when the load was a pure resistor of 50 Ω, which consumed about 10% of the load power used in the previous subsection. The PWM block and the inverter were modelled as a saturation block. The output voltage is shown in Figure 4.9(a) and the tracking errors, with and without the disturbance current shown in Figure 4.8(a), are shown in Figure 4.9(b). No performance degradation can be observed from these figures.

Another simulation explored the transient responses when the load was changed from the nominal load to a pure resistor of 50 Ω when i_d was set at 0. The tracking error and the current i_o are shown in Figure 4.10. The load was changed at $t = 0.301$ sec (when the load current was close to 0 so that the resulting spikes were small). The system reached the steady state within 4 mains cycles and the dynamic error was less than 1.5 V.

4.5.3 Response to Grid Distortions

A typical grid voltage is flattened at its peaks. Such a grid voltage of $u_g = 325 \sin \omega t - 32.5 \sin 3\omega t - 32.5 \sin 5\omega t$, as shown in Figure 4.11(a), was used as an example. The tracking

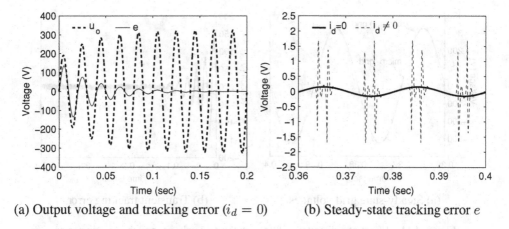

(a) Output voltage and tracking error ($i_d = 0$)   (b) Steady-state tracking error e

Figure 4.9 Output voltage and tracking error with a purely resistive load of 50 Ω

error decayed rapidly, as shown in Figure 4.11(b), and in the steady state it was very small (with ripples of about 7 V due to the effect of the PWM switching). Although the external grid was extremely distorted, the microgrid was very clean with a THD of about 1.20%, mostly due to the switching noise. In this simulation, the disturbance i_d was set to 0 and the load was the nominal load.

Another simulation was done when there was a shallow sag in the grid. The sag was −10% from 0.4 sec to 0.6 sec, as shown in Figure 4.12(a). The maximum dynamic error, as shown in Figure 4.12(b), was less than 6 V (peak) and the microgrid reached the steady state within 5 mains cycles. In this simulation, $i_d = 0$ and the PWM block and the inverter were modelled as a saturation.

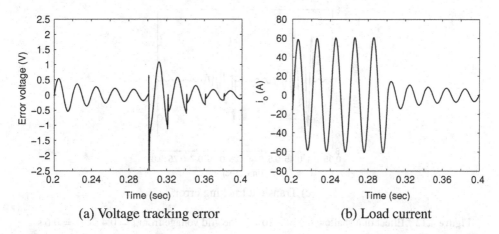

(a) Voltage tracking error   (b) Load current

Figure 4.10 Transient response when the load was changed at $t = 0.301$ s from the nominal load (as in Figure 4.1) to a resistor of 50 Ω

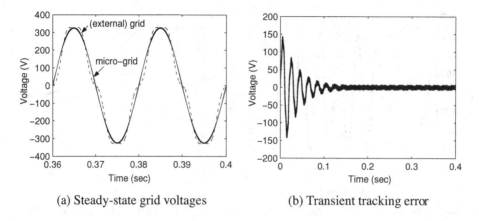

(a) Steady-state grid voltages

(b) Transient tracking error

Figure 4.11 Effect of a distorted public grid: the grid voltages and the tracking error

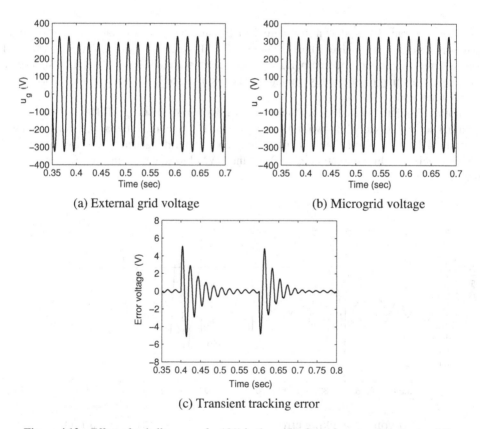

(a) External grid voltage

(b) Microgrid voltage

(c) Transient tracking error

Figure 4.12 Effect of a shallow sag of -10% in the grid voltage from $t = 0.4$ s to $t = 0.6$ s

4.6 Summary

Based on (Weiss *et al.* 2004; Zhong *et al.* 2002b), a voltage controller is designed for an inverter connected to a microgrid with local loads and a public grid interface. The controller adopts repetitive control on a per-phase basis in order to reject harmonic disturbances from non-linear loads or the public grid. An H^∞ design method is used to ensure that the controller performs effectively with a range of local load impedances. The system has been modelled and tested in MATLAB®/Simulink®. The inverter model includes a realistic switching frequency filter and a full model of the inverter PWM process. The results have shown that, apart from the switching noise, the tracking of voltage references is very accurate. When the load changes, the repetitive control loop converges and the tracking error remains very small after approximately 5 mains cycles.

5

Voltage H^∞ Repetitive Control with a Frequency-adaptive Mechanism

In the previous chapter, an H^∞ repetitive controller is designed with the feedback of the output voltage and the output current. In this chapter, this is further developed after simplifying the model of the inverter and taking the voltage only as the feedback for the repetitive controller. The feedback of the inductor current is adopted to form an inner-loop controller. The control plant for the repetitive controller is then reduced to a single-input-single-output (SISO) one and the complexity of the control design is considerably reduced. The stability evaluation of the system becomes easier as well. Moreover, a frequency-adaptive mechanism is embedded into the internal model and, hence, the controller is able to cope with grid frequency variations in the grid-connected mode. This mechanism allows the controller to maintain very good tracking performance over a wide range of grid frequencies. Extensive experimental results are included to demonstrate the control strategy.

5.1 System Description

The general idea of the control strategy is to adopt an individual controller for each phase in the natural frame, under the assumption that the system is implemented with a neutral point controller (Zhong *et al.* 2006). The overall control structure of the system is shown in Figure 5.1. It consists of three cascaded controllers: a power controller to generate the voltage reference u_{ref} while regulating the power exchanged with the grid, a voltage controller to track u_{ref}, and a current controller to generate the PWM signals to drive the power switches while tracking the current reference generated by the voltage controller. The current controller is chosen as a simple proportional controller K_c with the inductor current as feedback. It can be equipped with current protection as well. It does not affect the tracking accuracy of the outer voltage loop (Li *et al.* 2006; Vilathgamuwa *et al.* 2006). Actually, according to Chapter 7, this

---

Control of Power Inverters in Renewable Energy and Smart Grid Integration, First Edition.
Qing-Chang Zhong and Tomas Hornik.
© 2013 John Wiley & Sons, Ltd. Published 2013 by John Wiley & Sons, Ltd.

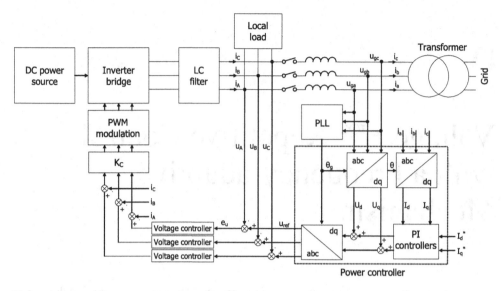

Figure 5.1 Block diagram of a voltage-controlled VSI with a voltage H^∞ repetitive controller in the natural frame

is equivalent to increasing the inductor resistance. Hence, it does not affect the design of the voltage controller and will be omitted when designing the voltage controller.

A PLL is adopted to provide the phase information of the grid voltage so that the power controller can generate the voltage reference u_{ref} while regulating the active power P and the reactive power Q according to the grid current references I_d^* and I_q^*. The inverter is assumed to be powered by a constant DC power source and, hence, no controller is needed to regulate the DC-link voltage. Otherwise, a controller can be introduced to regulate the DC-link voltage and to generate I_d^*. The main objective in this chapter is to design the voltage controller so the power controllers for P and Q are simply chosen as PI controllers. More details about power flow control can be found in Part III.

5.2 Controller Design

In this section, a voltage controller based on the H^∞ and repetitive control techniques is designed. Its main objective is to maintain a clean and balanced local load voltage in the presence of non-linear loads and/or grid distortion. The block diagram of the H^∞ repetitive control scheme is shown in Figure 5.2, where P is the transfer function of the plant, C is the transfer function of the stabilising compensator and M is the transfer function of the internal model. The stabilising compensator C and internal model M are the two components of the controller. The stabilising compensator C, designed by solving a weighted sensitivity H^∞ problem (Weiss et al. 1998), assures the exponential stability of the entire system, which implies that the tracking error $e = e_u$ between the voltage reference and the inverter output voltage converges to a small steady-state error (Weiss and Hafele 1999). The external signal

Voltage H^∞ Repetitive Control with a Frequency-adaptive Mechanism

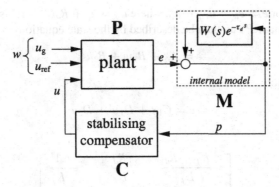

Figure 5.2 Block diagram of the voltage H^∞ repetitive control scheme

w contains both the grid voltage u_g and the voltage reference u_{ref}, which are assumed to be periodic with a fundamental frequency of 50 Hz.

5.2.1 State-space Model of the Control Plant P

The single-phase diagram of the system is shown in Figure 5.3. It consists of the inverter bridge, an LC filter (L_f and C_f) and a grid interface inductor L_g. The LC filter and the grid interface inductor form an LCL filter. The series winding resistance of the inductors is considered but the parallel resistance considered in Chapter 4 is not taken into account. The circuit breaker S_C is provided to facilitate the synchronisation and shut-down procedure and is considered to be closed (i.e. the synchronisation process is omitted). The PWM block together with the inverter are modelled by using an average voltage approach with the limits of the available DC-link voltage (Weiss et al. 2004) so that the fundamental component of u_f is equal to u. As a result, the PWM block and the inverter bridge can be ignored when designing the controller.

The currents of the two inductors and the voltage of the capacitor are chosen as state variables $x = [\,i_1 \quad i_2 \quad u_c\,]^T$. The external input $w = [\,u_g \quad u_{ref}\,]^T$ consists of the grid voltage u_g and the reference voltage u_{ref}, and the control input is u. The output signal from the plant

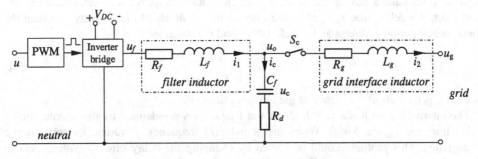

Figure 5.3 Single phase representation of the plant P

P is the tracking error $e = e_u = u_{ref} - u_o$, where $u_o = u_c + R_d(i_1 - i_2)$ is the output voltage of the inverter. The plant P can then be described by the state equation

$$\dot{x} = Ax + B_1 w + B_2 u \tag{5.1}$$

and the output equation

$$y = e = C_1 x + D_1 w + D_2 u \tag{5.2}$$

with

$$A = \begin{bmatrix} -\dfrac{R_f + R_d}{L_f} & \dfrac{R_d}{L_f} & -\dfrac{1}{L_f} \\ \dfrac{R_d}{L_g} & -\dfrac{R_g + R_d}{L_g} & \dfrac{1}{L_g} \\ \dfrac{1}{C_f} & -\dfrac{1}{C_f} & 0 \end{bmatrix},$$

$$B_1 = \begin{bmatrix} 0 & 0 \\ -\dfrac{1}{L_g} & 0 \\ 0 & 0 \end{bmatrix}, \quad B_2 = \begin{bmatrix} \dfrac{1}{L_f} \\ 0 \\ 0 \end{bmatrix},$$

$$C_1 = \begin{bmatrix} -R_d & R_d & -1 \end{bmatrix},$$

$$D_1 = \begin{bmatrix} 0 & 1 \end{bmatrix}, \quad D_2 = 0.$$

The corresponding plant transfer function is then

$$P = \begin{bmatrix} D_1 & D_2 \end{bmatrix} + C_1(sI - A)^{-1} \begin{bmatrix} B_1 & B_2 \end{bmatrix} = \left[\begin{array}{c|cc} A & B_1 & B_2 \\ \hline C_1 & D_1 & D_2 \end{array} \right]. \tag{5.3}$$

5.2.2 Frequency-adaptive Internal Model M

The internal model M, shown in Figure 5.2, is infinite dimensional and consists of a low-pass filter $W(s) = \frac{\omega_c}{s + \omega_c}$ cascaded with a delay line $e^{-\tau_d s}$. It is capable of generating periodic signals of a given fundamental period τ_d so it is capable of tracking periodic references and rejecting periodic disturbances having the same period. In order to improve the performance of the controller, the delay time τ_d used in the internal model M should be slightly less than the fundamental period τ (Weiss and Hafele 1999), and is chosen as

$$\tau_d = \tau - \dfrac{1}{\omega_c}, \tag{5.4}$$

where ω_c is the cut-off frequency of the low-pass filter W.

The internal model has a very high gain at frequencies pre-defined by the internal model delay line; see Figure 5.4(a). When the actual grid frequency f varies, its performance is degraded. This problem could be solved by changing the delay time τ_d with respect to the grid frequency. However, following the discrete-time implementation and low sampling frequency used (e.g. 5 kHz), it is impossible to implement the adaptive delay time without

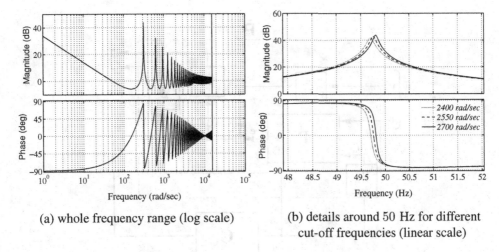

(a) whole frequency range (log scale)　　(b) details around 50 Hz for different cut-off frequencies (linear scale)

Figure 5.4 Bode plots of the discretised internal model

further degrading the controller performance. Alternatively, in order to maintain good tracking performance of the controller, the cut-off frequency of the low-pass filter ω_c can be changed according to the grid frequency variations. This adaptive mechanism is based on the formula

$$\omega_c = \frac{1}{\tau_d(1 - \tau_d f)}, \qquad (5.5)$$

which is derived from (2.9). This is to make the poles of the internal model close to the multiples of the fundamental frequency on the $j\omega$-axis. After several rounds of trial-and-error, τ_d has been chosen as 0.0196 s (98 out of 100 samples at 5 kHz). The frequency provided by the PLL can be adopted to change ω_c of the low-pass filter in the internal model. When the variations of the frequency are wide, τ_d might need to be changed. Figure 5.4(a) shows the Bode plots of the discretised internal model M for different ω_c, with details around 50 Hz shown in Figure 5.4(b).

5.2.3 Formulation of the Standard H^∞ Problem

In order to guarantee the stability of the system, the H^∞ control problem, as shown in Figure 5.5, is formulated to minimise the H^∞ norm of the transfer function $T_{z\tilde{w}} = \mathcal{F}_l(\tilde{P}, C)$ from $\tilde{w} = [\,v \ \ w\,]^T$ to $\tilde{z} = [\,z_1 \ \ z_2\,]^T$, after opening the local positive feedback loop of the internal model and introducing weighting parameters ξ and μ. The closed-loop system can be represented as

$$\begin{bmatrix} \tilde{z} \\ \tilde{y} \end{bmatrix} = \tilde{P} \begin{bmatrix} \tilde{w} \\ u \end{bmatrix}, \qquad (5.6)$$

$$u = C\tilde{y},$$

where $\tilde{P}$ is the generalised plant and C is the controller to be designed. The generalised plant $\tilde{P}$ consists of the original plant P together with the low-pass filter W, the processing delay

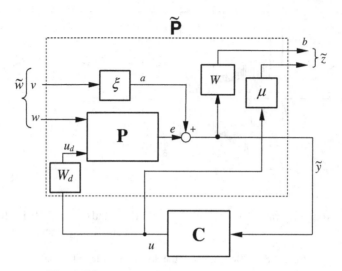

Figure 5.5 Formulation of the H^∞ control problem

represented by W_d and weighting parameters ξ and μ. The additional parameters ξ and μ are added to provide additional freedom in the design.

Assume that W is realised as

$$W = \left[\begin{array}{c|c} A_w & B_w \\ \hline C_w & 0 \end{array}\right] = \left[\begin{array}{c|c} -\omega_c & \omega_c \\ \hline 1 & 0 \end{array}\right]$$

and W_d is realised (Wang 2008) as

$$W_d = \left[\begin{array}{c|c} A_d & B_d \\ \hline C_d & D_d \end{array}\right] = \left[\begin{array}{c|c} -\frac{2}{T_s} & \frac{4}{T_s} \\ \hline 1 & -1 \end{array}\right],$$

where T_s is the sampling period. From Figure 5.5, the following equations can be deduced:

$$\tilde{y} = e + \xi v = \xi v + \left[\begin{array}{c|cc} A & B_1 & B_2 \\ \hline C & D_1 & D_2 \end{array}\right]\left[\begin{array}{c} w \\ u_d \end{array}\right]$$

$$= \left[\begin{array}{ccc|ccc} A & B_2 C_d & 0 & B_1 & B_2 D_d \\ 0 & A_d & 0 & 0 & B_d \\ \hline C & D_2 C_d & \xi & D_1 & D_d D_2 \end{array}\right]\left[\begin{array}{c} v \\ w \\ u \end{array}\right]. \qquad (5.7)$$

$$\tilde{z}_1 = W\tilde{y} = \left[\begin{array}{ccc|ccc} A & B_2 C_d & 0 & 0 & B_1 & B_2 D_d \\ 0 & A_d & 0 & 0 & 0 & B_d \\ B_w C_1 & B_w D_2 C_d & A_w & B_w \xi & B_w D_1 & B_w D_2 D_d \\ D_w C_1 & D_w D_2 C_d & C_w & D_w \xi & D_w D_1 & D_w D_2 D_d \end{array}\right]\left[\begin{array}{c} v \\ w \\ u \end{array}\right], \qquad (5.8)$$

$$\tilde{z}_2 = \mu u. \qquad (5.9)$$

Combining equations from (5.7) to (5.9), the generalised plant is then realised as

$$\tilde{P} = \left[\begin{array}{ccc|ccc|c} A & B_2C_d & 0 & 0 & B_1 & B_2D_d \\ 0 & A_d & 0 & 0 & 0 & B_d \\ B_wC_1 & B_wD_2C_d & A_w & B_w\xi & B_wD_1 & B_wD_2D_d \\ D_wC_1 & D_wD_2C_d & C_w & D_w\xi & D_wD_1 & D_wD_2D_d \\ 0 & 0 & 0 & 0 & 0 & \mu \\ \hline C_1 & D_2C_d & 0 & \xi & D_1 & D_2D_d \end{array}\right], \quad (5.10)$$

for which the stabilising controller C can be calculated using the well-known results on H^∞ controller design (Zhou et al. 1996).

5.2.4 Evaluation of System Stability

According to (Weiss and Hafele 1999; Weiss et al. 2004), the closed-loop system in Figure 5.2 is exponentially stable if the closed-loop system from Figure 5.5 is stable and its transfer function from a to b, denoted T_{ba}, satisfies $\|T_{ba}\|_\infty < 1$.

Assume that the state-space realisation of the controller is

$$C = \left[\begin{array}{c|c} A_c & B_c \\ \hline C_c & 0 \end{array}\right].$$

Note that the optimal controller obtained from the H_∞ design is always strictly proper. The realisation of the transfer function from a to b, assuming that $w = 0$, can be found as follows:

$$T_{ba} = \left(1 - \left[\begin{array}{c|c} A & B_2 \\ \hline C_1 & D_2 \end{array}\right]W_dC\right)^{-1}W$$

$$= \left(1 - \left[\begin{array}{c|c} A & B_2 \\ \hline C_1 & D_2 \end{array}\right]\left[\begin{array}{c|c} A_d & B_d \\ \hline C_d & D_d \end{array}\right]\left[\begin{array}{c|c} A_c & B_c \\ \hline C_c & 0 \end{array}\right]\right)^{-1}\left[\begin{array}{c|c} A_w & B_w \\ \hline C_w & 0 \end{array}\right]$$

$$= \left[\begin{array}{ccc|c} A & B_2C_d & B_2D_dC_c & 0 \\ 0 & A_d & B_dC_c & 0 \\ 0 & 0 & A_c & -B_c \\ \hline C_1 & D_2C_d & D_2D_dC_c & 1 \end{array}\right]^{-1}\left[\begin{array}{c|c} A_w & B_w \\ \hline C_w & 0 \end{array}\right]$$

$$= \left[\begin{array}{ccc|c} A & B_2C_d & B_2D_dC_c & 0 \\ 0 & A_d & B_dC_c & 0 \\ B_cC_1 & B_cD_2C_d & A_c+B_cD_2D_dC_c & B_c \\ \hline C_1 & D_2C_d & D_2D_dC_c & 1 \end{array}\right]\left[\begin{array}{c|c} A_w & B_w \\ \hline C_w & 0 \end{array}\right]$$

$$= \left[\begin{array}{cccc|c} A & B_2C_d & B_2D_dC_c & 0 & 0 \\ 0 & A_d & B_dC_c & 0 & 0 \\ B_cC_1 & B_cD_2C_d & A_c+B_cD_2D_dC_c & B_cC_w & 0 \\ 0 & 0 & 0 & A_w & B_w \\ \hline C_1 & D_2C_d & D_2D_dC_c & C_w & 0 \end{array}\right].$$

Since $D_2 = 0$,

$$T_{ba} = \begin{bmatrix} A & B_2C_d & B_2D_dC_c & 0 & 0 \\ 0 & A_d & B_dC_c & 0 & 0 \\ B_cC_1 & 0 & A_c & B_cC_w & 0 \\ 0 & 0 & 0 & A_w & B_w \\ \hline C_1 & 0 & 0 & C_w & 0 \end{bmatrix}.$$

Once the controller C is designed, the stability of the system can be verified by checking $\|T_{ba}\|_\infty$. Moreover, according to (Weiss and Hafele 1999; Weiss *et al.* 2004), a small value for $\frac{\gamma_0}{1-\gamma}$, where $\gamma_0 = \|T_{ew}\|_\infty$ and $\gamma = \|T_{ba}\|_\infty$, results in a small steady-state error. Hence, the weighting parameters ξ and μ can be chosen to minimise $\frac{\gamma_0}{1-\gamma}$ while keeping $\gamma < 1$. The realisation of the transfer function from w to $e = e_u$, assuming that $v = 0$, can be found as:

$$T_{ew} = \left(1 - \left[\begin{array}{c|c} A & B_2 \\ \hline C_1 & 0 \end{array}\right]\left[\begin{array}{c|c} A_d & B_d \\ \hline C_d & D_d \end{array}\right]\left[\begin{array}{c|c} A_c & B_c \\ \hline C_c & 0 \end{array}\right]\right)^{-1}\left[\begin{array}{c|c} A & B_1 \\ \hline C_1 & D_1 \end{array}\right]$$

$$= \begin{bmatrix} A & B_2C_d & B_2D_dC_c & 0 \\ 0 & A_d & B_dC_c & 0 \\ B_cC_1 & 0 & A_c & B_c \\ \hline C_1 & 0 & 0 & 1 \end{bmatrix}\left[\begin{array}{c|c} A & B_1 \\ \hline C_1 & D_1 \end{array}\right]$$

$$= \begin{bmatrix} A & B_2C_d & B_2D_dC_c & 0 & 0 \\ 0 & A_d & B_dC_c & 0 & 0 \\ B_cC_1 & 0 & A_c & B_cC_1 & B_cD_1 \\ 0 & 0 & 0 & A & B_1 \\ \hline C_1 & 0 & 0 & C_1 & D_1 \end{bmatrix}$$

$$= \begin{bmatrix} A & B_2C_d & B_2D_dC_c & 0 & B_1 \\ 0 & A_d & B_dC_c & 0 & 0 \\ B_cC_1 & 0 & A_c & 0 & B_cD_1 \\ 0 & 0 & 0 & A & B_1 \\ \hline C_1 & 0 & 0 & 0 & D_1 \end{bmatrix}$$

$$= \begin{bmatrix} A & B_2C_d & B_2D_dC_c & B_1 \\ 0 & A_d & B_dC_c & 0 \\ B_cC_1 & 0 & A_c & B_cD_1 \\ \hline C_1 & 0 & 0 & D_1 \end{bmatrix}.$$

5.3 Design Example

The parameters of the inverter are given in Table 3.1. The nominal low-pass filter W is chosen according to (5.5) as

$$W = \left[\begin{array}{c|c} -2550 & 2550 \\ \hline 1 & 0 \end{array}\right].$$

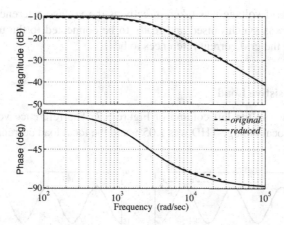

Figure 5.6 Bode plots of the original and reduced controllers

for $f = 50$ Hz and the processing delay is realised as

$$W_d = \begin{bmatrix} -10000 & 20000 \\ 1 & -1 \end{bmatrix}$$

for $f_s = 5$ kHz. The weighting parameters are chosen to be $\xi = 24$ and $\mu = 1.6$ after some trial-and-error. Using the MATLAB® *hinfsyn* algorithm, the H^∞ controller C which nearly minimises the H^∞ norm of the transfer matrix from $\tilde{w}$ to $\tilde{z}$ is obtained as

$$C(s) = \frac{864.6214(s + 10^4)(s^2 + 9189s + 4.04 \times 10^8)}{(s + 1.118 \times 10^4)(s + 2550)(s^2 + 9047s + 4.198 \times 10^8)}.$$

The resulting $\gamma = \|T_{ba}\|_\infty$ is 0.8198 and $\gamma_0 = \|T_{ew}\|_\infty$ is 1.2083. The controller can be reduced as

$$C(s) = \frac{864.6214(s + 10^4)}{(s + 1.118 \times 10^4)(s + 2550)}$$

without causing noticeable performance degradation, after cancelling the poles and zeros that are close to each other. The Bode plots of the original and reduced controllers in the continuous time domain are shown in Figure 5.6 for comparison. This leads to $\|T_{ba}\|_\infty = 0.815$ and $\|T_{ew}\|_\infty = 1.2030$, which still maintains the system stability.

5.4 Experimental Results

5.4.1 Steady-state Performance in the Stand-alone Mode

Experiments were carried out in both stand-alone and grid-connected modes. In the stand-alone mode, the experiments were carried out for a balanced resistive load $R_A = R_B = R_C = 12\ \Omega$, a non-linear three-phase uncontrolled rectifier loaded with an LC filter $L = 150\ \mu H$, $C = 1000\ \mu F$ and a resistor $R = 20\ \Omega$, and an unbalanced linear load with $R_A = R_C = 12\ \Omega$

and $R_B = \infty$. The grid voltage was taken directly as the voltage reference without filtering so that the inverter was synchronised and ready to be connected to the utility grid. This is equivalent to setting the grid current references to be 0.

5.4.1.1 With a Resistive Load

The results are shown in the left column of Figure 5.7. The reference voltage was tracked well. The recorded local voltage THD was 1.05% and the local load current THD was 1.06%.

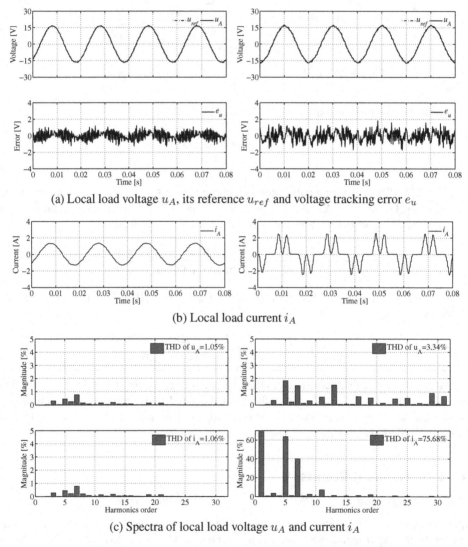

(a) Local load voltage u_A, its reference u_{ref} and voltage tracking error e_u

(b) Local load current i_A

(c) Spectra of local load voltage u_A and current i_A

Figure 5.7 Stand-alone mode: with a resistive load (left column) and with a non-linear load (right column)

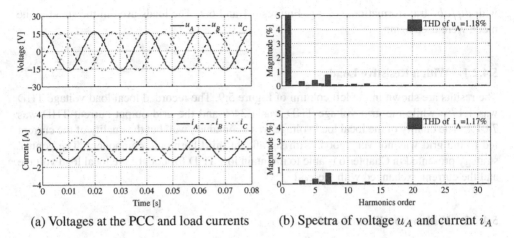

Figure 5.8 Stand-alone mode: with an unbalanced load

Although the utility grid voltage was used as the reference without any filtering, the quality of the local voltage is actually better than that of the grid voltage, with the THD of 1.29%.

5.4.1.2 With a Non-linear Load

The results are shown in the right column of Figure 5.7. Again, the reference voltage was tracked very well without noticeable fundamental components in the tracking error. The recorded local load voltage THD was 3.34%, while the grid voltage THD was 1.33% and the local load current THD was 75.68%. The voltage controller performed very well with this highly non-linear load.

5.4.1.3 With an Unbalanced Load

The results are shown in Figure 5.8. The Phase-B current was zero as there was no load connected. The recorded local load voltage THD was 1.18% while the grid voltage THD was 1.25% and the local load current THD was 1.17%. Since the control structure adopted a separate controller for each phase, the load imbalance had no influence on the controller performance and the local load voltages remained well balanced.

5.4.2 Steady-state Performance in the Grid-connected Mode

In addition to the three experiments carried out in the stand-alone mode, the case without a local load was tested in the grid-connected mode as well. Moreover, two other experiments were carried out to demonstrate the transient responses to a step change in the grid output current I_d^* reference and to the variations of the grid frequency.

In the grid-connected mode, the current reference of the grid output current I_d^* was set at 2 A (1.41 A RMS), after connecting the inverter to the grid. The reactive power was set at 0 Var ($I_q^* = 0$). The instantaneous grid current reference i_{ref} could be obtained from I_d^* and I_q^*

with dq/abc transformation; see e.g. Figures 3.1 and 6.7. The loads used in the stand-alone mode were used again.

5.4.2.1 With a Resistive Load

The results are shown in the left column of Figure 5.9. The recorded local load voltage THD was 1.17% while the grid voltage THD was 1.34% and the grid output current THD was 7.01%. The quality of the local load voltage was maintained well but the quality of the current fed to the grid was not very good, with significant third harmonic component. The control strategy presented in Chapter 6 is able to maintain low THD for both of the local load voltage and the current exchanged with the grid.

5.4.2.2 With a Non-linear Load

The results are shown in the right column of Figure 5.9. The recorded local load voltage THD was 2.05% while the grid voltage THD was 1.38% and the grid output current THD was 10.58%. Again, the local load voltage was controlled very well but the current exchanged with the grid was not. There were significant third, fifth and seventh harmonic components.

5.4.2.3 With an Unbalanced Load

The results are shown in Figure 5.10. The recorded local load voltage THD was 1.32% while the grid voltage THD was 1.30% and the grid output current THD was 6.92%. The load imbalance had no influence on the control performance and the voltages remained well balanced. However, the grid currents were not balanced for the reason explained above.

5.4.2.4 Without a Local Load

The results are shown in Figure 5.11. The recorded local voltage THD was 1.16% while the grid voltage THD was 1.32% and the grid output current THD was 6.67%. Again, the quality of the local voltage was better than that of the grid voltage but the quality of the grid current was not so good.

5.4.3 Transient Response: without a Local Load

In this experiment, a step change in the grid current reference I_d^* from 2 A (1.41 A RMS) to 3 A (2.12 A RMS) was applied while keeping $I_q^* = 0$. The results are shown in Figure 5.12. It took about 5 cycles to settle down.

5.4.4 Response to Variations of the Grid Frequency

All the above experiments were carried out with the frequency-adaptive mechanism enabled. In order to see the improvement of the frequency-adaptive mechanism, several other experiments were carried out at different grid frequencies.

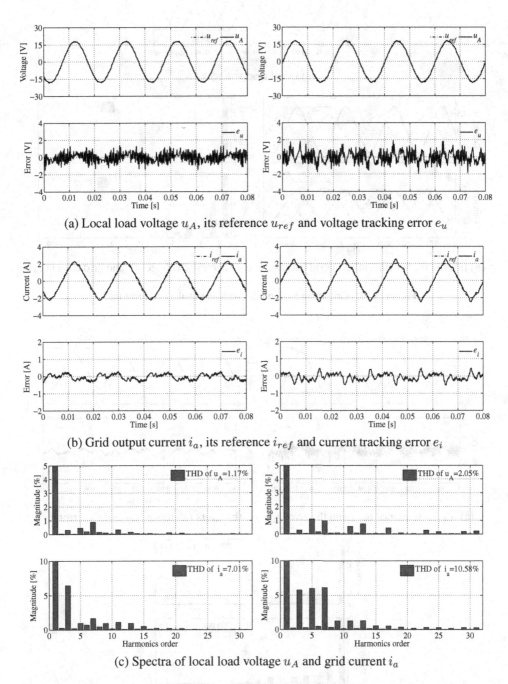

(a) Local load voltage u_A, its reference u_{ref} and voltage tracking error e_u

(b) Grid output current i_a, its reference i_{ref} and current tracking error e_i

(c) Spectra of local load voltage u_A and grid current i_a

Figure 5.9 Grid-connected mode: with a resistive load (left column) and with a non-linear load (right column)

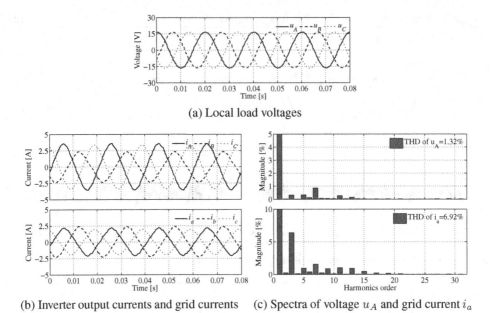

Figure 5.10 Grid-connected mode: with an unbalanced local load

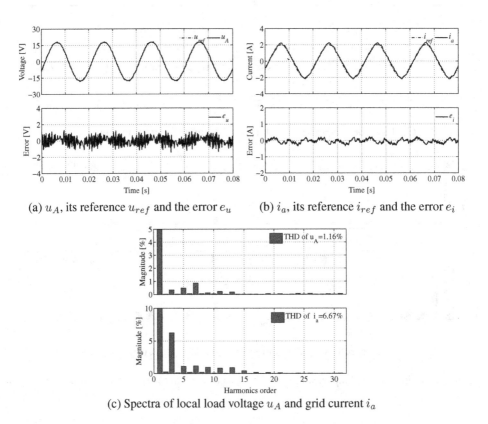

Figure 5.11 Grid-connected mode: without a local load

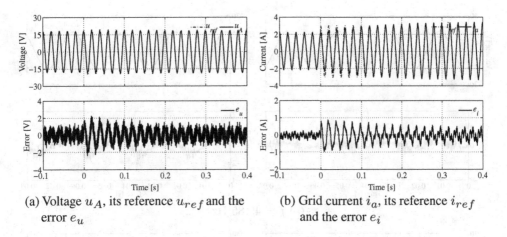

(a) Voltage u_A, its reference u_{ref} and the error e_u

(b) Grid current i_a, its reference i_{ref} and the error e_i

Figure 5.12 Transient response to a 1 A step change of I_d^* in the grid-connected mode without a local load

When the grid frequency was at 50.00 Hz, the results when the frequency-adaptive mechanism was enabled and was not enabled are shown in Figure 5.13. Because the frequency was the same as the nominal value, the results are comparable.

When the grid frequency was not 50.00 Hz, the results when the frequency-adaptive mechanism was enabled and was not enabled are shown in Figure 5.14 for the grid frequency of 49.85 Hz, 49.90 Hz and 49.95 Hz and in Figure 5.15 for the grid frequency of 50.05 Hz, 50.10 Hz and 50.15 Hz, respectively. The improvement in the mechanism can be clearly seen from the figures. A phase shift between the voltage output u_A and the reference voltage u_{ref} was noticeable, which increased the steady-state tracking error. Since the controller was tuned for the grid frequency at 50.00 Hz, the generated voltage u_A was leading the reference voltage when the grid frequency was below 50.00 Hz, as shown in Figure 5.14, and the generated

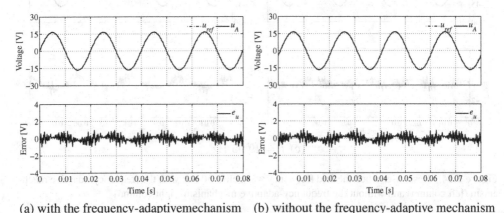

(a) with the frequency-adaptive mechanism   (b) without the frequency-adaptive mechanism

Figure 5.13 Responses when the grid frequency was at $f = 50.00$ Hz

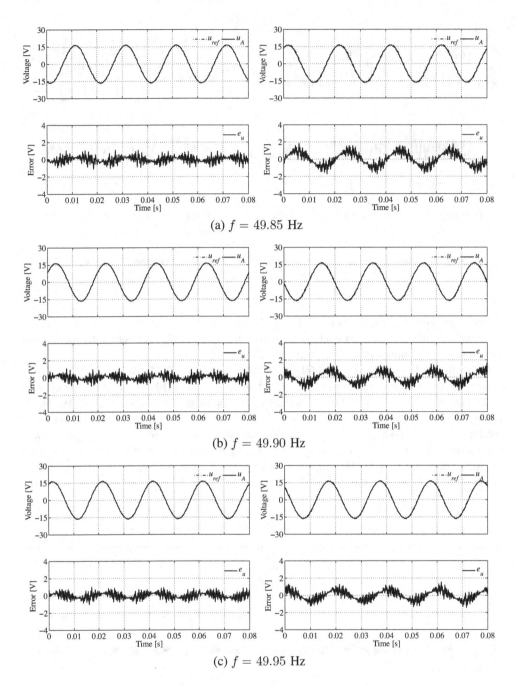

Figure 5.14 Responses when the grid frequency was below 50 Hz: with the frequency-adaptive mechanism (left column) and without the frequency-adaptive mechanism (right column)

Voltage H^∞ Repetitive Control with a Frequency-adaptive Mechanism

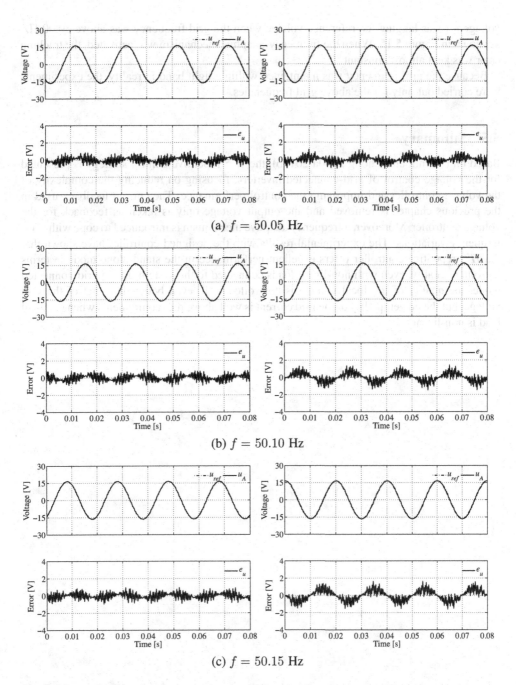

(a) $f = 50.05$ Hz

(b) $f = 50.10$ Hz

(c) $f = 50.15$ Hz

Figure 5.15 Responses when the grid frequency was above 50 Hz: with the frequency-adaptive mechanism (left column) and without the frequency-adaptive mechanism (right column)

voltage u_A was lagging the reference voltage when the grid frequency was above 50.00 Hz, as shown in Figure 5.15. When the frequency-adaptive mechanism was activated, the tracking error was kept almost constant.

Because the grid frequency was maintained within a tight band in reality, the experiments were carried out only for the above grid frequencies.

5.5 Summary

Based on (Hornik and Zhong 2010a 2010b), the H^∞ repetitive control strategy is presented for the voltage control of grid-connected inverters, focusing on reducing the complexity of the voltage controller and improving the control performance. The current feedback used in the previous chapter is removed and the output voltage only is taken as feedback for the voltage controller. Moreover, a frequency adaptive mechanism is introduced to cope with grid-frequency variations. The experimental results with the designed controller have shown that the H^∞ repetitive controller offers excellent performance in the stand-alone mode in terms of voltage quality, even with non-linear loads connected to the system. In the grid-connected mode, the quality of the local voltage is controlled well, even better than that of the grid voltage, but the power quality of the grid current is well above the requirements when the local load is non-linear.

6

Cascaded Current-Voltage H^∞ Repetitive Control

Both the THD of the microgrid voltage and the THD of the current exchanged with the grid need to be kept low according to industrial regulations. It has been shown that it is not a problem to obtain low THD for either the microgrid voltage (Chapters 4 and 5) or for the current exchanged with the grid (Chapter 3). However, it has been a challenge to obtain low THD for both the voltage and the current simultaneously. In this chapter, the advantages of the voltage and current controllers based on H^∞ and repetitive control techniques, presented in Chapters 3 and 5 respectively, are brought together to achieve low THD for both the microgrid voltage and the current exchanged with the grid. It is not a simple combination of the two control strategies but a complete re-design after realising that the inverter LCL filter can be split into two separate parts to design the controllers. The LC part can be used for the voltage controller design and the grid inductor can be used for the current controller design. The voltage controller is responsible for the power quality of the local load voltage (microgrid voltage), power distribution within the microgrid and synchronisation with the grid. The current controller is responsible for the power exchanged between the grid and the microgrid, the power quality of the current exchanged with the grid, and the over-current protection. When the microgrid is connected to the grid, both controllers are active; when the microgrid is not connected to the grid, the current controller works with zero current reference. Hence, no extra effort is needed when changing the operation mode of the microgrid, which considerably facilitates the smooth operation of grid-connected inverters (Yao et al. 2010). It also shows that there is no need to consider the processing delay W_d considered in Chapter 5.

6.1 Operation Modes in Microgrids

Microgrids are normally operated in the grid-connected mode. However, it is also expected to provide sufficient generation capacity, controls, and operational strategies to supply at least a part of the load after being disconnected from the distribution system and to remain operational as a stand-alone (islanded) system (Chen et al. 2010; Hatziargyriou et al. 2007; Katiraei et al.

Control of Power Inverters in Renewable Energy and Smart Grid Integration, First Edition.
Qing-Chang Zhong and Tomas Hornik.
© 2013 John Wiley & Sons, Ltd. Published 2013 by John Wiley & Sons, Ltd.

2008; Li and Kao 2009; Mohamed and El-Saadany 2008; Xiarnay *et al.* 2008). Traditionally, the inverters used in microgrids behave like current sources when they are connected to the grid and as voltage sources when they work autonomously (Chen *et al.* 2010). This involves the change of the controller when the operational mode is changed from stand-alone to grid-connected or vice versa (Yao *et al.* 2010). It is advantageous to operate inverters as voltage sources because there is no need to change the controller when the operation mode is changed. A parallel control structure consisting of an output voltage controller and a grid current controller was proposed in (Yao *et al.* 2010) to achieve seamless transfer via changing the references to the controller without changing the controller.

As mentioned before, since non-linear and/or unbalanced loads can represent a high proportion of the total load in small-scale systems, the problem with power quality is a particular concern in microgrids (Prodanovic and Green 2006). The THD of the inverter local load voltage and the current exchanged with the grid (referred to as the grid current) needs to be kept low according to industrial regulations. It has been known that it is not a problem to obtain low THD for either the inverter local load voltage (Hornik and Zhong 2010b; Weiss *et al.* 2004) or for the grid current (Hornik and Zhong 2009, 2011). However, it is a challenge to obtain low THD for both the inverter local load voltage and the grid current simultaneously. This may even have been believed impossible because there may be non-linear local loads. In this chapter, a cascaded control structure consisting of an inner-loop voltage controller and an outer-loop current controller is presented to achieve this, after recognising that the inverter LCL filter can be split into two separate parts. The LC part can be used to design the voltage controller and the grid interface inductor can be used to design the current controller. The voltage controller is responsible for the power quality of the inverter local load voltage, power distribution and synchronisation with the grid, and the current controller is responsible for the power quality of the grid current, the power exchanged with the grid and over-current protection. With the help of the H^∞ repetitive control (Hornik and Zhong 2009, 2010b, 2011), the control strategy is able to maintain low THD in both the inverter local load voltage and the grid current at the same time. When the inverter is connected to the grid, both controllers are active; when the inverter is not connected to the grid, the current controller works with zero current reference. Hence, no extra effort is needed when changing the operation mode of the inverter, which considerably facilitates the seamless mode transfer for grid-connected inverters. For three-phase inverters, the same individual controller can be used for each phase in the natural frame when the system is implemented with a neutral-point controller, e.g. the one described in (Zhong *et al.* 2006) and Part II, to cope with unbalanced utility grid voltages and utility voltage sags, etc, which are the two most common utility voltage quality problems (Li *et al.* 2006; Vilathgamuwa *et al.* 2006). As a result, the inverter can cope with unbalanced local loads as well as three-phase applications. In other words, the harmonic currents and unbalanced local load currents are all contained locally and do not affect the grid.

It is worth stressing that the cascaded current–voltage control structure improves the quality of both the inverter local load voltage and the grid current at the same time. Moreover, it is able to achieve the seamless transfer of the operation mode. The outer-loop current controller is to provide a reference for the inner-loop voltage controller, which is the key to allow the simultaneous improvement of the THD in the grid current and the inverter local load voltage and to achieve the seamless transfer of operation mode. This is different from the conventional voltage-current control scheme (Li *et al.* 2006), where the (inner) current loop is to regulate the filter inductor current of the inverter (not the grid current), so it is impossible to achieve simultaneous improvement of the THD in the grid current and the inverter local

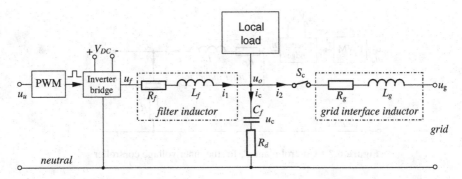

Figure 6.1 Sketch of a grid-connected single-phase inverter with local loads

load voltage. An inner current loop can still be added to the presented structure inside the voltage loop without any difficulty to perform the conventional function, if needed. The H^∞ repetitive control strategy (Hornik and Zhong 2009, 2010b, 2011) is adopted here to design the controllers but this is not a must and other approaches can be used as well.

The multiloop control strategies analysed in (Loh and Holmes 2005) indicated that it was impossible to stabilise an inverter with a proportional feedback of the capacitor voltage and that the performance with an inner-loop proportional-derivative voltage controller is not good either. It is demonstrated in this chapter that excellent performance can be achieved with an inner-loop repetitive controller.

6.2 Control Scheme

Figure 6.1 depicts the structure of a single-phase inverter connected to the grid. It consists of an inverter bridge, an LC filter and a grid interface inductor connected with a circuit breaker. It is worth noting that the local loads are connected in parallel with the filter capacitor. The current i_1 flowing through the filter inductor is called the filter inductor current and the current i_2 flowing through the grid interface inductor is called the grid current. The control objective is to maintain low THD for the inverter local load voltage u_o and, simultaneously, for the grid current i_2.

As a matter of fact, the system can be regarded as two parts, as shown in Figures 6.2 and 6.3, cascaded together. Hence, a cascaded controller, as shown in Figure 6.4, can be adopted and designed naturally. It consists of two loops: an inner voltage loop to regulate the inverter local load voltage u_o and an outer current loop to regulate the grid current i_2. According to the basic principles of control theory of cascaded control, if the dynamics of the outer loop is designed to be slower than that of the inner loop, then the two loops can be designed separately. As a result, the outer-loop controller can be designed under the assumption that the inner-loop is already in the steady state, that is, $u_o = u_{ref}$. It is also worth stressing that the current controller is in the outer loop and the voltage controller is in the inner loop. This is contrary to what is normally done. In this chapter, both controllers are designed using the H^∞ repetitive control strategy because of its excellent performance in reducing THD.

The main functions of the voltage controller are: to deal with power quality issues of the inverter local load voltage even under unbalanced and/or non-linear local loads, to generate

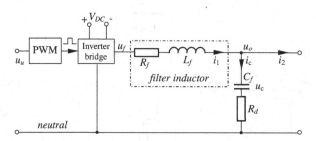

Figure 6.2 Control plant P_u for the inner voltage controller

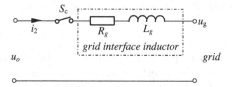

Figure 6.3 Control plant P_i for the outer current controller

and dispatch power to the local load and to synchronise the inverter with the grid. When the inverter is synchronised and connected with the grid, the voltage and the frequency are determined by the grid.

The main function of the outer-loop current controller is to exchange a clean current with the grid even in the presence of grid voltage distortion and/or non-linear (and/or unbalanced for three-phase applications) local loads connected to the inverter. The current controller can be used for over-current protection but, normally, it is included in the driving circuits of the inverter bridge. A PLL can be used to provide the phase information of the grid voltage, which is needed to generate the current reference i_{ref} (see Section 6.5 for an example). As the control structure described here uses just one inverter connected to the system and the inverter is assumed to be powered by a constant DC voltage source, no controller is needed to regulate the DC-link voltage. Otherwise, a controller can be introduced to regulate the DC-link voltage.

Another important feature is that the grid voltage u_g is feed-forwarded and added to the output of the current controller. This is used as a synchronisation mechanism and does not affect the design of the controller, as will be seen later.

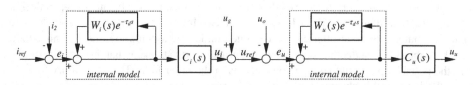

Figure 6.4 Cascaded current-voltage controller for inverters, where both controllers adopt the H^∞ repetitive strategy

6.3 Design of the Voltage Controller

The design of the voltage controller is outlined below, following the detailed procedures presented in (Hornik and Zhong 2010b) and Chapters 4 and 5. A prominent feature different from what is known is that the control plant of the voltage controller is no longer the whole LCL filter but just the LC filter, as shown in Figure 6.2.

The block diagram of the voltage H^∞ repetitive control scheme is shown in Figure 6.5, where P_u is the transfer function of the plant (i.e. the LC filter here), C_u is the transfer function of the stabilising compensator and M_u is the transfer function of the internal model.

6.3.1 State-space Model of the Plant P_u

The corresponding control plant shown in Figure 6.2 for the voltage controller consists of the inverter bridge and the LC filter (L_f and C_f). The filter inductor is modelled with a series winding resistance. The PWM block together with the inverter are modelled by using an average voltage approach with the limits of the available DC-link voltage (Weiss *et al.* 2004) so that the average value of u_f over a sampling period is equal to u_u. As a result, the PWM block and the inverter bridge can be ignored when designing the controller.

The filter inductor current i_1 and the capacitor voltage u_c are chosen as state variables $x_u = \begin{bmatrix} i_1 & u_c \end{bmatrix}^T$. The external input $w_u = \begin{bmatrix} i_2 & u_{ref} \end{bmatrix}^T$ consists of the grid current i_2 and the reference voltage u_{ref}. The control input is u_u. The output signal from the plant P_u is the tracking error $e_u = u_{ref} - u_o$, where $u_o = u_c + R_d(i_1 - i_2)$ is the inverter local load voltage. The plant P_u can be described by the state equation

$$\dot{x}_u = A_u x_u + B_{u1} w_u + B_{u2} u_u \tag{6.1}$$

and the output equation

$$y_u = e_u = C_{u1} x_u + D_{u1} w_u + D_{u2} u_u \tag{6.2}$$

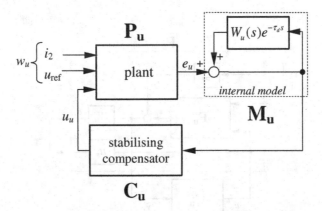

Figure 6.5 Block diagram of the H^∞ repetitive voltage control scheme

with

$$A_u = \begin{bmatrix} -\dfrac{R_f + R_d}{L_f} & -\dfrac{1}{L_f} \\ \dfrac{1}{C_f} & 0 \end{bmatrix},$$

$$B_{u1} = \begin{bmatrix} \dfrac{R_d}{L_f} & 0 \\ -\dfrac{1}{C_f} & 0 \end{bmatrix}, \quad B_{u2} = \begin{bmatrix} \dfrac{1}{L_f} \\ 0 \end{bmatrix},$$

$$C_{u1} = \begin{bmatrix} -R_d & -1 \end{bmatrix},$$

$$D_{u1} = \begin{bmatrix} R_d & 1 \end{bmatrix}, \quad D_{u2} = 0.$$

The corresponding plant transfer function is then

$$P_u = \left[\begin{array}{c|cc} A_u & B_{u1} & B_{u2} \\ \hline C_{u1} & D_{u1} & D_{u2} \end{array} \right]. \tag{6.3}$$

6.3.2 Formulation of the Standard H^∞ Problem

In order to guarantee the stability of the inner voltage loop, an H^∞ control problem as shown in Figure 6.6 is formulated to minimise the H^∞ norm of the transfer function $T_{\tilde{z}_u \tilde{w}_u} = \mathcal{F}_l(\tilde{P}_u, C_u)$ from $\tilde{w}_u = [v_u \quad w_u]^T$ to $\tilde{z}_u = [z_{u1} \quad z_{u2}]^T$, after opening the local positive feedback loop of the internal model and introducing weighting parameters ξ_u and μ_u. The closed-loop system can be represented as

$$\begin{bmatrix} \tilde{z}_u \\ \tilde{y}_u \end{bmatrix} = \tilde{P}_u \begin{bmatrix} \tilde{w}_u \\ u_u \end{bmatrix}, \tag{6.4}$$

$$u_u = C_u \tilde{y}_u,$$

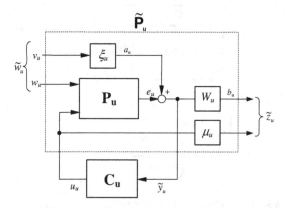

Figure 6.6 Formulation of the H^∞ control problem for the voltage controller

where $\tilde{P}_u$ is the generalised plant and C_u is the voltage controller to be designed. The generalised plant $\tilde{P}_u$ consists of the original plant P_u together with the low-pass filter $W_u = \begin{bmatrix} A_{w_u} & B_{w_u} \\ C_{w_u} & D_{w_u} \end{bmatrix}$, which is part of the internal model for repetitive control. The details of how to select W_u can be found in (Hornik and Zhong 2010b, 2011). A weighting parameter ξ_u is added to adjust the relative importance of v_u with respect to w_u and another weighting parameter μ_u is added to adjust the relative importance of u_u with respect to b_u. The parameters ξ_u and μ_u also play a role in guaranteeing the stability of the system; see more details in (Hornik and Zhong 2010b, 2011). It can be found out that the generalised plant $\tilde{P}_u$ is realised as

$$\tilde{P}_u = \left[\begin{array}{ccc|cc} A_u & 0 & 0 & B_{u1} & B_{u2} \\ B_{w_u}C_{u1} & A_{w_u} & B_{w_u}\xi_u & B_{w_u}D_{u1} & B_{w_u}D_{u2} \\ D_{w_u}C_{u1} & C_{w_u} & D_{w_u}\xi_u & D_{w_u}D_{u1} & D_{w_u}D_{u2} \\ 0 & 0 & 0 & 0 & \mu_u \\ \hline C_{u1} & 0 & \xi_u & D_{u1} & D_{u2} \end{array} \right]. \tag{6.5}$$

The controller C_u can then be found according to the generalised plant $\tilde{P}_u$ using the H^∞ control theory, e.g. by using the function *hinfsyn* provided in MATLAB®.

6.4 Design of the Current Controller

As explained before, when designing the outer-loop current controller, it can be assumed that the inner voltage loop tracks the reference voltage perfectly, that is $u_o = u_{ref}$. Hence, the control plant for the current loop is simply the grid inductor, as shown in Figure 6.3. The formulation of the H^∞ control problem to design the H^∞ compensator C_i is similar to that in the case of the voltage control loop shown in Figure 6.6, but with a different plant P_i and the subscript $_u$ replaced with $_i$.

6.4.1 State-space Model of the Plant P_i

Since it can be assumed that $u_o = u_{ref}$, then $u_o = u_g + u_i$ or $u_i = u_o - u_g$ from Figures 6.3 and 6.4, that is, u_i is actually the voltage dropped on the grid inductor. The feed-forwarded grid voltage u_g provides a base local load voltage for the inverter. The same voltage u_g appears on both sides of the grid interface inductor L_g and does not affect the controller design. Hence, the feed-forwarded voltage path can be ignored during the design process. This is a very important feature. The only contribution that needs to be considered during the design process is the output u_i of the repetitive current controller.

The grid current i_2 flowing through the grid interface inductor L_g is chosen as the state variable $x_i = i_2$. The external input is $w_i = i_{ref}$ and the control input is u_i. The output signal from the plant P_i is the tracking error $e_i = i_{ref} - i_2$, i.e. the difference between the current reference and the grid current. The plant P_i can then be described by the state equation

$$\dot{x}_i = A_i x_i + B_{i1} w_i + B_{i2} u_i$$

and the output equation

$$y_i = e_i = C_{i1}x_i + D_{i1}w_i + D_{i2}u_i$$

with

$$A_i = -\frac{R_g}{L_g}, \quad B_{i1} = 0, \quad B_{i2} = \frac{1}{L_g},$$
$$C_{i1} = -1, \quad D_{i1} = 1, \quad D_{i2} = 0.$$

The corresponding transfer function of P_i is

$$P_i = \left[\begin{array}{c|cc} A_i & B_{i1} & B_{i2} \\ \hline C_{i1} & D_{i1} & D_{i2} \end{array}\right].$$

6.4.2 Formulation of the Standard H^∞ Problem

Similarly, a standard H^∞ problem can be formulated as in the case of the voltage controller shown in Figure 6.6, replacing the subscript $_u$ with $_i$. The resulting generalised plant can be obtained as

$$\tilde{P}_i = \left[\begin{array}{cc|ccc|c} A_i & 0 & 0 & B_{i1} & & B_{i2} \\ B_{w_i}C_{i1} & A_{w_i} & B_{w_i}\xi_i & B_{i1}D_{i1} & & B_{w_i}D_{i2} \\ D_{w_i}C_{i1} & C_{w_i} & D_{w_i}\xi_i & D_{w_i}D_{i1} & & D_{w_i}D_{i2} \\ 0 & 0 & 0 & 0 & & \mu_i \\ \hline C_{i1} & 0 & \xi_i & D_{i1} & & D_{i2} \end{array}\right], \qquad (6.6)$$

with weighting parameters ξ_i and μ_i and low-pass filter $W_i = \left[\begin{array}{c|c} A_{w_i} & B_{w_i} \\ \hline C_{w_i} & D_{w_i} \end{array}\right]$, which can be selected similarly as the corresponding ones for the voltage controller.

The controller C_i can then be found according to the generalised plant $\tilde{P}_i$ using the H^∞ control theory, e.g. by using the function *hinfsyn* provided in MATLAB®.

6.5 Design Example

As an example, the controllers will be designed in this section for an experimental set-up, which consists of an inverter board, a three-phase LC filter, a three-phase grid interface inductor, a board consisting of voltage and current sensors, a step-up Wye-Wye transformer (12V/230V/50Hz), a dSPACE DS1104 R&D controller board with ControlDesk software, and a MATLAB®/Simulink® SimPower software package. The inverter board consists of two independent three-phase inverters and has the capability to generate PWM voltages from a constant 42V DC voltage source. One inverter was used to generate a stable neutral line for the three-phase inverter. The generated three-phase voltage was connected to the grid via a controlled circuit breaker and a step-up transformer. The PWM switching frequency was 12 kHz. A Yokogawa power analyser WT1600 was used to measure the THD. The parameters of the inverter are given in Table 3.1. Three sets of identical controllers were used for the three

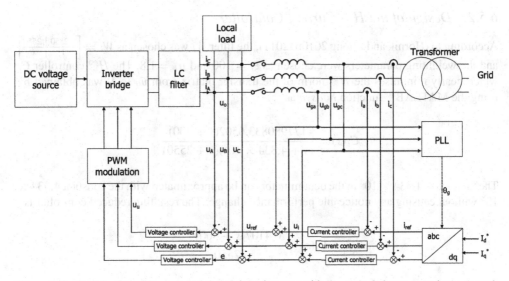

Figure 6.7 Sketch of a grid-connected three-phase inverter with the cascaded current-voltage control strategy

phases because of the stable neutral line available. The control structure for the three-phase system is shown in Figure 6.7. A traditional dq phase-locked-loop was used to provide the phase information needed to generate the three-phase grid current references via a dq/abc transformation from the current references I_d^* and I_q^*. The internal model was implemented with the capability to adapt to the frequency change in the grid described in Chapter 5, according to (Hornik and Zhong 2010b).

6.5.1 Design of the H^∞ Voltage Controller

According to (Hornik and Zhong 2010b, 2011), the weighting function was chosen as $W_u = \begin{bmatrix} -2550 & 2550 \\ 1 & 0 \end{bmatrix}$ for $f = 50$ Hz and the weighting parameters were chosen as $\xi_u = 100$ and $\mu_u = 1.85$. For the parameters of the plant given in Table 3.1, the H^∞ controller C_u which nearly minimises the H^∞ norm of the transfer matrix from $\tilde{w}_u$ to $\tilde{z}_u$ was obtained by using the MATLAB® function *hinfsyn* as

$$C_u(s) = \frac{748.649(s^2 + 6954s + 3.026 \times 10^8)}{(s + 2550)(s^2 + 7969s + 3.043 \times 10^8)}.$$

It can be reduced to

$$C_u(s) = \frac{748.649}{s + 2550},$$

without causing noticeable performance degradation, after cancelling the poles and zeros that are close to each other.

6.5.2 Design of the H^∞ Current Controller

According to (Hornik and Zhong 2010b, 2011), the filter W_i was chosen as $W_i = \begin{bmatrix} -2550 & 2550 \\ 1 & 0 \end{bmatrix}$ and the weighting parameters were chosen as $\xi_i = 100$ and $\mu_i = 1.8$. The H^∞ controller C_i which nearly minimises the H^∞ norm of the transfer matrix from $\tilde{w}_i$ to $\tilde{z}_i$ was obtained by using the MATLAB® function *hinfsyn* as

$$C_i(s) = \frac{177980833.6502(s + 300)}{(s + 4.334 \times 10^8)(s + 2550)}.$$

The factor $s + 4.334 \times 10^8$ in the denominator can be approximated with the constant 4.334×10^8 without causing any noticeable performance change. The resulting reduced controller is

$$C_i(s) = \frac{0.4107(s + 300)}{s + 2550}.$$

6.6 Experimental Results

The above-designed controller was implemented to evaluate its performance in both stand-alone and grid-connected modes with different loads. The seamless transfer of the operation mode was also carried out. The H^∞ repetitive current controller was replaced with a PR current controller for comparison in the grid-connected mode. In the stand-alone mode, since the grid current reference was set to zero and the circuit breaker was turned off (which means the current controller was not functioning), the experimental results with both the repetitive current controller and the PR current controller are similar and hence no comparative results are provided for the stand-alone mode. The PR controller was designed according to (Timbus et al. 2009) with the plant used in Section 6.4.1 as

$$C_{i-PR}(s) = 0.735 + \frac{20s}{s^2 + 10000\pi^2}.$$

6.6.1 Steady-state Performance in the Stand-alone Mode

The voltage reference was set to the grid voltage (so the inverter was synchronised and ready to be connected to the utility grid). The evaluation of the controller was made for a resistive load ($R_A = R_B = R_C = 12\ \Omega$), a non-linear load (a three-phase uncontrolled rectifier loaded with an LC filter $L = 150\ \mu H$, $C = 1000\ \mu F$ and a resistor $R = 20\ \Omega$), and an unbalanced load ($R_A = R_C = 12\ \Omega$ and $R_B = \infty$).

6.6.1.1 With a Resistive Load

The local load voltage u_A, voltage reference u_{ref} and filter inductor current i_A are shown in Figure 6.8(a). Figure 6.8(b) shows the spectra of the inverter local load voltage and the local

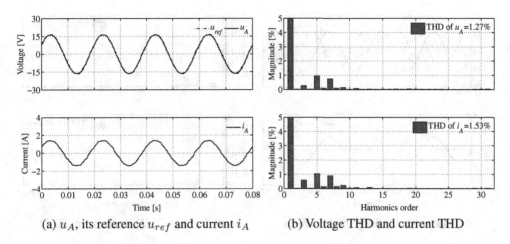

Figure 6.8 Experimental results in the stand-alone mode with a resistive load

load current. The recorded local voltage THD was 1.27% while the grid voltage THD was 1.8%. Since the utility grid voltage was used as the reference, it is worth mentioning that the quality of the inverter local load voltage was better than that of the grid voltage, even without using an active filter.

6.6.1.2 With a Non-linear Load

The local load voltage u_A, voltage reference u_{ref} and filter inductor current i_A are shown in Figure 6.9(a). The spectra of the inverter local load voltage and the local load current are shown in Figure 6.9(b). The recorded local load voltage THD was 4.73% while the grid voltage THD

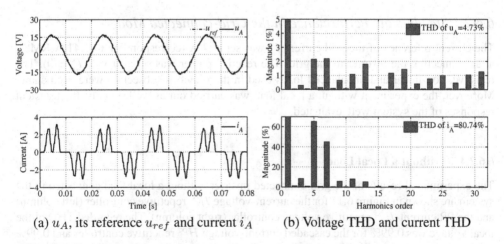

Figure 6.9 Experimental results in the stand-alone mode with a non-linear load

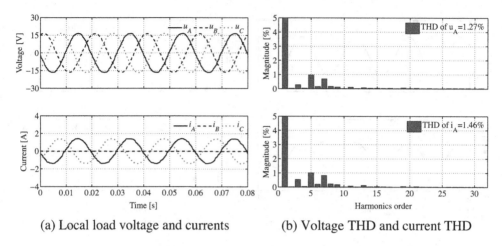

(a) Local load voltage and currents   (b) Voltage THD and current THD

Figure 6.10 Experimental results in the stand-alone mode with an unbalanced load

was 1.78%. The experimental results demonstrate satisfactory performance of the voltage controller for non-linear loads.

6.6.1.3 With an Unbalanced Load

The inverter local load voltage and the local load currents are shown in Figure 6.10(a) with their spectra shown in Figure 6.10(b). The recorded local load voltage THD was 1.27% while the grid voltage THD was 1.77%. Since the control structure adopts separate controllers for each phase, the unbalanced loads had no influence on the voltage controller performance and the inverter local load voltages remained balanced.

6.6.2 Steady-state Performance in the Grid-connected Mode

The current reference of the grid current I_d^* was set at 2 A (corresponding to 1.41 A RMS), after connecting the inverter to the grid. The reactive power was set at 0 Var ($I_q^* = 0$). The resistive, non-linear and unbalanced loads used in the previous subsection were used again. Moreover, the experiment without a local load was carried out as well. Finally, the transient responses of the system were evaluated.

6.6.2.1 Without a Local Load

The experimental results of the grid-connected inverter without a local load connected to the system are shown in Figure 6.11 for the current-voltage H^∞ repetitive controller (left column) and the PR current-H^∞ repetitive voltage controller (right column). The recorded THD of the local voltage was 0.99% for the cascaded current-voltage H^∞ repetitive controller and 0.99% for the PR controller, while the grid voltage THD was 1.58% and 0.96% respectively. The THD

Cascaded Current-Voltage H^∞ Repetitive Control

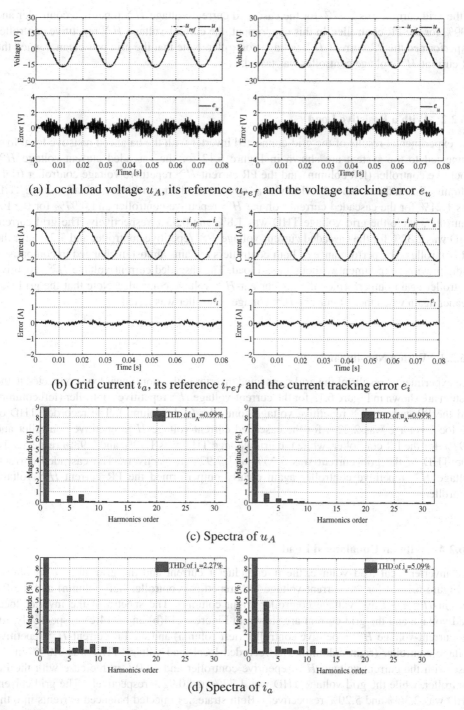

(a) Local load voltage u_A, its reference u_{ref} and the voltage tracking error e_u

(b) Grid current i_a, its reference i_{ref} and the current tracking error e_i

(c) Spectra of u_A

(d) Spectra of i_a

Figure 6.11 Experimental results in the grid-connected mode without a load: current-voltage H^∞ repetitive controller (left column) and PR current-H^∞ repetitive voltage controller (right column)

of the grid current was 2.27% for the cascaded current-voltage H^∞ repetitive controller and 5.09% for the PR controller. In this experiment, the current-voltage H^∞ repetitive controller outperformed the PR current-H^∞ voltage controller. Note that the grid was cleaner when the PR current-H^∞ voltage controller was tested.

6.6.2.2 With a Resistive Load

The experimental results of the grid-connected inverter with the balanced resistive local load connected to the system are shown in Figure 6.12 for the cascaded current-voltage H^∞ repetitive controller (left column) and the PR current-H^∞ repetitive voltage controller (right column). When the resistive local load was connected, the recorded local load voltage THD was 1.21% for the cascaded current-voltage H^∞ repetitive controller and 0.97% for the PR controller, while the grid voltage THD was 1.8% and 0.95% respectively. The grid current THD was 2.32% for the cascaded current-voltage H^∞ repetitive controller and 5.24% for the PR controller. The performance of both controllers remains almost unchanged in comparison to the previous experiment without a local load. The cascaded current-voltage H^∞ repetitive controller again outperformed the PR current-H^∞ voltage controller. Note that the grid was cleaner again when the PR current-H^∞ voltage controller was tested.

6.6.2.3 With a Non-linear Load

The experimental results of the grid-connected inverter with a non-linear load connected to the system are shown in Figure 6.13 for the current-voltage H^∞ repetitive controller (left column) and the PR current-H^∞ repetitive voltage controller (right column). The recorded THD of the local voltage was 2.22% for the cascaded current-voltage H^∞ repetitive controller and 2.97% for the PR controller, while the grid voltage THD was 1.72% and 0.93% respectively. The THD of the grid current was 5.35% and 7.97% respectively. The cascaded current-voltage H^∞ repetitive controller again clearly outperformed the PR current-H^∞ voltage controller.

6.6.2.4 With an Unbalanced Load

The inverter local load voltage, the filter inductor current and the grid current are shown in Figure 6.14(a) for the current-voltage H^∞ repetitive controller (left column) and the PR current-H^∞ repetitive voltage controller (right column). The spectra of the inverter local load voltage and the grid current are shown in Figures 6.14(b) and 6.14(c), respectively, for the current-voltage H^∞ repetitive controller (left column) and the PR current-H^∞ repetitive voltage controller (right column). The recorded local load voltage THD was 1.09% in the case with the current-voltage H^∞ repetitive controller and 1.00% in the case with the PR controller, while the grid voltage THD was 1.77% and 0.92% respectively. The grid current THD was 2.36% and 5.20% respectively. Both strategies injected balanced currents into the grid although the local load was not balanced.

Cascaded Current-Voltage H^∞ Repetitive Control 141

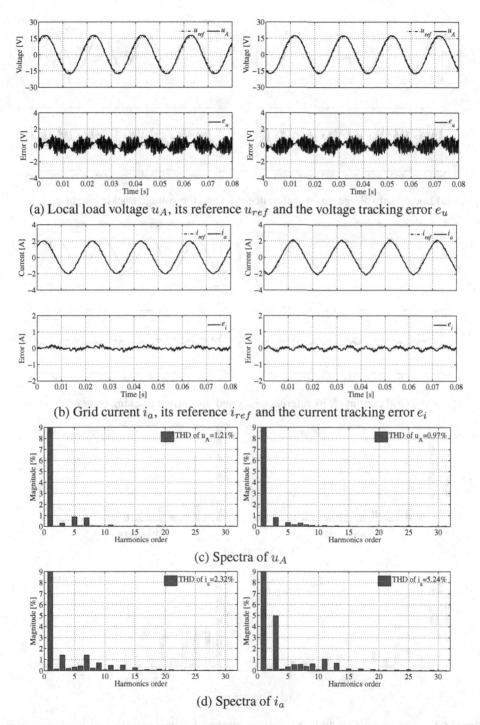

Figure 6.12 Experimental results in the grid-connected mode with a balanced linear load: current-voltage H^∞ repetitive controller (left column) and PR current-H^∞ repetitive voltage controller (right column)

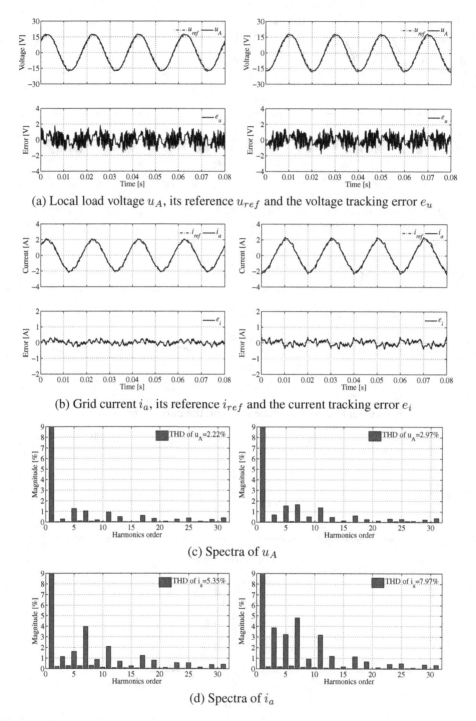

Figure 6.13 Experimental results in the grid-connected mode with a non-linear load: current-voltage H^∞ repetitive controller (left column) and PR current-H^∞ repetitive voltage controller (right column)

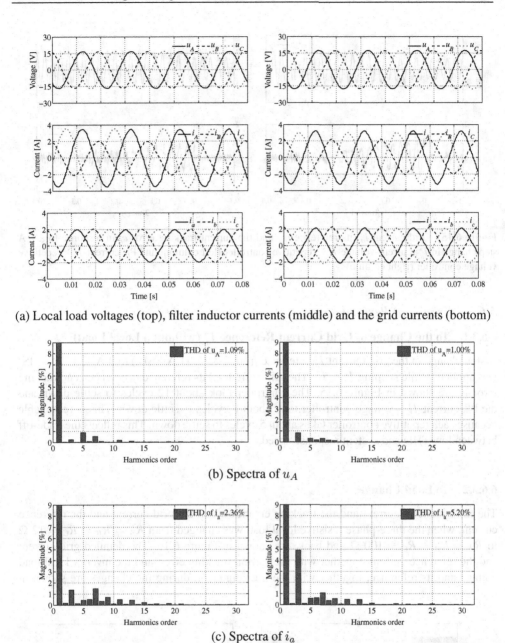

(a) Local load voltages (top), filter inductor currents (middle) and the grid currents (bottom)

(b) Spectra of u_A

(c) Spectra of i_a

Figure 6.14 Experimental results in the grid-connected mode with unbalanced loads: current-voltage H^∞ repetitive controller (left column) and PR current-H^∞ repetitive voltage controller (right column)

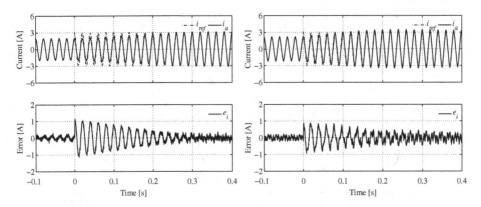

Figure 6.15 Grid current i_a, its reference i_{ref} and the current tracking error e_i in response to a 1 A step change in I_d^*: current-voltage H^∞ repetitive controller (left column) and PR current-H^∞ repetitive voltage controller (right column)

6.6.3 Transient Performance

6.6.3.1 To the Change of Grid Current Reference I_d^* (without a Local Load)

A step change in the grid current I_d^* reference from 2 A (1.41 A RMS) to 3 A (2.12 A RMS) was applied (while keeping $I_q^* = 0$). The grid current i_a, its reference i_{ref} and current tracking error e_i are shown in Figure 6.15. The controller took about 12 cycles to settle down and the PR current-H^∞ voltage controller took about 8 cycles to settle down. This is reasonable because each repetitive controller takes about 5 cycles to settle down. This reflects the trade-off between low THD and system response speed.

6.6.3.2 To Load Changes

The filter inductor current and the tracking error between the grid current and the reference current, when the three-phase resistive local load was changed from $R_A = R_B = R_C = 12\,\Omega$ to $R_A = R_B = R_C = 100\,\Omega$ and back, are shown in Figure 6.16. The detailed grid current and the reference current, together with the current tracking error, and the inverter local load voltage and its reference, together with the tracking error, during the changes are shown in

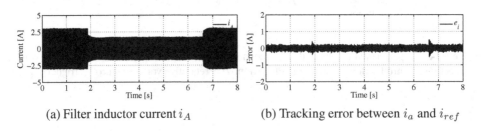

(a) Filter inductor current i_A (b) Tracking error between i_a and i_{ref}

Figure 6.16 Transient responses of the inverter when the local load was changed

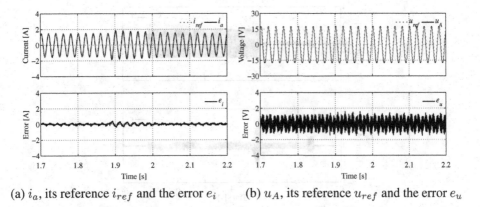

(a) i_a, its reference i_{ref} and the error e_i (b) u_A, its reference u_{ref} and the error e_u

Figure 6.17 Detailed responses when the local load was changed from 12 Ω to 100 Ω at $t = 1.88$ s

Figure 6.17 for the change from 12 Ω to 100 Ω at $t = 1.88$ s and in Figure 6.18 for the change from 100 Ω to 12 Ω at $t = 6.61$ s. The current controller took about 5 cycles to settle down, which is in line with the findings from the previous experiment. There was no noticeable change in the inverter local load voltage.

6.6.4 Seamless Transfer of the Operation Mode

The transient response of the grid current when the inverter was changed from the stand-alone mode to the grid-connected mode and back is shown in Figure 6.19. The detailed responses during the transfers are shown in Figures 6.20, 6.21 and 6.22.

At $t = 1$ s, the inverter was connected to the grid. The details of the transfer from the stand-alone mode to the grid-connected mode are shown in Figure 6.20. There was not much

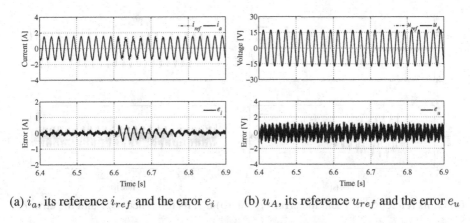

(a) i_a, its reference i_{ref} and the error e_i (b) u_A, its reference u_{ref} and the error e_u

Figure 6.18 Detailed responses when the local load was changed from 100 Ω to 12 Ω at $t = 6.61$ s

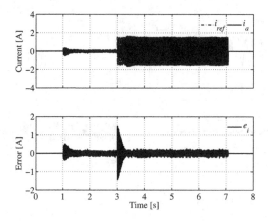

Figure 6.19 Transient response of the inverter when transferred from the stand-alone mode to the grid-connected mode and then back

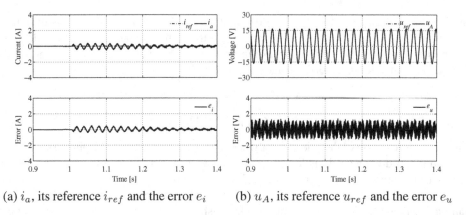

(a) i_a, its reference i_{ref} and the error e_i (b) u_A, its reference u_{ref} and the error e_u

Figure 6.20 Detailed responses when transferred from the stand-alone mode to the grid-connected mode at $t = 1$ s

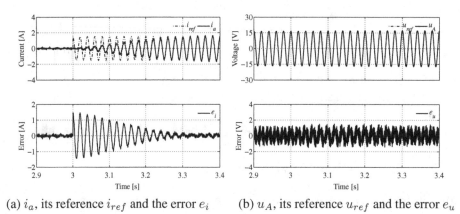

(a) i_a, its reference i_{ref} and the error e_i (b) u_A, its reference u_{ref} and the error e_u

Figure 6.21 Detailed responses to a step change in the grid current reference I_d^* from 0 A to 1.5 A at $t = 3$ s

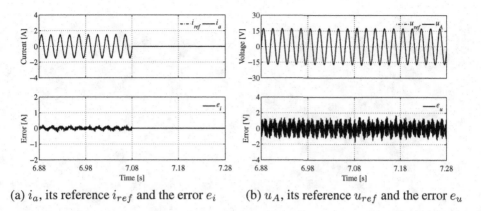

(a) i_a, its reference i_{ref} and the error e_i (b) u_A, its reference u_{ref} and the error e_u

Figure 6.22 Detailed responses when transferred from the grid-connected mode to the stand-alone mode at $t = 7.08$ s

dynamics in the current. There was no noticeable change in the inverter local load voltage, either. Hence, seamless grid connection was achieved.

A step change in the grid current reference I_d^* from 0 A to 1.5 A (1.06 A RMS) was applied at time $t = 3$ s and the responses are shown in Figure 6.21. The system took about 12 cycles to settle down, which is consistent with the test done in Section 6.6.3.1.

At $t = 7.08$ s, the inverter was disconnected from the grid and the details of the responses are shown in Figure 6.22(a) and (b). There was no noticeable transients in the inverter local load voltage and seamless disconnection from the grid was achieved.

In summary, the control strategy is able to achieve seamless transfer of operation modes from stand-alone to grid-connected or vice versa.

6.7 Summary

Based on (Hornik and Zhong 2010a; Zhong and Hornik 2012), a cascaded current-voltage control strategy is presented for inverters in microgrids. It consists of an inner voltage loop and an outer current loop and offers excellent performance in terms of THD for both the inverter local load voltage and the grid current. In particular, when non-linear and/or unbalanced loads are connected to the inverter in the grid-connected mode, this strategy significantly improves the THD of the inverter local load voltage and the grid current at the same time. The controllers are designed using the H^∞ repetitive control but can be designed with other approaches as well. This strategy also achieves seamless transfer between the stand-alone mode and the grid-connected mode. It can be used for both single-phase systems and three-phase systems. As a result, non-linear harmonic currents and unbalanced local load currents are all contained locally and do not affect the grid. Experimental results under various scenarios are presented to demonstrate the performance.

7

Control of Inverter Output Impedance

The filter capacitor of an inverter can be regarded as a part of the load and, as a result, the output impedance of the inverter is inductive. Such inverters are referred to as L-inverters. As mentioned in Chapter 2, the output impedance of an inverter plays an important role in the THD of the output voltage. In this chapter, control strategies are presented to design the output impedance to be resistive and capacitive, respectively. The corresponding inverters are referred to as R-inverters and C-inverters. Moreover, it is shown that C-inverters can be designed to significantly improve the THD of the output voltage. Both simulation and experimental results are presented.

7.1 Inverters with Inductive Output Impedances (L-inverters)

The circuit of a single-phase inverter under consideration is shown in Figure 7.1(a). It consists of a single-phase H-bridge inverter powered by a DC source, and an LC filter. The inverter is connected to the AC bus via a circuit breaker CB and the load is assumed to be connected to the AC bus. The control signal u is converted to a PWM signal to drive the H-bridge so that the average of u_f over a switching period is the same as u, i.e. $u \approx u_f$. Hence, the PWM block and the H-bridge can be ignored in the controller design (Zhong 2012a).

Although the inverter consists of an LC filter, the capacitor C can be regarded as a part of the load instead of a part of the inverter (Zhong 2012c). This reduces the control plant to an H-bridge and an inductor, as shown in Figure 7.1(b). The advantages of this are: (1) it reduces the order of the control plant to be 1; and (2) it considerably simplifies the design and analysis of the controller, which facilitates the understanding of the nature of inverter control.

Since the average of u_f over a switching period is the same as u, which is the same as the reference voltage v_r, there is approximately

$$u = v_r \approx u_f = v_o + sLi,$$

Control of Power Inverters in Renewable Energy and Smart Grid Integration, First Edition.
Qing-Chang Zhong and Tomas Hornik.
© 2013 John Wiley & Sons, Ltd. Published 2013 by John Wiley & Sons, Ltd.

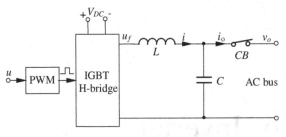

(a) Used for physical implementation

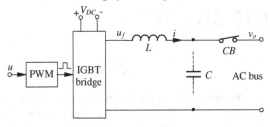

(b) Used for controller design

Figure 7.1 Single-phase inverter

which gives

$$v_o = v_r - Z_o(s) \cdot i$$

with

$$Z_o(s) = sL.$$

That is, the output impedance $Z_o(s)$ is inductive. Inverters with inductive output impedance are called L-inverters.

7.2 Inverters with Resistive Output Impedances (R-inverters)

7.2.1 Controller Design

The inductor current i can be measured to construct a proportional controller K_i, as shown in Figure 7.2, so that the output impedance of the inverter is forced to be resistive and that it dominates the impedance between the inverter and the AC bus. This is also widely referred to as a virtual resistor (Dahono 2003, 2004; Dahono et al. 2001; Li 2009) to dampen the resonance of the LC filter.

The following two equations hold for the closed-loop system consisting of Figure 7.1(b) and Figure 7.2:

$$u = v_r - K_i i,$$
$$u_f = sLi + v_o.$$

Control of Inverter Output Impedance

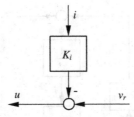

Figure 7.2 Controller to achieve a resistive output impedance

Since the average of u_f over a switching period is the same as u, there is approximately

$$v_r - K_i i = sLi + v_o,$$

which gives

$$v_o = v_r - Z_o(s) \cdot i$$

with

$$Z_o(s) = sL + K_i.$$

If the gain K_i is chosen big enough, the effect of the inductance is not significant and the output impedance can be made nearly purely resistive over a wide range of frequencies. Then, the output impedance is roughly

$$Z_o(s) \approx K_i,$$

which is independent of the inductance. However, a big K_i causes considerable harmonic components in the output voltage for non-linear loads and, hence, small K_i is preferred in order to achieve low THD in the output voltage. How to reduce the THD of the output voltage while using large K_i will be discussed in Chapter 20. Note that when $K_i = 0$, the output impedance becomes inductive and the inverter becomes an L-inverter.

With the above control strategy, the inverter can be approximated as a controlled ideal voltage supply v_r cascaded with a resistive output impedance R_o described as

$$v_o = v_r - R_o i \qquad (7.1)$$

with

$$R_o = K_i.$$

Such inverters are called R-inverters. Note that $v_o \approx u = v_r$ if no load is connected.

7.2.2 Stability Analysis

If the controller is implemented using analogue electronic circuits, then the control loop is stable for a very large gain K_i. However, if it is implemented with a digital controller,

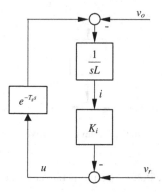

Figure 7.3 Approximate block diagram for the current loop

then the proportional gain is limited. The effect of computation and PWM conversion can be approximated by a one-step delay $e^{-T_s s}$, where T_s is the sampling period. Then, an approximate block diagram for the control loop is shown in Figure 7.3.

The characteristic equation of the loop is

$$1 + \frac{K_i}{sL} e^{-T_s s} = 0.$$

If the gain is chosen to satisfy

$$K_i < \frac{\pi}{2} \frac{L}{T_s}, \tag{7.2}$$

then the loop is stable. When $K_i = \frac{\pi L}{4T_s}$, the phase margin of the loop is $\frac{\pi}{4}$ rad. Note that the analysis done here is to demonstrate that there is a limit, sometime very strict, on the current feedback gain. The gain K_i should be further decreased if a low-pass filter is involved in the measurement of the current i.

7.3 Inverters with Capacitive Output Impedances (C-inverters)

The inductor current i can be measured to construct an integral controller $\frac{1}{sC_o}$, as shown in Figure 7.4, so that the output impedance of the inverter is forced to be capacitive and that it dominates the impedance between the inverter and the AC bus. This is equivalent to having a virtual capacitor C_o connected in series with the filter inductor L, as will be shown later.

The following two equations hold for the closed-loop system consisting of Figure 7.1(b) and Figure 7.4:

$$u = v_r - \frac{1}{sC_o} i \quad \text{and} \quad u_f = sLi + v_o.$$

Control of Inverter Output Impedance

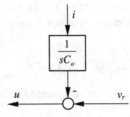

Figure 7.4 Controller to achieve a capacitive output impedance

Since the average of u_f over a switching period is the same as u, there is approximately

$$v_r - \frac{1}{sC_o}i = sLi + v_o,$$

which leads to

$$v_o = v_r - Z_o(s) \cdot i,$$

with the output impedance $Z_o(s)$ of the inverter given as

$$Z_o(s) = sL + \frac{1}{sC_o}. \tag{7.3}$$

If the capacitor C_o is chosen small enough, the effect of the inductor is not significant and the output impedance can be made nearly purely capacitive at the fundamental frequency, i.e., roughly

$$Z_o(s) \approx \frac{1}{sC_o}.$$

Such inverters are called C-inverters.

7.4 Design of C-inverters to Improve the Voltage THD

As discussed in Section 2.1.2, the THD of the output voltage depends on the output impedance at the harmonic frequencies. If the output impedance of an inverter is designed to be capacitive, then it is possible to improve the THD of the output voltage by selecting appropriate C_o.

7.4.1 General Case

According to (7.3), there is

$$|Z_o(jh\omega^*)|^2 = \left(h\omega^* L - \frac{1}{h\omega^* C_o}\right)^2,$$

where ω^* is the rated fundamental frequency and h is the harmonic order. In order to minimise the THD of the output voltage, the virtual capacitor C_o should be chosen, according to (2.1), to minimise

$$\Sigma_{h=2}^{\infty} I_h^2 |Z_o(jh\omega^*)|^2$$

because the fundamental component V_1 can be assumed to be almost constant. This is equivalent to

$$\min_{C_o} \Sigma_{h=2}^{\infty} i_{1h}^2 \left(h\omega^* L - \frac{1}{h\omega^* C_o} \right)^2, \quad (7.4)$$

where $i_{1h} = \frac{I_h}{I_1}$ is the normalised h-th harmonic current I_h with respect to the fundamental current I_1. Depending on the distribution of the harmonic current components, different strategies can be obtained.

Assume that the harmonic current is negligible for the harmonics higher than the N-th order (with an odd number N). Then C_o can be found via solving (7.4). Define

$$f(C_o) = \Sigma_{h=2}^{N} i_{1h}^2 \left(h\omega^* L - \frac{1}{h\omega^* C_o} \right)^2.$$

Then C_o needs to satisfy

$$\frac{\mathrm{d}f(C_o)}{\mathrm{d}C_o} = 2\Sigma_{h=2}^{N} i_{1h}^2 \left(h\omega^* L - \frac{1}{h\omega^* C_o} \right) \frac{1}{h\omega^* C_o^2} = 0,$$

which is equivalent to

$$\Sigma_{h=2}^{N} i_{1h}^2 \left(L - \frac{1}{(h\omega^*)^2 C_o} \right) = 0.$$

Hence,

$$\Sigma_{h=2}^{N} i_{1h}^2 L = \frac{1}{(\omega^*)^2 C_o} \Sigma_{h=2}^{N} \frac{i_{1h}^2}{h^2},$$

and the optimal capacitance can be solved as

$$C_o = \frac{1}{(\omega^*)^2 L} \frac{\Sigma_{h=2}^{N} \frac{i_{1h}^2}{h^2}}{\Sigma_{h=2}^{N} i_{1h}^2}, \quad (7.5)$$

which is applicable for any known harmonic distribution of current i. The corresponding $f(C_o)$ is

$$f_{min}(C_o) = \Sigma_{h=2}^{N} i_{1h}^2 \left(h\omega^* L - \frac{\omega^* L}{h} \frac{\Sigma_{h=2}^{N} i_{1h}^2}{\Sigma_{h=2}^{N} \frac{i_{1h}^2}{h^2}} \right)^2$$

$$= (\omega^* L)^2 \Sigma_{h=2}^{N} i_{1h}^2 \left(h - \frac{1}{h} \frac{\Sigma_{h=2}^{N} i_{1h}^2}{\Sigma_{h=2}^{N} \frac{i_{1h}^2}{h^2}} \right)^2.$$

Control of Inverter Output Impedance

Hence, the THD of v_o is in proportion to the inductance L of the inverter LC filter. A small L not only reduces the cost but also improves the voltage quality. Moreover, since $\frac{1}{C_o} \sim L$, a small L leads to a small gain for the integrator, which improves the stability of the current loop. However, a small L leads to a high $\frac{di}{dt}$ for the switches and large current ripples.

If the distribution of the harmonic components is not known, then it can be assumed that the even harmonics are zero (which is normally the case) and the odd harmonics are equally distributed. As a result, the optimal C_o can be chosen, according to (7.5), as

$$C_o = \frac{1}{(\omega^*)^2 L} \frac{\Sigma_{h=3,5,7,\ldots,N} \frac{1}{h^2}}{\Sigma_{h=3,5,7,\ldots,N}^{N} 1}$$

$$= \frac{1}{(\omega^*)^2 L} \frac{\Sigma_{h=3,5,7,\ldots,N} \frac{1}{h^2}}{(N-1)/2}.$$

This can be written as

$$C_o = \frac{1}{(\omega^*)^2 L} \frac{1}{(N-1)/2} \left(\frac{1}{3^2} + \frac{1}{5^2} + \cdots + \frac{1}{N^2} \right),$$

where $(N-1)/2$ is the number of terms in the summation. The corresponding $f(C_o)$ is

$$f_{min}(C_o) = (\omega^* L)^2 \Sigma_{h=3,5,7,\ldots,N} \left(h - \frac{1}{h} \frac{(N-1)/2}{\Sigma_{h=3,5,7,\ldots,N} \frac{1}{h^2}} \right)^2.$$

If a single h-th harmonic component is concerned, then the optimal C_o is

$$C_o = \frac{1}{(h\omega^*)^2 L}.$$

This forces the impedance at the h-th harmonic frequency nearly 0 and hence no voltage at this frequency is caused.

7.4.2 Special Case I: to Minimise the 3rd and 5th Harmonic Components

In most cases, it is enough to consider the 3rd and 5th harmonics only. This gives the optimal capacitance

$$C_o = \frac{17}{225(\omega^*)^2 L}.$$

As a result, the output impedance of the inverter is

$$Z_o(j\omega) = j \left(\omega L - \frac{1}{\omega C_o} \right)$$

$$= j\omega^* L \left(\frac{\omega}{\omega^*} - \frac{225}{17} \frac{\omega^*}{\omega} \right).$$

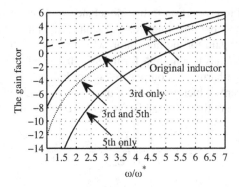

Figure 7.5 Gain factors to meet different criteria

The gain factor $\frac{\omega}{\omega^*} - \frac{225}{17}\frac{\omega^*}{\omega}$ of the imaginary part with respect to the normalised frequency $\frac{\omega}{\omega^*}$ is shown in Figure 7.5. It changes from negative to positive at around $\frac{\omega}{\omega^*} = 3.638$. At the fundamental frequency, i.e., when $\omega = \omega^*$, the output impedance is

$$Z_o = -j\frac{208}{17}\omega^* L \approx -j12.23\omega^* L.$$

It is capacitive as expected.

7.4.3 Special Case II: to Minimise the 3rd Harmonic Component

In this case, the optimal C_o is

$$C_o = \frac{1}{(3\omega^*)^2 L}$$

and the corresponding impedance is

$$Z_o(j\omega) = j\omega^* L \left(\frac{\omega}{\omega^*} - \frac{9\omega^*}{\omega}\right).$$

The gain factor $\frac{\omega}{\omega^*} - \frac{9\omega^*}{\omega}$ of the imaginary part with respect to the normalised frequency $\frac{\omega}{\omega^*}$ is also shown in Figure 7.5. It changes from negative to positive at $\omega = 3\omega^*$. At the fundamental frequency, i.e., when $\omega = \omega^*$, the output impedance is

$$Z_o = -j8\omega^* L \approx -j8\omega^* L,$$

which is capacitive as well.

7.4.4 Special Case III: to Minimise the 5th Harmonic Component

In this case, the optimal C_o is

$$C_o = \frac{1}{(5\omega^*)^2 L}$$

and the corresponding impedance is

$$Z_o(j\omega) = j\omega^* L \left(\frac{\omega}{\omega^*} - \frac{25\omega^*}{\omega} \right).$$

The gain factor $\frac{\omega}{\omega^*} - \frac{25\omega^*}{\omega}$ of the imaginary part with respect to the normalised frequency $\frac{\omega}{\omega^*}$ is also shown in Figure 7.5. It changes from negative to positive at $\omega = 5\omega^*$. At the fundamental frequency, i.e., when $\omega = \omega^*$, the output impedance is

$$Z_o = -j24\omega^* L \approx -j24\omega^* L.$$

This is capacitive as well.

7.5 Simulation Results for R-, L- and C-inverters

Simulations were carried out on a single-phase inverter powered by a 42 V DC voltage supply. The inverter was equipped with an outer-loop controller, as shown in Figure 7.6, to regulate the output voltage. This outer-loop controller is actually the robust droop controller (Zhong 2012c) for C-inverters to be discussed in detail in Chapter 19. The parameters were $n_i = 2.2$, $m_i = 0.14$ and $K_e = 20$. The switching frequency was 7.5 kHz and the frequency of the system was 50 Hz. The rated voltage was 12 V. The filter capacitance was 22 μF

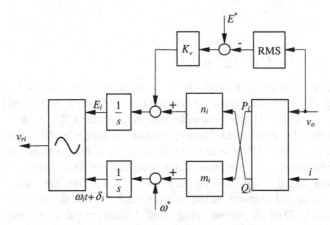

Figure 7.6 Outer-loop controller to generate the voltage reference v_r for C-inverters. See Chapter 19 for more details

Table 7.1 Steady-state performance of the inverter with $L = 2.35$ mH

Type of Z_o	L	R	$C_o = 325\ \mu F$ (3rd + 5th)	$C_o = 479\ \mu F$ (3rd)
THD of v_o	39.4%	25.9%	19.8%	25.9%
V_o	11.92	11.02	11.42	11.63

and the load was a full-bridge rectifier loaded with an LC filter $L = 150\ \mu H$, $C = 1000\ \mu F$ and $R_L = 9\ \Omega$.

7.5.1 The Case with $L = 2.35$ mH

The parasitic resistance of the inductor is assumed to be $0.1\ \Omega$. According to the analysis in the previous section, the optimal capacitance C_o can be chosen as $479\ \mu F$ to minimise the effect of the 3rd harmonics in v_o and $325\ \mu F$ to minimise the effect of the 3rd and 5th harmonics in v_o. The steady-state performance of the inverters with these controllers is shown in Table 7.1. The results of the system when the inverters were designed to have resistive output impedances (with $K_i = 4$) and inductive output impedances (with $K_i = 0$) using the robust droop controller proposed in (Zhong 2012c) are also shown in Table 7.1 for comparison. The C-inverters considerably improved the THD of the output voltage: from 39.4% obtained by the L-inverters and 25.9% obtained by the R-inverters to 19.8%. The lowest THD (19.8%) was obtained when the capacitor was designed as $C_o = 325\ \mu F$ to minimise the effect of the 3rd and 5th harmonics. The output voltage, the THD of the output voltage and the current curves for the inverters with different output impedance is shown in Figure 7.7. Apparently, the C-inverters offered the best THD and the L-inverters offered the worst THD. It is worth noting that the purpose of these simulations is to demonstrate that C-inverters can improve the THD of the output voltage because of the extra freedom introduced to optimise the THD but not to reduce the overall THD to meet industrial regulations. Other techniques, e.g. the ones proposed in (Shen et al. 2010; Zhong et al. 2011), can be applied to further decrease the THD.

7.5.2 The Case with $L = 0.25$ mH

According to the above analysis, a small filter inductor helps reduce the cost and the THD of the output voltage. In order to demonstrate this, the filter inductor $L = 2.35$ mH was replaced with an inductor $L = 0.25$ mH and the simulations were repeated. The parasitic resistance of the inductor is assumed to be $0.045\ \Omega$. The steady-state performance of the system is shown in Table 7.2. As expected, the THD was reduced significantly with comparison to the case with $L = 2.35$ mH. Moreover, the C-inverters again improved the THD from 8.2% to 7.0% for the R-inverters and from 9.2% to 7.0% for the L-inverters. The best THD was obtained when the capacitor was designed to minimise the effect of the 3rd harmonics as $C_o = 4500\ \mu F$. The output voltage, the THD of the output voltage and the current curves for the inverters with different output impedance is shown in Figure 7.8. Note that there are some variations in the THD of the output voltage. The values in Table 7.2 are the mean values.

Control of Inverter Output Impedance

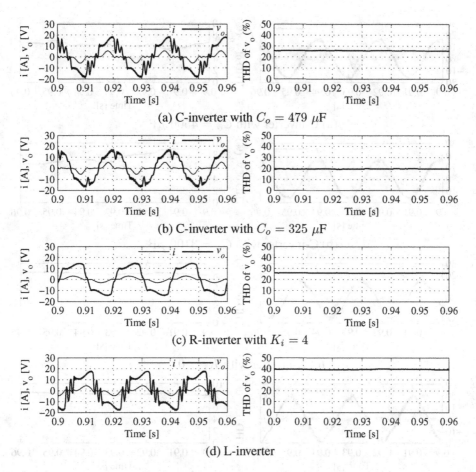

Figure 7.7 Simulation results for the case with $L = 2.35$ mH: output voltage and current (left column) and THD of the output voltage (right column)

7.6 Experimental Results for R-, L- and C-inverters

Experiments were also carried out on a test rig, of which the parameters are the same as those in the simulations, to further demonstrate the analysis. The load, which is highly non-linear, remained the same as well. Two cases were tested: one with $L = 2.35$ mH and the other with $L = 0.25$ mH.

Table 7.2 Steady-state performance of the inverter with $L = 0.25$ mH

Type of Z_o	L	R	$C_o = 3100\ \mu F$ (3rd + 5th)	$C_o = 4500\ \mu F$ (3rd)
THD of v_o	9.2%	8.2%	7.4%	7.0%
V_o	12.73	10.97	11.36	11.33

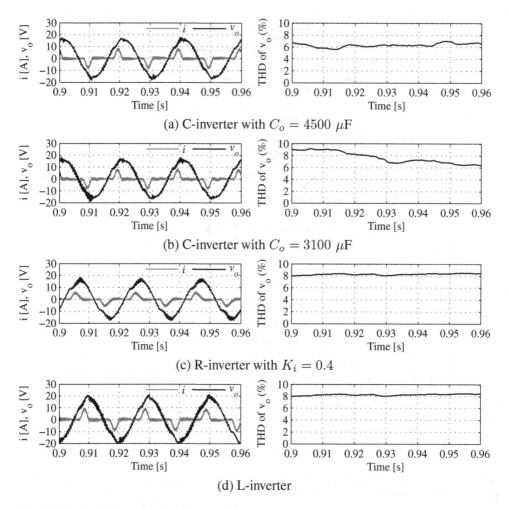

Figure 7.8 Simulation results for the case with $L = 0.25$ mH: output voltage and current (left column) and THD of the output voltage (right column)

7.6.1 The Case with $L = 2.35$ mH

The experimental results when the inverter was designed to have different types of output impedance is shown in Figure 7.9. When the inverter was designed to have a capacitive output impedance to minimise the effect of the 3rd harmonics, the THD was improved by about 5% from the case with an inductive output impedance and by about 3% from the case with a resistive output impedance (with $K_i = 4$). When the inverter was designed to have a capacitive output impedance to minimise the effect of both 3rd and 5th harmonics, the THD was improved by 3% and 1%, respectively.

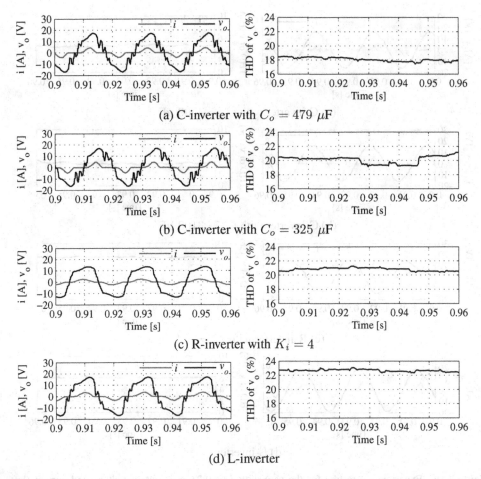

Figure 7.9 Experimental results for the case with $L = 2.35$ mH: output voltage and current (left column) and THD of the output voltage (right column)

7.6.2 The Case with $L = 0.25$ mH

The experimental results when the inverter was designed to have different types of output impedance is shown in Figure 7.10. When the inverter was designed to have a capacitive output impedance to minimise the effect of both 3rd and 5th harmonics, the THD was improved by 2% from the case with an inductive output impedance and by nearly 4% from the case with a resistive output impedance (with $K_i = 0.4$). When the inverter was designed to have a capacitive output impedance to minimise the effect of the 3rd harmonics, the THD was improved by about 3% and 1%, respectively. It is worth noting that when the inverter was designed to have a capacitive output impedance, the THD of the output voltage dropped below 5%.

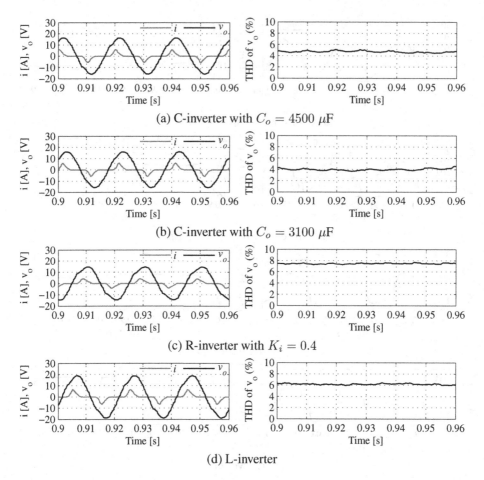

Figure 7.10 Experimental results for the case with $L = 0.25$ mH: output voltage and current (left column) and THD of the output voltage (right column)

7.7 Impact of the Filter Capacitor

It has been demonstrated that the filter capacitor can be regarded as a part of the load instead of a part of the inverter. The impact of the filter capacitor on the inverter can be analysed as below. The no-load voltage is

$$v_o = \frac{\frac{1}{sC}}{\frac{1}{sC} + Z_o(s)} v_r = \frac{1}{1 + sCZ_o(s)} v_r$$

and the equivalent output impedance after considering the capacitor is

$$\tilde{Z}_o(s) = \frac{\frac{1}{sC} Z_o(s)}{\frac{1}{sC} + Z_o(s)} = \frac{Z_o(s)}{1 + sCZ_o(s)}.$$

Since the capacitance C is often small, $\frac{1}{1+sCZ_o(s)}$ is a low-pass filter with a unity gain for a wide range of frequencies. As a result,

$$v_o \approx v_r \quad \text{and} \quad \tilde{Z}_o(s) \approx Z_o(s)$$

at low frequencies. Indeed, the role of the filter capacitor is negligible.

The impact of the the filter capacitor on the stability is discussed in Chapter 8.

7.8 Summary

The fact that the output impedance of an inverter plays an important role on the THD of the output voltage offers an alternative to improve the quality of output voltage of an inverter. Partially based on (Zhong 2012c; Zhong and Zeng 2011), control strategies are presented to change the output impedance of an inverter from inductive to resistive and capacitive, respectively. It is shown that when the output impedance is designed as capacitive, it is possible to minimise the THD of the output voltage. The capacitance of a C-inverter can be selected to eliminate certain harmonics. Both simulation and experimental results have demonstrated that C-inverters are able to offer the best voltage quality among R-, L- and C-inverters.

8

Bypassing Harmonic Current Components

As discussed in the previous chapter, the filter capacitor C can be regarded as part of the load instead of part of the inverter and the output impedance of an inverter can be changed to resistive or capacitive with a simple proportional or integral controller. In general, C-inverters offer better voltage quality than R-inverters and L-inverters. Simulation and experimental results have indicated that the voltage THD is high for R-inverters and L-inverters when a non-linear load is connected. It can be improved via reducing the filter inductor but the inductor cannot be too small. In this chapter, a voltage controller is introduced to improve the voltage quality. Moreover, it is found that the physical interpretation of the strategy is to bypass harmonic components of the load current.

8.1 Controller Design

For the single-phase inverter shown in Figure 8.1, the inductor current i is measured to construct a proportional controller so that the output impedance of the inverter is forced to be resistive and so that it dominates the impedance between the inverter and the AC bus, as shown in Chapter 7 to design R-inverters. This is also widely referred to as a virtual resistor (Dahono 2003, 2004; Dahono et al. 2001; Li 2009) to dampen the resonance of the LC filter. Here, it is used to force the output impedance of the inverter to be predominantly resistive with a large enough K_i. However, a large K_i results in high THD in the output voltage if non-linear loads are present. In order to improve the voltage quality, the output voltage v_o can be measured and compared with the voltage reference v_r to form a voltage loop, as shown in Figure 8.2. The following two equations hold for the closed-loop system consisting of Figure 8.1 and Figure 8.2:

$$u = v_r - K_i i + K_R(s)(v_r - v_o),$$
$$u_f = sLi + v_o.$$

Control of Power Inverters in Renewable Energy and Smart Grid Integration, First Edition.
Qing-Chang Zhong and Tomas Hornik.
© 2013 John Wiley & Sons, Ltd. Published 2013 by John Wiley & Sons, Ltd.

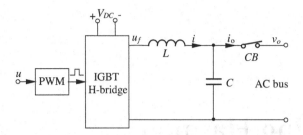

Figure 8.1 Single-phase inverter with a PWM block and an LC filter

Since the average of u_f over a switching period is the same as u, there is approximately

$$v_r - K_i i + K_R(s)(v_r - v_o) = sLi + v_o,$$

which gives

$$v_o = v_r - Z_o(s) \cdot i, \qquad (8.1)$$

where the output impedance $Z_o(s)$ is

$$Z_o(s) = \frac{sL + K_i}{1 + K_R(s)}. \qquad (8.2)$$

The resistive and inductive part of the output impedance are all decreased when the real part of K_R is positive, which is able to improve the THD of the output voltage. In general, there are harmonics in the current i because of non-linear loads and/or the pulse width modulation, which cause harmonic voltage drops on the output impedance Z_o. Since the reference v_r is often purely sinusoidal and does not contain any harmonic voltage components, the harmonic voltage drop on the output impedance Z_o appears in the output voltage, which degrades the voltage quality and causes a high THD. It is hence worth noting that the main factor that affects the voltage THD is the value of the output impedance instead of the type (resistive or inductive). The THD can be small even if the output impedance is inductive as long as it is small. The output impedance does not have to be resistive and the voltage THD can be large if the output impedance is resistive and large. Another observation is that the inductor current feedback increases the output impedance, which increases the voltage THD, and the voltage feedback decreases the output impedance, which decreases the voltage THD.

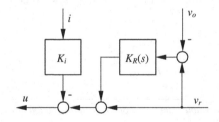

Figure 8.2 Controller to improve voltage THD

Bypassing Harmonic Current Components

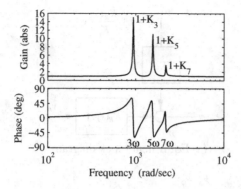

Figure 8.3 Bode plot of a typical $1 + K_R(s)$ with $\xi = 0.01$

The block $K_R(s)$ in Figure 8.2 can be designed to have small gains at low frequencies and high gains at high frequencies so that the output impedance is resistive at low frequencies (in particular at the fundamental frequency) for other purposes, e.g. power sharing, and is small at harmonic frequencies for the purpose of improving the voltage THD. There are many ways to achieve this. One option is to adopt the resonant harmonic compensators (Castilla *et al.* 2009; Shen *et al.* 2010)

$$K_R(s) = \sum_{h=3,5,\ldots} \frac{2\xi h\omega s}{s^2 + 2\xi h\omega s + (h\omega)^2} \times K_h, \qquad (8.3)$$

of which the gain at frequency $h\omega$ is K_h with zero phase; see the Bode plot of a typical $1 + K_R(s)$ with $\xi = 0.01$ shown in Figure 8.3. It is more or less 1 everywhere apart from at the frequencies around the harmonics. This is equivalent to reducing the inductance L to $\frac{L}{1+K_h}$ at the frequency $h\omega$ (and also the output resistance to $\frac{K_i}{1+K_h}$), which is able to improve the THD of the output voltage v_o. The damping factor ξ can be chosen as $\xi = 0.01$ to accommodate frequency variations and h can be chosen to cover the major harmonic components in the current, e.g. the 3rd, 5th and 7th harmonics.

8.2 Physical Interpretation of the Controller

As explained in the previous section, the strategy is able to reduce the output impedance of the inverter at the h-th harmonic frequency by $\frac{1}{1+K_h}$. As a matter of fact, this inner-loop controller is equivalent to the block diagram shown in Figure 8.4(a), that is, it is equivalent to the feedback of current $i_L = i + \frac{K_R(s)}{K_i} v_o$, which is the sum of the original current i and the current flowing through the admittance $\frac{K_R(s)}{K_i}$ that is connected across the output terminals. Since the reference voltage v_r only contains the fundamental component, it is the same after passing through the block $K_R(s) + 1$. Hence, the structure is in principle the same as a proportional current feedback adopted to achieve R-inverters discussed in Chapter 7 but the inductor current i_L has an additional component $\frac{K_R(s)}{K_i} v_o$. The corresponding equivalent circuit diagram is shown in Figure 8.4(b), where the admittance block $\frac{K_R(s)}{K_i}$ is connected in parallel with the filter capacitor

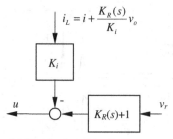

(a) Equivalent block diagram

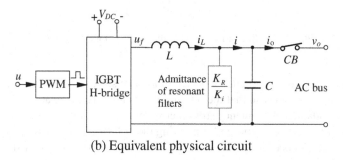

(b) Equivalent physical circuit

Figure 8.4 Equivalent structures of the controller shown in Figure 8.2

C. In other words, the inner-loop controller shown in Figure 8.2 is equivalent to adding the admittance block $\frac{K_R(s)}{K_i}$ to parallel with the filter C.

The admittance block $\frac{K_R(s)}{K_i}$ with the $K_R(s)$ given in (8.3) is

$$\frac{K_R(s)}{K_i} = \sum_{h=3,5,\ldots} \frac{2\xi h\omega s}{s^2 + 2\xi h\omega s + (h\omega)^2} \times \frac{K_h}{K_i},$$

which is the parallel connection of the admittance branches of

$$\frac{2\xi h\omega s}{s^2 + 2\xi h\omega s + (h\omega)^2} \times \frac{K_h}{K_i} = \frac{1}{\frac{K_i}{2\xi h\omega K_h}s + \frac{K_i}{K_h} + \frac{K_i}{K_h}\frac{h\omega}{2\xi s}}$$

for $h = 3, 5, \ldots$. Each branch is equivalent to the series connection of resistance $\frac{K_i}{K_h}$, inductance $\frac{K_i}{2\xi h\omega K_h}$ and capacitance $\frac{2\xi K_h}{h\omega K_i}$, which resonate at the h-th harmonic frequency $h\omega$. Hence, the controller shown in Figure 8.2 is equivalent to adding shunt series resonant filters at individual harmonic frequencies to the output to bypass harmonic current components. It is interesting to see that $\frac{K_i}{K_h}$ is the remaining impedance at the h-th harmonic frequency. In order to completely eliminate the h-th harmonic voltage component, K_i should be zero because K_h is finite. This indicates that forcing the output impedance of the inverter resistive via the current feedback increases certain harmonic voltage components. However, it is able to dampen high frequency oscillations (Dahono et al. 2001; Li 2009), which may reduce some harmonic voltage components. Hence, there exists a trade-off when determining K_i.

8.3 Stability Analysis

8.3.1 *Without Consideration of the Sampling Effect*

Assume that the load is passive and its impedance is $Z(s)$. Furthermore, assume that the ESR of the filter capacitor is $r > 0$. Then

$$i = \left(\frac{1}{Z(s)} + \frac{1}{r + \frac{1}{sC}}\right)v_o = \frac{1}{Z_c(s)}v_o, \tag{8.4}$$

where

$$Z_c(s) = \frac{Z(s)\left(r + \frac{1}{sC}\right)}{Z(s) + r + \frac{1}{sC}} = \frac{\frac{1 + srC}{sC}}{1 + \frac{1 + srC}{sC}\frac{1}{Z(s)}}$$

is the impedance of the augmented load after regarding the filter capacitor as part of the load and can be represented as the feedback loop shown in Figure 8.5. Since the impedance $Z(s)$ of any passive network must be positive real (Smith 2002; van der Schaft 1996), the Nyquist plot of the loop transfer function $\frac{1+srC}{sC}\frac{1}{Z(s)}$ does not encircle the critical point $(-1,0)$ on the s-plane, which means $Z_c(s)$ is always stable for any passive $Z(s)$. It is worth noting that the non-zero ESR of the capacitor r plays an important role in guaranteeing the stability. If $r = 0$, then the parallel connection of the load and the capacitor might be oscillatory. For example, a purely inductive load forms an oscillator with the capacitor when $r = 0$.

Substituting (8.4) into (8.1), then the closed-loop transfer function $G(s)$ from v_r to v_o is

$$G(s) = \frac{Z_c(s)}{Z_o(s) + Z_c(s)}$$

$$= \frac{(1 + K_R(s))Z_c(s)}{(sL + K_i) + (1 + K_R(s))Z_c(s)}$$

$$= \frac{H(s)}{1 + H(s)},$$

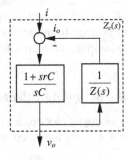

Figure 8.5 Augmented load $Z_c(s)$

where $H(s) = \frac{1+K_R(s)}{sL+K_i} Z_c(s)$. Since $1 + K_R(s)$ is stable for any positive ξ and ω, and $\frac{1}{sL+K_i}$ is stable for any positive L and K_i, $H(s)$ is stable for any passive $Z(s)$. The stability of the system can then be checked by plotting the Nyquist diagram of $H(s)$ and the range of the parameters can be determined by plotting the root locus of $H(s)$.

A sufficient condition is given below. The transfer function $G(s)$ or the system is stable if

$$\left\| \frac{1+K_R(s)}{sL+K_i} Z_c(s) \right\|_\infty < 1 \qquad (8.5)$$

according to the well-known small gain theorem (Zhou *et al.* 1996). The norm $\left\| \frac{1+K_R(s)}{sL+K_i} Z_c(s) \right\|_\infty$ is the maximum amplitude obtained from the Bode plot of $H(s)$. It can also be calculated iteratively, e.g. via using the MATLAB® routine *hinfnorm*. It is worth noting that this condition is sufficient only and it can be very conservative in some cases. A more conservative condition, which can be derived from (8.5), is

$$\|Z_c(s)\|_\infty < \frac{K_i}{1+K_{hm}},$$

where $K_{hm} = \max\{K_h\}$. For a resistive load $Z(s) = R$, there is

$$\|Z_c(s)\|_\infty = R$$

and, hence, the system is stable if

$$\frac{K_i}{1+K_{hm}} > R$$

or, equivalently, if

$$K_{hm} < \frac{K_i}{R} - 1.$$

This condition is likely to be very conservative because of the approximations applied to obtain it and also because of the inherent conservativeness of the small-gain theorem. What is clear from this is that there is an upper bound on K_h and, hence, K_h cannot be too big. The system is more stable if the load is heavier (with a smaller R). A large current feedback gain K_i also helps stabilise the system. However, K_i should not be too large because, as discussed before, a large K_i leads to high THD. It will be seen from the next subsection that a large K_i also destabilises the current loop when the sampling effect is taken into consideration.

8.3.2 With Consideration of the Sampling Effect

If the controller is implemented with analogue electronic circuits, then the current feedback gain K_i can be very large. However, if it is implemented with a digital controller, then it is limited, as will be illustrated below.

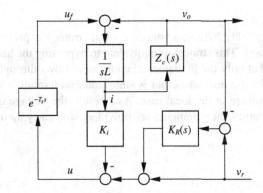

Figure 8.6 Approximate block diagram of the complete system

There are various ways to approximate the effect of sampling, computation and PWM conversion. One of them is to use a one-step delay $e^{-T_s s}$, where T_s is the sampling period. With this approximation, the complete system block diagram is shown in Figure 8.6 and the characteristic equation of the closed-loop system is

$$1 + \frac{K_i + K_R(s)Z_c(s)}{sL + Z_c(s)} e^{-T_s s} = 0,$$

or, equivalently,

$$sL + K_i e^{-T_s s} + (1 + K_R(s)e^{-T_s s})Z_c(s) = 0.$$

It is the same as that of the system with the open-loop transfer function of $\frac{1}{sL+K_i e^{-T_s s}}(1 + K_R(s)e^{-T_s s})Z_c(s)$. The terms $Z_c(s)$ and $1 + K_R(s)e^{-T_s s}$ remain stable under the aforementioned conditions. However, the term $\frac{1}{sL+K_i e^{-T_s s}}$ is no longer stable for very large K_i. In order to make sure that $\frac{1}{sL+K_i e^{-T_s s}}$ is stable, the gain K_i should satisfy

$$K_i < \frac{\pi}{2} \frac{L}{T_s}. \tag{8.6}$$

This is consistent with (7.2) and clearly shows that there is an upper bound on the current feedback gain K_i. The gain K_i should be further decreased if a low-pass filter is involved in the measurement of the current because the filter introduces extra phase lag.

8.4 Experimental Results

This control strategy can be combined with other strategies, e.g. the robust droop control for parallel operation of inverters discussed in Chapter 19. Extensive experimental results with the combined strategy to achieve accurate power sharing, tight voltage regulation and low voltage THD are presented in Chapter 20.

8.5 Summary

Based on (Zhong et al. 2011, 2012c), a voltage control strategy is presented to improve the voltage quality of inverters. This strategy is equivalent to bypassing the harmonic components in the load current so that only the fundamental component flows through the inverter. Since the reference voltage for the inverter output is sinusoidal, no harmonic voltage components appear on the output voltage in the ideal case. As a result, the voltage quality is improved. Some guidelines about tuning the parameters are provided, after carrying out stability analysis.

9
Power Quality Issues in Traction Power Systems

There are serious power quality issues in traction power systems, including negative-sequence currents, current harmonics and low power factor, in addition to voltage harmonics. In this chapter, these issues are dealt with. A topology for traction power systems with a single feeding wire is implemented with a three-phase V/V transformer and a three-phase converter. Compared to the traditional scheme with two feeding wires (in two phases) with three-phase V/V transformers, this topology improves the system reliability and has the potential for the traction of high-speed trains. Compared to the co-phase system proposed in the literature, this topology adopts a simple normal transformer instead of a complicated YNvd transformer and a three-phase converter instead of a back-to-back single-phase converter, which saves one converter leg. A strategy is then presented to control the three-phase converter so that all harmonic, negative-sequence and reactive currents generated by the non-linear single-phase load of locomotives are compensated. As a result, only balanced real power is drawn from the grid. Simulation results are provided to illustrate the performance of the system.

9.1 Introduction

In recent years, high-speed electrified trains have been rapidly developed all over the world and this is the future trend for railway transport. But for traditional traction power systems, there are some power quality problems such as low power factor, a significant amount of harmonics and negative-sequence currents caused by locomotives, which are present as single-phase non-linear loads (Chen et al. 1998). As a result, the grid currents are unbalanced and contain a lot of harmonics and reactive power (Chang et al. 2004; Ledwich and George 1994; Lee et al. 2006; Tan et al. 2003).

The problem of reactive power and harmonics is partially solved nowadays because high-speed locomotives are driven by four-quadrant PWM converters (Brenna et al. 2011; Busco et al. 2003; Chen et al. 2004). However, the problem with negative-sequence currents becomes more and more serious because the power of locomotives is increasing. An overview

---

Control of Power Inverters in Renewable Energy and Smart Grid Integration, First Edition.
Qing-Chang Zhong and Tomas Hornik.
© 2013 John Wiley & Sons, Ltd. Published 2013 by John Wiley & Sons, Ltd.

of the imbalance in traction power systems is presented in (Kneschke 1985) and the unbalanced currents in different kinds of power supply schemes are compared. The amount of negative-sequence currents are determined by the topology of the traction power system, especially the type of transformers adopted, and the power of locomotives. In traditional traction power systems, the topology with two-phase feeding wires is widely used (Chen et al. 2004). There are two main schemes in this category according to the transformers used: (i) with three-phase V/V transformers; and (ii) with some balancing transformers such as Scott transformers (Horita et al. 2010; Ming-Li et al. 2008), Woodbridge transformers (Morimoto et al. 2009) and Le Blanc transformers (Huang and Chen 2002). The detailed evaluation of negative-sequence currents injected into the grid from different traction substations equipped with different transformers is given in (Chen and Guo 1996; Wang et al. 2009). If a balancing transformer is used, the two-phase secondary currents result in balanced three phase currents on the grid side under some specific load conditions. However, since the speed and load of locomotives change frequently, the grid currents are normally unbalanced. In order to solve this problem, some active power compensators (APC) can be adopted on the three-phase grid side or on the two-phase track side. For example, an APC was proposed in (Sun et al. 2004) for a traction power system equipped with a Scott transformer to compensate for the negative-sequence currents.

Compared to balancing transformers, three-phase V/V transformers have a simple structure. However, since the V/V connection scheme is inherently unbalanced, its performance in reducing the three-phase imbalance is essentially worse than that of balancing transformers. In order to deal with this problem, a three-phase V/V transformer with a railway static power conditioner (RPC) is proposed in (Luo et al. 2011) and a strategy to compensate the negative-sequence and harmonic currents is explained. Some improved strategies are then proposed in (Wu et al. 2012), where a three-phase converter is used to replace the single-phase back-to-back converter.

In comparison with the traditional two-phase systems, the topology with a single feeding wire has some obvious advantages. First of all, the neutral section needed for each substation to separate the two phases can be removed. Secondly, the voltages on the two adjacent sections are nearly of the same phase so the insulation requirement between two adjacent sections is considerably reduced and the neutral sections needed by two-phase systems can be replaced with section insulators. Thirdly, the insulation/neutral sections for two-phase systems are quite long and the speed loss of locomotives when passing through neutral sections is quite significant. Compared to two-phase schemes, the number and length of neutral sections in single-feeder systems are considerably reduced. Therefore, the topology with a single feeding wire is more appropriate to provide power to high-speed trains.

Systems with a single feeding wire are explored in the literature. Such a system can be achieved by using a Steinmetz transformer (Driesen and Craenenbroeck 2002), which is a three-phase transformer with an extra power balancing load composed of a capacitor and an inductor rated proportional to the single-phase load. An obvious drawback is that the capacitor and the inductor need to be changed when the load changes. As a result, the capacitor and inductor can be replaced with static var compensators (SVC) (ABB 2010). A co-phase power traction system is proposed in (Shu et al. 2011; Zhao et al. 2010), where a complicated YNvd balancing transformer and an APC are used. After compensation, only the active power, including the load active power and system losses, is provided by the grid and in a balanced manner. All the active power is provided through the YNvd transformer and half of it flows through the APC. Some further optimised design and performance evaluation of the co-phase system are reported in (Chen et al. 2009).

Compared with the three-phase V/V scheme in (Luo et al. 2011), the YNvd transformer in the co-phase system is much more complicated than the single-phase transformers connected in the three-phase V/V scheme. Moreover, the APC adopted in the co-phase system is a single-phase back-to-back converter, which requires one more converter leg than the three converter legs needed by the three-phase V/V scheme.

In order to combine the advantages of the co-phase system and the three-phase V/V scheme, a new topology for traction power systems is presented in this chapter. It adopts a three-phase V/V transformer and a three-phase converter running as a static power conditioner (SPC). Moreover, it provides a single feeding wire without the need for a neutral section at each substation. The three-phase SPC is controlled to balance the three-phase grid currents and to compensate for the reactive and harmonic currents. The case when the power factor of the load is not unity is discussed in detail and the case with a unity power factor is presented as a special case. The harmonic components are compensated without any extra cost. A compensation strategy is presented so that all the harmonics and reactive power caused by the load are injected into the SPC. Hence, the grid currents are balanced and in phase with the corresponding phase voltages. It is worth noting that the SPC also maintains the DC-bus voltage and there is no need for an external power supply. The ripple voltage at the double frequency in the controller maintaining the DC-bus voltage is removed to make sure that the reference currents generated for the SPC are purely sinusoidal, which improves the THD of the grid currents.

9.2 Description of the Topology

The topology for traction power systems with a single feeding wire is shown in Figure 9.1. It adopts a three-phase V/V transformer to reduce the grid-side three-phase high voltage, e.g. 220 kV, to the track-side voltage, e.g. 27.5 kV, for traction. The turns ratio of the transformer is K_V. One open end of the secondary V windings, Terminal b in Figure 9.1, is connected to the track (earth) and the other open end of the secondary V windings, Terminal a in Figure 9.1, is connected to the catenary. A three-phase converter (called the static power conditioner, SPC) is connected to the two open ends of the secondary V windings and the common point, via two step-down single-phase transformers with a turns ratio of K_D in Figure 9.1. The SPC maintains the DC-bus voltage by itself and there is no need to provide an external power supply. The leakage inductances of the step-down transformers on the SPC side are denoted as L_a and L_b, respectively. Since the traction voltage is the grid line voltage divided by K_V, which is the same as the one in the conventional two-phase systems equipped with V/V transformers, an SPC can be easily retrofitted into existing two-phase traction power systems to improve power quality.

9.3 Compensation of Negative-sequence Currents, Reactive Power and Harmonic Currents

9.3.1 Grid-side Currents before Compensation

Assume the RMS value of the grid voltage is U and the phase angle of Phase A grid voltage is 0. Then the three phase grid voltages can be denoted as

$$\begin{cases} \dot{U}_A = U \angle 0, \\ \dot{U}_B = U \angle -\frac{2}{3}\pi, \\ \dot{U}_C = U \angle \frac{2}{3}\pi. \end{cases} \tag{9.1}$$

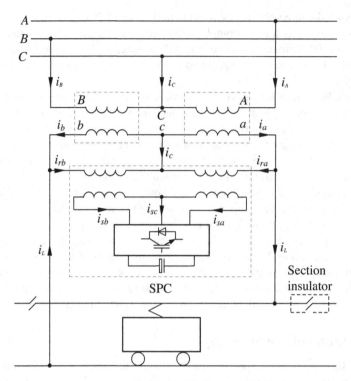

Figure 9.1 Traction power system with a single feeding wire equipped with a V/V transformer

The load current i_L is assumed to be purely sinusoidal without any harmonics for the moment (the case with harmonic currents will be discussed later) and the power factor is $\cos\theta$. Then, the load current can be expressed as

$$\dot{I}_L = I_{L1} \angle \left(\frac{\pi}{6} - \theta\right) \tag{9.2}$$

where I_{L1} is the RMS value of the fundamental load current. The three-phase grid currents without any compensation are

$$\begin{cases} \dot{I}_A = \dfrac{I_{L1}}{K_V} \angle \left(\dfrac{\pi}{6} - \theta\right), \\ \dot{I}_B = \dfrac{I_{L1}}{K_V} \angle \left(-\dfrac{5}{6}\pi - \theta\right), \\ \dot{I}_C = 0, \end{cases} \tag{9.3}$$

as shown in Figure 9.2(a). It is obvious that the three-phase grid currents are unbalanced. Both reactive and active current components are included in $\dot{I}_A$ and $\dot{I}_B$ as they are not in phase with $\dot{U}_A$ and $\dot{U}_B$, respectively. Phase-A current leads its voltage by $\frac{\pi}{6} - \theta$ and Phase-B current leads its voltage by $-\frac{\pi}{6} - \theta$. A significant amount of negative-sequence currents exists in the grid currents.

Power Quality Issues in Traction Power Systems

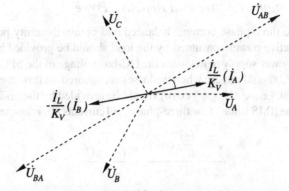

(a) At the grid side before compensation

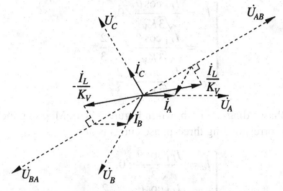

(b) At the grid side after compensation

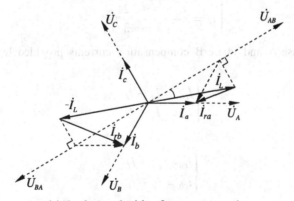

(c) At the track side after compensation

Figure 9.2 Phasor diagram of the system

9.3.2 Compensation of Active and Reactive Power

In order to make the three-phase currents balanced and obtain the unity power factor at the grid side, all the reactive power consumed by the load should be provided by the SPC. Since there is no external power supply to maintain the DC-bus voltage of the SPC, the active power consumed by the SPC should be zero when the losses are ignored. In this case, the active power consumed by the load, i.e., $\sqrt{3}U \times \frac{I_{L1}}{K_V} \cos\theta$, should be provided by the grid currents in a balanced way. Hence, the RMS value of the three-phase grid currents after compensation should be

$$\frac{\sqrt{3}U \frac{I_{L1}}{K_V} \cos\theta}{3U} = \frac{I_{L1}\cos\theta}{\sqrt{3}K_V},$$

which means the three-phase grid currents should be

$$\begin{cases} i_A = \dfrac{I_{L1}\cos\theta}{\sqrt{3}K_V}\angle 0, \\ i_B = \dfrac{I_{L1}\cos\theta}{\sqrt{3}K_V}\angle -\dfrac{2}{3}\pi, \\ i_C = \dfrac{I_{L1}\cos\theta}{\sqrt{3}K_V}\angle \dfrac{2}{3}\pi. \end{cases} \quad (9.4)$$

The corresponding phasor diagram is shown in Figure 9.2(b). Mapping the grid-side currents to the track side, the corresponding three-phase currents are

$$\begin{cases} i_a = \dfrac{I_{L1}\cos\theta}{\sqrt{3}}\angle 0, \\ i_b = \dfrac{I_{L1}\cos\theta}{\sqrt{3}}\angle -\dfrac{2}{3}\pi, \\ i_c = \dfrac{I_{L1}\cos\theta}{\sqrt{3}}\angle \dfrac{2}{3}\pi. \end{cases} \quad (9.5)$$

Therefore, the Phase-A and Phase-B compensation currents provided by the three-phase converter are

$$\begin{cases} i_{ra} = i_a - i_L, \\ i_{rb} = i_b + i_L. \end{cases}$$

or

$$\begin{cases} i_{ra} = i_a - i_L, \\ i_{rb} = i_b + i_L. \end{cases} \quad (9.6)$$

The Phase-C compensation current is

$$i_{rc} = i_c = -i_{ra} - i_{rb}.$$

The phasor diagram of the system at the track side after compensation is shown in Figure 9.2(c). It can be seen that I_{ra} and I_{rb} are not the same unless $\cos\theta = 1$. Because of the high voltage of the traction power system, two single-phase step-down transformers with turns ratio of K_D can be used.

It is worth noting that, after compensation, the loss in the grid transmission line is reduced by

$$1 - \frac{3 \times \left(\frac{I_{L1}\cos\theta}{\sqrt{3}K_V}\right)^2}{2 \times \left(\frac{I_{L1}}{K_V}\right)^2} = 1 - \frac{\cos^2\theta}{2}.$$

That is, the loss is reduced by at least 50%.

9.3.3 Compensation of Harmonic Currents

The above analysis is based on the assumption that there is no harmonics in the load current. As a matter of fact, according to (9.6), all the harmonic current components, if any, are automatically diverted into the compensation currents i_{ra} and i_{rb} since i_a and i_b only contain the fundamental current component. Therefore, no extra effort is needed to suppress the harmonics. It is worth noting that the current i_{rc} (i_c) only contains fundamental components even if the load current contains harmonics.

9.3.4 Regulation of the DC-bus Voltage

A stable DC-link voltage is required in order for the SPC to work properly. This can be achieved by introducing a PI controller to maintain the DC bus voltage V_c at the DC-bus reference voltage V_{cref}. The output of the DC-bus voltage controller is added on to the required RMS value of the track-side currents so that the right amount of active power can be injected into the SPC. Because of the double frequency ripple component in the DC-bus voltage, a low-pass filter, such as the hold filter

$$H(s) = \frac{1 - e^{-Ts/2}}{Ts/2},$$

where T is the fundamental period of the system, can be adopted to measure the DC component of V_c for feedback.

9.3.5 Implementation of the Compensation Strategy

The above compensation strategy can be implemented as shown in Figure 9.3. The sinusoidal tracking algorithm (STA) (Ziarani and Konrad 2004) (see also Chapter 22) is adopted to calculate the phase of the fundamental component of the grid line voltage u_{ab}. The phase of the voltage u_{ab} is $\omega t + \frac{\pi}{6}$ so it can be used to generate the signal $\sin(\omega t)$ and $\sin(\omega t - \frac{2\pi}{3})$ needed to form the reference compensation currents. The product of $\sin(\omega t + \frac{\pi}{6})$ with the fundamental load current is

$$\sin\left(\omega t + \frac{\pi}{6}\right) \times \sqrt{2}I_{L1}\sin\left(\omega t + \frac{\pi}{6} - \theta\right)$$
$$= \frac{\sqrt{2}I_{L1}}{2}\left(\cos\theta - \cos\left(2\omega t + \frac{\pi}{3} - \theta\right)\right),$$

of which the DC component $\frac{\sqrt{2}}{2}I_{L1}\cos\theta$ can be multiplied with $\frac{2}{\sqrt{3}}$ to obtain the required amplitude of the track-side currents, i.e., $\sqrt{2} \times \frac{1}{\sqrt{3}}I_{L1}\cos\theta$.

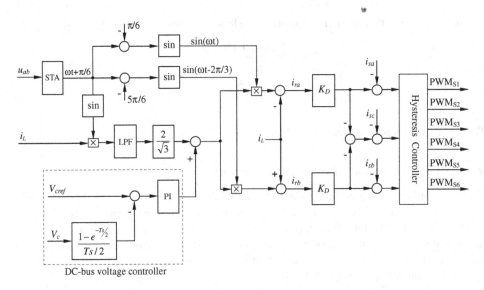

Figure 9.3 Control strategy to compensate negative-sequence, reactive and harmonic currents

The major control problem is for the SPC to track the calculated reference compensation currents i_{ra}, i_{rb} and i_{rc}. This can be done with many control strategies. For example, the repetitive controller discussed in other chapters is a very good candidate that works with a fixed switching frequency. In this chapter, three hysteresis controllers are adopted to generate PWM signals to drive the converter switches, as shown in Figure 9.3.

9.4 Special Case: $\cos\theta = 1$

Nowadays, many high-speed trains are equipped with four-quadrant converters and the power factor of the load is nearly 1. In this case, the load current is

$$i_L = I_{L1} \angle \frac{\pi}{6}$$

and the grid currents before compensation are

$$\begin{cases} i_A = \dfrac{I_{L1}}{K_V} \angle \dfrac{\pi}{6}, \\ i_B = \dfrac{I_{L1}}{K_V} \angle -\dfrac{5}{6}\pi, \\ i_C = 0. \end{cases}$$

After compensation, the grid currents are

$$\begin{cases} i_A = \dfrac{I_{L1}}{\sqrt{3}K_V} \angle 0, \\ i_B = \dfrac{I_{L1}}{\sqrt{3}K_V} \angle -\dfrac{2}{3}\pi, \\ i_C = \dfrac{I_{L1}}{\sqrt{3}K_V} \angle \dfrac{2}{3}\pi. \end{cases}$$

and the three-phase currents on the track side are

$$\begin{cases} i_a = \dfrac{I_{L1}}{\sqrt{3}} \angle 0, \\ i_b = \dfrac{I_{L1}}{\sqrt{3}} \angle -\dfrac{2}{3}\pi, \\ i_c = \dfrac{I_{L1}}{\sqrt{3}} \angle \dfrac{2}{3}\pi. \end{cases} \quad (9.7)$$

The corresponding phasor diagrams are shown in Figure 9.4. In this case, the compensation currents

$$\begin{cases} i_{ra} = \dfrac{I_{L1}}{\sqrt{3}} \angle -\dfrac{2}{3}\pi, \\ i_{rb} = \dfrac{I_{L1}}{\sqrt{3}} \angle 0, \\ i_{rc} = \dfrac{I_{L1}}{\sqrt{3}} \angle \dfrac{2}{3}\pi, \end{cases} \quad (9.8)$$

are balanced but in the negative sequence, with the same amplitude as that of the currents on the secondary side of the V/V transformer.

9.5 Simulation Results

Simulations were carried out in MATLAB®/Simulink® to verify the strategy. The solver used was ode23tb with a maximum step size of 1 μs.

The model of the high-speed train is shown in Figure 9.5. The turns ratio of the locomotive transformer is 27.5:1.5 and the parameters are: $R_1 = 0.5 \, \Omega$, $L_1 = 1.5$ mH, $C_1 = 1000 \, \mu F$, $R = 6 \, \Omega$, $L = 20$ mH and $C = 460 \, \mu F$. The capacitor C_1 is chosen as $1000 \, \mu F$ to obtain a power factor of 0.6. The parameters of the PI controller for the DC-bus voltage are $K_p = 0.1$ and $K_i = 0.005$. The parameters of the traction power system are given in Table 9.1.

9.5.1 The Case when $\cos\theta \neq 1$

The THD of the load current was 30%. The system was started at 0 s and the SPC was turned on at 0.08 s. The load current is shown in Figure 9.6(a) and the three-phase grid-side currents are shown in Figure 9.6(b). It can be seen that the Phase-C current before compensation is 0. In addition, there is a significant amount of harmonics in i_A and i_B. After compensation, the three-phase grid currents are balanced and clean. Moreover, the amplitude of the grid currents is reduced considerably. As shown in Figure 9.6(c), the DC-bus voltage was initially set at 4600 V and was maintained close to the reference voltage after the SPC was turned on. The compensation currents generated by the three-phase converter are shown in Figure 9.6(d). The THD of Phase-A current at the grid-side dropped from 30% to 1%, as shown in Figure 9.6(e) and its zoomed version in Figure 9.6(f).

9.5.2 The Case when $\cos\theta = 1$

In this case, the train model is changed to a purely resistive load of 5 Ω connected to the locomotive transformer. The corresponding curves are shown in Figure 9.7 when the SPC was turned on at 0.08 s. Similar performance was obtained.

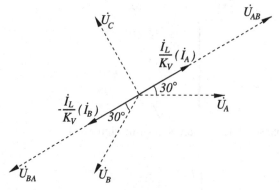

(a) At the grid side before compensation

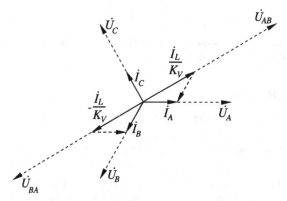

(b) At the grid side after compensation

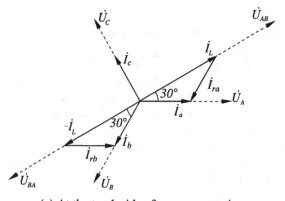

(c) At the track side after compensation

Figure 9.4 Phasor diagrams of the system when $\cos\theta = 1$

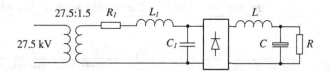

Figure 9.5 Load model of a high-speed train

Table 9.1 Parameters of the traction system

Parameters	Values
Grid-side line voltage	220 kV
K_V	220 : 27.5
K_D	27.5 : 1
L_a and L_b	1.5 mH
DC-bus capacitor	30000 μF
Initial voltage of the DC-bus capacitor	4600 V

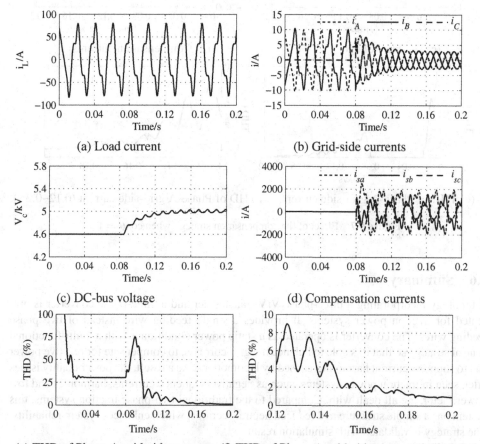

(a) Load current
(b) Grid-side currents
(c) DC-bus voltage
(d) Compensation currents
(e) THD of Phase-A grid-side current
(f) THD of Phase-A grid-side current (0.12~0.2 s)

Figure 9.6 Effect of the compensation strategy when $\cos\theta \neq 1$

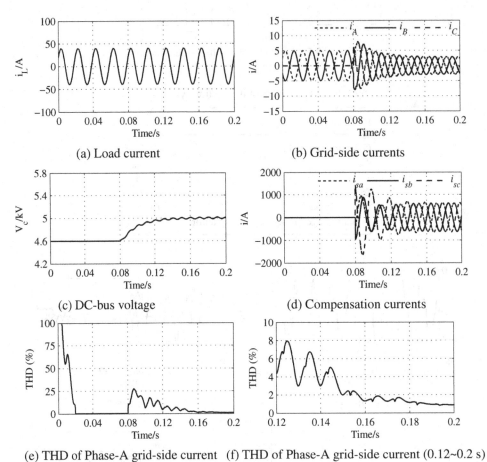

(e) THD of Phase-A grid-side current   (f) THD of Phase-A grid-side current (0.12~0.2 s)

Figure 9.7 Effect of the compensation strategy when $\cos\theta = 1$

9.6 Summary

A topology incorporating a three-phase V/V transformer and a three-phase converter is presented for traction power systems. It provides a single feeding wire instead of two phase feeding wires. The converter is operated as a static power conditioner with a multi-functional control strategy so that it is able to balance the grid currents, to compensate for reactive power and to suppress current harmonics caused by locomotives. As a result, the power quality issues often seen in traction power systems, such as negative-sequence currents, harmonics and low power factor, are all dealt with. Compared to the traditional two-phase traction systems, this system has a simple structure and reduced neutral sections, which enhances system reliability. The strategy is validated with simulation results.

Part II
Neutral Line Provision

10

Topology of a Neutral Leg

For applications in renewable energy, distributed generation and smart grids, there is often a need to have a neutral line to work with inverters so that a current path is provided for unbalanced loads. A neutral line is also needed when the phases of an inverter need to be independently operated so that the coupling effect among the phases is minimised. Indeed, as demonstrated in Part I, the power quality of the phase voltages generated by an inverter is best when the phases are independently controlled. In this chapter, some topologies to provide a neutral line are discussed.

10.1 Introduction

The introduction of microgrids and smart grids improves power quality, reduces transmission line congestion, decreases emission and energy losses, and effectively facilitates the utilisation of renewable energy resources (Chen *et al.* 2010; Lee and Cheng 2007; Li and Kao 2009; Li *et al.* 2007; Nikkhajoei and Lasseter 2009; Xiarnay *et al.* 2008). In some circumstances, the inverters used in microgrids and/or other applications must supply a mixture of single- and three-phase loads via a four-wire three-phase distribution network. A neutral line is often needed to provide a current path for unbalanced loads and the traditional six-switch inverter must be supplemented with a neutral connection (De and Ramanarayanan 2010; Li *et al.* 2006; Zhong *et al.* 2006). Figure 10.1 shows a microgrid, where a neutral line is provided for the local loads. If the neutral point of the inverter that provides the neutral line is not well balanced, then the neutral-point voltage may deviate severely from the midpoint of the DC source of the inverter. This may result in an unbalanced or variable output voltage, DC voltage/current components, large neutral current or even more serious problems. Thus, the generation of a balanced neutral point in a simple and effective manner has become an important issue (De and Ramanarayanan 2010; Li *et al.* 2006; Zhong *et al.* 2006). As usual, a microgrid equipped with a neutral line can be operated in the grid-connected mode or the stand-alone mode. If there are multiple inverters, then one of them can be equipped with the neutral line and the rest of them can be operated under the balanced three-phase mode. In this part, the case when

---

Control of Power Inverters in Renewable Energy and Smart Grid Integration, First Edition.
Qing-Chang Zhong and Tomas Hornik.
© 2013 John Wiley & Sons, Ltd. Published 2013 by John Wiley & Sons, Ltd.

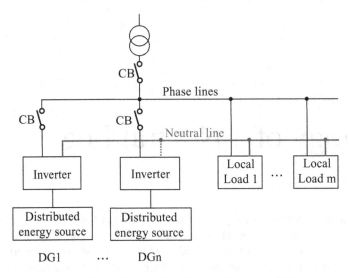

Figure 10.1 Microgrid equipped with a neutral line

one inverter provides the neutral line is discussed. If multiple inverters are fitted with a neutral line in a microgrid, then appropriate strategies should be developed to share the neutral current among them, in addition to the widely-studied load sharing among parallel-operated inverters (Guerrero et al. 2011; Yao et al. 2011; Zhong 2012c).

Another situation where a balanced neutral point is required is in multi-level pulse width modulation (PWM) converters, which have the potential to improve the total harmonic distortion and to reduce the voltage stress across the switches; see (Celanovic and Boroyevich 2000; Nabae et al. 1981; Soto-Sanchez and Green 2001; Wong et al. 2001) and the survey paper (Rodriguez et al. 2002).

10.2 Split DC Link

A straightforward way to provide a neutral point is to use two capacitors in parallel with large balancing resistors, as shown in Figure 10.2(a), with the neutral point clamped at half of the DC link voltage. Such a configuration is called a split DC link. This topology is widely adopted in active power filters because the neutral current in active power filters does not have a DC component and the fundamental component is relatively small; see, for example, (Quinn and Mohan 1992; Verdelho and Marques 1998). It has also been widely adopted in three-level converters because there is no need to provide a neutral line on the AC side and the aim is to maintain a balanced mid-point of the DC link. Some special PWM generating schemes are available to balance the charging of the capacitors; see e.g. (Celanovic and Boroyevich 2000; Lee et al. 1998; Ratnayake et al. 1999).

However, for applications with a large neutral current, the neutral current flows through the capacitors and, hence, bulky capacitors are needed to make the voltage ripple on the capacitors small. Another drawback is that the neutral point usually drifts and becomes unbalanced, in

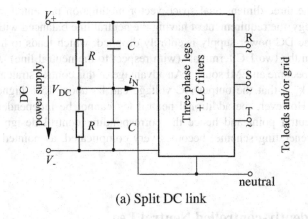

(a) Split DC link

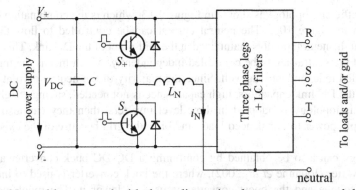

(b) Neutral leg as used with three-dimensional space vector modulation

Figure 10.2 Conventional topologies to generate a neutral line

particular, when the neutral current has a DC component. In order to improve the performance of the split DC link, different neutral point balancing strategies are reported, usually using redundant states of the Space Vector Pulse Width Modulation (SVPWM) (Bendre et al. 2006; Salaet et al. 2006; Zhou and Rouaud 2001) or varied voltage-balancing modulation techniques (Busquets-Monge et al. 2008a; Dai et al. 2008b; Ghennam et al. 2010; Lewicki et al. 2011; Wang and Li 2010; Zaragoza et al. 2009). However, this means the control of the neutral point is not decoupled from the control of the inverter, which may cause problems as well. Hence, this topology is not suitable for DC-AC converters which supply power to possibly unbalanced loads or to the grid.

10.3 Conventional Neutral Leg

A better way to provide a neutral line is to add an additional fourth leg, called a neutral leg, to the conventional three-leg converter (Jahns et al. 1993), as shown in Figure 10.2(b). It can be controlled by the three-dimensional space vector modulation presented in (Zhang *et al.* 1997). In this strategy, the requirement of having the neutral line balanced with respect to the two terminals of the DC power supply is entirely dropped, which leads to high voltages at high frequencies on the two DC terminals (with respect to the neutral line). As a result, the DC terminals may become an EMI source. An advantage of this control strategy, as compared to the split DC link, is that the output AC voltage can be about 15% higher (for a given DC link voltage). However, the additional neutral leg cannot be independently controlled to maintain the neutral point and hence the corresponding control design (including the PWM waveform generating scheme) becomes very complicated, as pointed out in (Zhang *et al.* 1997).

10.4 Independently-controlled Neutral Leg

A simple and efficient topology is shown in Figure 10.3, which is the combination of the two circuits shown in Figure 10.2. The neutral current can be controlled to flow through the inductor so that the neutral point is maintained at the mid-point of the DC link. This allows the phase legs and the neutral leg to be controlled independently, which means the three phases can be controlled independently as well. Since the majority of the neutral current no longer flows through the DC link capacitors, high capacitance is not needed, saving weight, volume and cost. If this topology is adopted in three-level inverters, then they can also be easily applied to supply power to unbalanced loads and/or to the grid because of the existence of a neutral line.

This topology can also be obtained by combining a DC-DC buck converter and a DC-DC boost converter (Jouanne *et al.* 2002), where the buck converter is used to increase the neutral point voltage and the boost converter is used to lower it. The topology can also be obtained from a class-B chopper (Chen *et al.* 2000). In both (Jouanne *et al.* 2002) and (Chen *et al.* 2000), a comparator with hysteresis is used to generate pulses to operate these two switches. Hence, the controller is non-linear. In Chapters 11–13, several linear control strategies are discussed.

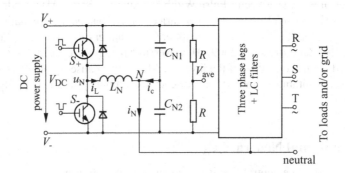

Figure 10.3 Independently-controlled neutral leg

10.5 Summary

Based on (Hornik and Zhong 2011; Zhong *et al.* 2002a, 2005a, 2005b, 2006), three different topologies to provide a neutral line are discussed. These are a split DC link, a conventional neutral leg and an independently-controlled neutral leg, which is the combination of the former two. It allows the neutral leg to be controlled independently from the inverter phase legs. Moreover, it allows the phase legs of the inverter to be controlled independently as well.

11

Classical Control of a Neutral Leg

In this chapter, a controller is developed for the independently-controlled neutral leg by using classical control techniques to provide a neutral line for a 3-phase 4-wire inverter. The neutral point can be steadily regulated at the mid-point of the DC link by a simple controller, involved in voltage and current feedback, even when the three-phase converter system (mostly, the load) is extremely unbalanced. The achievable performance is analysed quantitatively and the parameters of the neutral leg are determined accordingly.

11.1 Mathematical Modelling

Here, the linear model of the neutral leg is developed in a slightly different way from that in (Zhong et al. 2006) in order to gain some insightful understanding about the neutral leg, by ignoring the resistance of the inductor and assuming that the capacitors are the same.

Assume that the voltages across the capacitors C_N with respect to the neutral point N are V_+ and V_-, respectively. Then the DC link voltage is

$$V_{DC} = V_+ - V_-,$$

and the shift of the neutral point is[1]

$$\varepsilon = V_+ + V_-.$$

According to Kirchhoff's laws, the following equations are satisfied:

$$\begin{cases} u_N = L_N \dfrac{di_L}{dt}, \\ i_N = i_L + i_c, \\ i_c = C_N \dfrac{dV_+}{dt} + C_N \dfrac{dV_-}{dt} = C_N \dfrac{d\varepsilon}{dt}. \end{cases} \quad (11.1)$$

---

[1] Actually, ε is twice the shift of the neutral point with respect to the mid-point of the DC link. In order to simplify the exposition, it is called the shift.

Control of Power Inverters in Renewable Energy and Smart Grid Integration, First Edition.
Qing-Chang Zhong and Tomas Hornik.
© 2013 John Wiley & Sons, Ltd. Published 2013 by John Wiley & Sons, Ltd.

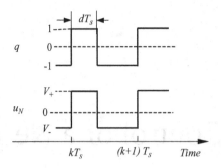

Figure 11.1 Firing pulse q and inductor voltage u_N

Assume that the average of the firing pulse during a switching period is p (with the magnitude of ± 1), as shown in Figure 11.1, then the duty cycle of the firing pulse is

$$d = \frac{1+p}{2}$$

and the average of the voltage u_N (i.e. the inductor voltage), as shown in Figure 11.1, is

$$\begin{aligned} u_N &= dV_+ + (1-d)V_- \\ &= \frac{1+p}{2}V_+ + \frac{1-p}{2}V_- \\ &= \frac{p}{2}V_{DC} + \frac{1}{2}\varepsilon. \end{aligned} \quad (11.2)$$

Combining the Laplace transformation of (11.1) and (11.2), the block diagram of the neutral leg can be obtained as shown in Figure 11.2.

The amplifying effect of the IGBT bridge (the block $\frac{V_{DC}}{2}$ in Figure 11.2) can be cancelled/scaled by $\frac{2}{V_{DC}}$ in the controller to be designed and, hence, ignoring this block does not affect the control design. Moreover, the block $\frac{1}{2}$ in Figure 11.2 can be cancelled by the proportional gain of the controller to be designed because the setpoint for the shift ε is 0.

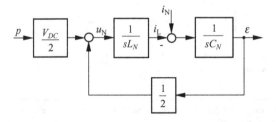

Figure 11.2 Block diagram of the independently-controlled neutral leg

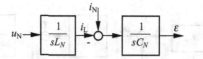

Figure 11.3 Simplified model of the neutral leg

Based on these two facts, the problem is equivalently changed from controlling the average p of the firing pulse as shown in Figure 11.2 to controlling the inductor voltage u_N, as shown in Figure 11.3. As can be seen later, the proportional gain from the shift ε to u_N is designed to be much larger than $\frac{1}{2}$. Hence, when implementing the controller, this block is usually ignored and only the scaling block $\frac{2}{V_{DC}}$ is needed to cascade to the designed controller.

According to (11.1),

$$\varepsilon = \int_0^t \frac{1}{C_N}(i_N - i_L)\mathrm{dt}. \tag{11.3}$$

Since the control objective is to maintain the point N as a neutral point (i.e. the shift $\varepsilon \approx 0$), then

$$\int_0^t (i_N - i_L)\mathrm{dt} = C_N \varepsilon(t)$$

is expected to be very close to 0 all the time. As a result, i_L is expected to be almost equal to i_N. This means the majority of the neutral current should flow through the inductor L_N but not through the capacitors C_N. The smaller the i_c, the smaller the shift ε. Hence, the capacitors are not necessarily very large like those in the conventional split DC link topology. This is a very important fact.

11.2 Controller Design

The system shown in Figure 11.3 is a double integrator system. A proportional-integral (PI) controller or even a proportional (P) controller cannot stabilise it. A possible option is to adopt a lead compensator

$$C(s) = K_p \frac{bs+1}{as+1},$$

with $K_p > 0$ and $b > a > 0$, taking the measurement of ε. Three control parameters need to be determined.

Since the neutral leg is effectively a second-order system, another option is to adopt a controller that has only two parameters to be determined. The states of the neutral leg can be chosen as the shift ε and the current i_c based on the fact that the smaller the current i_c, the smaller the shift ε. The corresponding control scheme is shown in Fig. 11.4, where an inner current loop is introduced to attenuate the effect of the neutral current i_N on the current i_c and an outer voltage loop is used to regulate the shift ε. Because there exists an integrator in

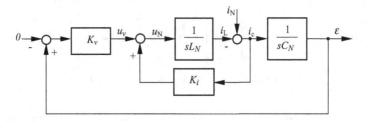

Figure 11.4 Voltage-current control scheme for the neutral leg

both the voltage loop and the current loop, the two loop controllers can simply be designed as proportional controllers K_v and K_i, respectively.

11.2.1 Design of the Current Controller K_i

The objective of the current loop is to attenuate the effect of neutral current i_N on i_c as much as possible, at least for those frequency components under consideration. For systems with a fundamental frequency of 50 Hz, the highest frequency components under consideration can be up to the 31st harmonics, i.e., 1550 Hz. Approximately, the corresponding angular frequency is 10000 rad/s. As can be seen later, this is a crucial parameter for the performance of the neutral point (the level of the shift ε).

The transfer function from i_N to i_c in Figure 11.5(a) is $\frac{sL_N}{sL_N+K_i}$, of which the Bode plot is shown in Figure 11.5(b). The corner angular frequency, called the inner loop (angular) frequency, is $\omega_i = \frac{K_i}{L_N}$. As discussed in Section 11.1, the smaller the current i_c, the smaller the shift ε. In order to make i_c as small as possible, the inner loop frequency ω_i should be high enough while considering the physical limit.[2] Hence, the current-loop control parameter is

$$K_i = \omega_i L_N. \tag{11.4}$$

Here, ω_i is chosen as 10000 rad/s. Accordingly, the switching frequency f_s is chosen as 10 kHz, which is larger than 3 ~ 4 times of 1550 Hz.

11.2.2 Design of the Voltage Controller K_v

After the current loop is designed, the system shown in Figure 11.4 is equivalent to the one shown in Figure 11.6, based on which the voltage controller K_v can be designed.

The transfer function from i_N to ε is

$$T(s) = \frac{sL_N}{s^2 L_N C_N + s K_i C_N + K_v}. \tag{11.5}$$

---

[2] As can be seen later in Figure 11.8, the larger the ω_i, the higher the high frequency gain from i_N to u_N is required. But this is usually limited by the low-pass filtering effect of the system.

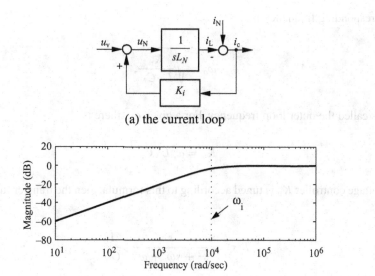

(a) the current loop

(b) the magnitude Bode plot from i_N to i_c

Figure 11.5 Design of the current loop

and its magnitude frequency response is

$$|T(j\omega)| = \frac{\omega L_N}{\sqrt{(K_v - \omega^2 L_N C_N)^2 + (\omega K_i C_N)^2}}$$

$$= \frac{L_N}{\sqrt{\left(\frac{K_v}{\omega} - \omega L_N C_N\right)^2 + (K_i C_N)^2}}. \tag{11.6}$$

It reaches its maximum when

$$\frac{K_v}{\omega} - \omega L_N C_N = 0. \tag{11.7}$$

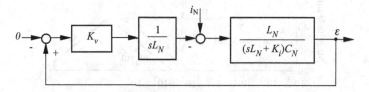

Figure 11.6 Equivalent structure for the voltage loop

The corresponding frequency is

$$\omega_o = \sqrt{\frac{K_v}{L_N C_N}},$$

which is called the outer loop frequency. For a given ω_o, there is

$$K_v = \omega_o^2 L_N C_N.$$

If the voltage controller K_v is tuned according to this formula, then the transfer function (11.5) becomes

$$T(s) = \frac{s}{s^2 + s\omega_i + \omega_o^2} \cdot \frac{1}{C_N}.$$

The Bode plot is shown in Figure 11.7 for different ω_o. It consists of three parts A, B and C. When K_v or ω_o is increased, the left part A moves towards right while the length of the middle part B is shortened and the right part C remains almost unchanged (when K_v or ω_o is not extremely large), and the shift ε becomes smaller. When C_N is decreased, parts B and C move up and part A extends while the length of part B is shortened, and the shift ε becomes larger. Hence, in order to reduce the shift ε, larger K_v (or ω_o) and larger C_N are needed. Moreover, the left part A of the Bode plot should move towards the right as close as possible to the right part C. The extreme case is to choose

$$\omega_o = \omega_i.$$

This makes use of the full physical capability of the controller. The corresponding Bode plot is shown as the thick line in Figure 11.7 and the voltage controller is obtained with

$$K_v = \omega_i^2 L_N C_N. \tag{11.8}$$

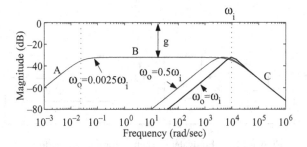

Figure 11.7 Bode plot for the voltage loop

11.3 Performance Evaluation

Substitute (11.7) into (11.6), the maximum value of the magnitude frequency response from the neutral current i_N to the shift ε is

$$20 \lg |T(j\omega)|_{max} = 20 \lg \frac{1}{\omega_i C_N},$$

which is illustrated as g in Figure 11.7. It is independent of the inductor L_N. This equation means that for any neutral current i_N containing any single frequency component (assume its peak value is I_N), the maximal shift ε_m is not larger than $\frac{I_N}{\omega_i C_N}$, i.e.

$$\varepsilon_m \leq \frac{I_N}{\omega_i C_N}.$$

This is equivalent to the voltage resulted by a sinusoidal current source I_N with a frequency equal to the inner loop frequency ω_i flowing through the DC link capacitor. In order to obtain a small shift ε, a small neutral current I_N and/or a large capacitance C_N and/or a high inner loop frequency ω_i are needed. For a sinusoidal neutral current i_N with a specific frequency component f_j and a peak value of I_j, the magnitude frequency response (dB), according to (11.6), is

$$20 \lg |T(j\omega)| = 20 \lg \frac{1}{\omega_i C_N} - 20 \left(\lg \omega_o - \lg(2\pi f_j) \right)$$
$$= 20 \lg \frac{2\pi f_j}{\omega_i \omega_o C_N}.$$

Thus, the corresponding shift ε_j is

$$\varepsilon_j = \frac{2\pi f_j I_j}{\omega_i \omega_o C_N}. \tag{11.9}$$

Moreover, when $\omega_o = \omega_i$,

$$\varepsilon_j = \frac{2\pi f_j I_j}{\omega_i^2 C_N}. \tag{11.10}$$

As is well known, any periodic neutral current i_N can be decomposed as a series

$$i_N = \Sigma_{j=0}^{+\infty} I_j \sin(2\pi j f t + \phi_j), \tag{11.11}$$

where f is the fundamental frequency. In practice, only a few components, e.g. up to the first 31 harmonics, need to be considered, bearing in mind that the DC component does not contribute to the shift according to (11.10). Hence, the root-mean-square value of the shift is

$$\varepsilon = \sqrt{\frac{1}{2}\Sigma_{j=1}^{31}\varepsilon_j^2}$$

$$= \sqrt{2 \cdot \Sigma_{j=1}^{31}\left(\frac{j\pi f I_j}{\omega_i^2 C_N}\right)^2}$$

$$= \frac{\pi f}{\omega_i^2 C_N}\sqrt{\Sigma_{j=1}^{31} 2 j^2 I_j^2}.$$

Another important issue in engineering is how the control signal u_N behaves. The transfer function from i_N to u_N is

$$T_u(s) = \frac{sL_N(sK_iC_N + K_v)}{s^2 L_N C_N + sK_i C_N + K_v}$$

$$= \frac{sL_N(s\omega_i + \omega_o^2)}{s^2 + s\omega_i + \omega_o^2}. \tag{11.12}$$

Moreover, when $\omega_o = \omega_i$,

$$T_u(s) = \frac{s(s + \omega_i)}{s^2 + s\omega_i + \omega_i^2}\omega_i L_N.$$

The Bode plot is shown in Figure 11.8. There is a small peak of about 3 dB near ω_o. This does not cause a large control signal u_N. In the frequency range under consideration, the control signal can be calculated as

$$u_N \approx 2\pi f_N L_N I_N, \tag{11.13}$$

where f_N is the frequency of the neutral current i_N with a peak value of I_N (assuming there is only one frequency component to simplify the exposition). This coincides with the analysis

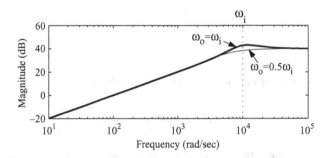

Figure 11.8 Frequency response from i_N to control signal u_N

in Section 11.1. It means that for a given system the allowable high frequency neutral current decreases when the frequency increases. The Bode plot with $\omega_o = \frac{1}{2}\omega_i$, corresponding to the case when the two poles of the closed-loop system are identical and $K_v = \frac{1}{4}\omega_i^2 L_N C_N$, is also shown in Figure 11.8. There is no peak anywhere, but as shown in (11.9) the shift will be doubled and, as can be seen later, the capacitor needed is doubled, too. Hence, it is worth using $\omega_o = \omega_i$.

11.4 Selection of the Components

11.4.1 Capacitor C_N

The capacitor C_N can be designed, according to (11.9), as

$$C_N \geq \frac{2\pi f I_N}{\omega_i \omega_o \varepsilon_m} \qquad (11.14)$$

where I_N is the maximal peak value of the main component of the neutral current i_N, f is the frequency of the main component of i_N, ω_i is the inner loop frequency, ω_o is the outer loop frequency and ε_m is the desired maximal ripple voltage.

If the other components in the neutral current form a large portion of i_N, a conservative design is to add all the capacitance for the different components together and then verify if the achievable shift meets the requirement.

Example. Assume the neutral current is $I_N = 100$ A (peak) at $f = 50$ Hz and $\omega_i = \omega_o = 10000$ rad/s. If the desired maximal shift is 1 V, then $C_N \geq 314$ μF.

11.4.2 Inductor L_N

Equation (11.9) shows that the inductor does not affect the shift (when the control signal u_N is not limited). However, it does affect the shift when the neutral current is too big and the neutral leg cannot supply the required voltage.

As discussed in Section 11.1, the current i_L needs to be almost equal to the neutral current i_N. Hence, for a sinusoidal neutral current i_N with a specific frequency f (of which the peak value is I_N), the control signal u_N is required to be

$$u_N \approx 2\pi f L_N I_N.$$

This has been verified in (11.13). Assume the possible peak voltage of u_N is about $\frac{k}{2}V_{DC}$ (k is a constant coefficient, which may often be $0.8 \sim 1$ depending on the PWM waveform generating scheme and V_{DC} is the DC link voltage), then

$$L_N \leq \frac{k}{4\pi f I_N} V_{DC}. \qquad (11.15)$$

This formula can also be used to verify if the inductor L_N makes the control signal u_N constrained by the DC link voltage. Since a large inductor does not improve the shift, a small inductor is desired because a smaller inductor is more cost-effective and it allows a higher neutral current with higher frequency (when V_{DC} is the same). However, too small an inductor results in very large surge of current flowing through the switches. Hence, the inductor L_N should be designed with more consideration on this point, i.e.

$$L_N \geq \frac{\frac{1}{2}V_{DC}}{\delta_m},$$

where δ_m is the maximum allowable $\frac{di}{dt}$ of the switches. Usually, 0.1 mH is quite enough (Rashid 1993). An extreme condition is that the current flowing through the inductor is a triangle waveform with a frequency equal to the switching frequency f_s. Then the following condition should be met:

$$L_N \geq \frac{V_{DC}}{8 I_m f_s}, \qquad (11.16)$$

where I_m is the maximum allowable current of the switch. Usually, this condition is stronger than the previous one and the inductor is chosen according to (11.15) and (11.16), i.e.

$$\frac{V_{DC}}{8 I_m f_s} \leq L_N \leq \frac{k V_{DC}}{4\pi f I_N}.$$

Another factor which should be taken into account is that too small a u_N means a weak control action. Hence, if the neutral current is normally large, then a small inductor is desired; if the neutral current is normally small, then a larger inductor is required.

11.5 Simulation Results

In the following simulations, a single-phase buck converter is used to simulate the neutral current i_N of a three-phase inverter. The fundamental frequency of this buck converter can be set at different frequencies to simulate different harmonic components in a neutral line. The parameters of the neutral leg used in these simulations are $L_N = 10$ mH, $C_N = 4000$ μF, $V_{DC} = 900$ V and $f_s = 10$ kHz. The controller is designed as $K_i = 100$ and $K_v = 4000$, i.e. with $\omega_i = \omega_o = 10000$ rad/s.

In order to see the signals clearly, the signals i_c and u_N are filtered by a hold filter

$$F(s) = \frac{\left(1 - e^{-\frac{1}{f_s}s}\right) f_s}{s}.$$

This is a notch filter, of which the notching frequencies are the switching frequency f_s and its integer multiples. Hence, the switching effects are eliminated.

11.5.1 With $i_N = 0$

This is an ideal situation. The simulation results are shown in Figure 11.9. The peak value of the shift is less than 0.01 V; the duty cycle of the firing pulse q is about 50% and u_N is about

Classical Control of a Neutral Leg

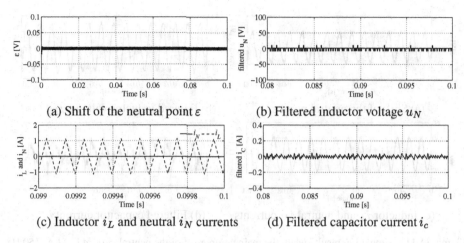

Figure 11.9 Simulation results when $i_N = 0$

0 V although there are some pulses left after filtered by the hold filter; the current flowing through the inductor is nearly a triangle wave with an amplitude of about 1.25 A; the current i_c is very small although there are some spikes, which are not shown in the figure because of the filtering effect of the hold filter $F(s)$. The average of the inductor current is 0, which is the same as the neutral current.

11.5.2 With a 50 Hz Neutral Current

When the buck converter has a single-phase load $R = 5\,\Omega$ in series with an inductor $L = 5$ mH and works at 50 Hz, the peak amplitude of the output voltage is about 360 V. This offers a neutral current of about 68 A peak. The simulation results are shown in Figure 11.10. The

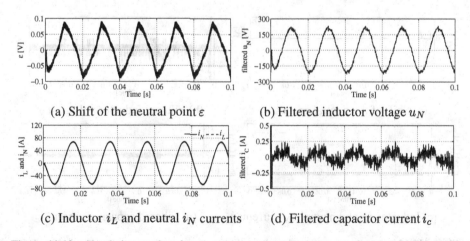

Figure 11.10 Simulation results when the main component of the neutral current is 68 A, 50 Hz

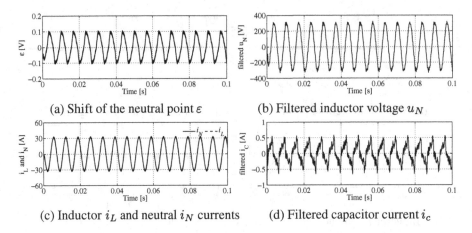

Figure 11.11 Simulation results when the main component of the neutral current is 34 A, 150 Hz

peak value of the shift is less than 0.1 V; the inductor current is almost the same as the neutral current and i_c is very small. The neutral leg provides a voltage of about 200 V peak.

11.5.3 With a 150 Hz Neutral Current

When the buck converter has a single phase load $R = 10\,\Omega$ in series with an inductor $L = 5$ mH and works at 150 Hz, the peak amplitude of the output voltage is about 360 V. This offers a neutral current of about 34 A peak at 150 Hz. The simulation results are shown in Figure 11.11. The peak value of the shift is about 0.1 V; the inductor current is almost the same as the neutral current and the current i_c is very small. The neutral leg provides a voltage of about 300 V peak.

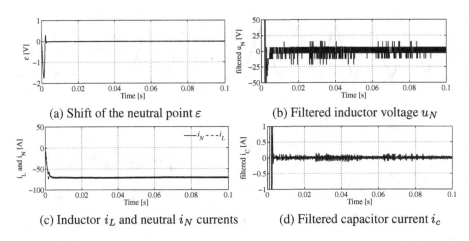

Figure 11.12 Simulation results when the main component of the neutral current is 72 A DC

Table 11.1 Frequency-related parameters

Frequency	Value
fundamental frequency f	50 Hz
harmonics up to 31st	100 ~ 1550 Hz
outer loop frequency ω_o	10000 rad/s
inner loop frequency ω_i	10000 rad/s
switching frequency f_s	10 kHz

11.5.4 With a DC Neutral Current

When the single-phase converter works at 0 Hz with an output voltage of 360 V and the load is $R = 5\ \Omega$ in series with an inductor $L = 5$ mH, the neutral line current is about 72 A. The simulation results are shown in Figure 11.12. The shift is almost 0 when the neutral current is stable.

11.6 Summary

Based on (Zhong *et al.*, 2005a, 2005b), the independently-controlled neutral leg is analysed in detail in this chapter. A controller is designed with classical control techniques, from which the inductor and the capacitor can be selected. Simulation results are provided to demonstrate the control performance.

There are several frequency-related parameters involved, which are summarised in Table 11.1. If the frequency $\omega_o = \omega_i = 10000$ rad/s leads to instability in real-time implementation, then it can be reduced.

12

H^∞ Voltage-Current Control of a Neutral Leg

In Chapter 11, a voltage-current controller is designed with classical control techniques to control an independently-controlled neutral leg. In this chapter, the advanced H^∞ control technique is applied to design a controller so that it leads to very small deviations of the neutral point from the mid-point of the DC link, in spite of possibly very large neutral currents.

12.1 Mathematical Modelling

The independently-controlled neutral leg shown in Figure 10.3 is re-drawn in Figure 12.1 for convenience. The objective of the neutral controller is to generate firing pulses to turn the switches S_+ and S_- ON or OFF, in a complementary way, so that the neutral point is very close to the mid-point of the DC link. This can be achieved by maintaining the inductor current i_L equal to the neutral current i_N so that no current flows through the capacitors. Thus, the capacitors maintain constant and equal voltages and, hence, the neutral point remains balanced. The mathematical model of the neutral leg is established in a slightly different way from that in Chapter 11, taking into account the resistance of the inductor and the mismatch of the capacitors.

Assume that all voltages are measured with respect to the neutral point N. The DC-link voltage is

$$V_{DC} = V_+ - V_-, \qquad (12.1)$$

and the mid-point (or average) of the DC link, which is desired to be 0, is

$$V_{ave} = \frac{V_+ + V_-}{2}. \qquad (12.2)$$

Control of Power Inverters in Renewable Energy and Smart Grid Integration, First Edition.
Qing-Chang Zhong and Tomas Hornik.
© 2013 John Wiley & Sons, Ltd. Published 2013 by John Wiley & Sons, Ltd.

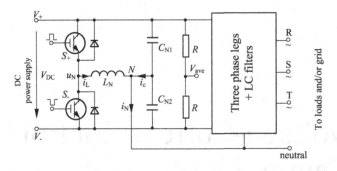

Figure 12.1 Independently-controlled neutral leg under consideration

Note that V_{ave} is half of the voltage shift ε defined in the previous chapter, i.e. $V_{ave} = \frac{1}{2}\varepsilon$. According to Kirchhoff's laws, there are

$$u_N = L_N \frac{di_L}{dt} + R_N i_L, \qquad i_N = i_L + i_c \qquad (12.3)$$

where R_N is the equivalent series resistance of the inductor, and

$$i_c = C_{N1} \frac{dV_+}{dt} + C_{N2} \frac{dV_-}{dt}. \qquad (12.4)$$

Actually, (12.1) and (12.2) determine the coordinate transformation

$$\begin{bmatrix} V_+ \\ V_- \end{bmatrix} = \begin{bmatrix} 0.5 & 1 \\ -0.5 & 1 \end{bmatrix} \begin{bmatrix} V_{DC} \\ V_{ave} \end{bmatrix}.$$

Hence, (12.4) can be re-formulated as

$$i_c = (C_{N1} + C_{N2}) \frac{dV_{ave}}{dt} + \frac{C_{N1} - C_{N2}}{2} \frac{dV_{DC}}{dt}.$$

Define a new voltage variable (actually, a disturbance signal)

$$V_0 = -\frac{C_{N1} - C_{N2}}{C_{N1} + C_{N2}} \frac{V_{DC}}{2}. \qquad (12.5)$$

This reflects the effect of the possibly mismatched capacitors C_{N1} and C_{N2} and the effect of the possible variations of V_{DC}. Then the last formula for i_c becomes

$$i_c = (C_{N1} + C_{N2}) \frac{d(V_{ave} - V_0)}{dt}. \qquad (12.6)$$

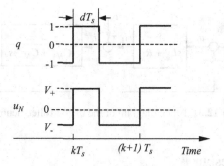

Figure 12.2 Signals q and u_N

Assume that the switching frequency is f_s and the switching period is $T_s = 1/f_s$. Define

$$q = \begin{cases} 1 & \text{when } S_+ \text{ is on and } S_- \text{ is off,} \\ -1 & \text{when } S_- \text{ is on and } S_+ \text{ is off.} \end{cases}$$

Then, the corresponding voltage u_N, together with its given q, are shown in Figure 12.2. Denote the average of q over a switching period by p. Then $p \in [-1, 1]$ and the duty cycle for S_+ is

$$d = \frac{1+p}{2}.$$

The average of u_N over a switching period (assuming that V_+ and V_- are constant within a switching period) is

$$dV_+ + (1-d)V_- = \frac{p}{2}V_{DC} + V_{ave}.$$

That is, approximately,

$$u_N = \frac{p}{2}V_{DC} + V_{ave}. \tag{12.7}$$

Combining (12.3), (12.6) and (12.7), the block diagram of the neutral leg is shown in Figure 12.3, where V_0 plays the role of an external disturbance and p is the control variable. Strictly speaking, even this simplified model is non-linear, because p is multiplied with the variable V_{DC}. However, since the variations of V_{DC} are relatively small, it is assumed that the V_{DC} appearing in the first block in Figure 12.3 is equal to its nominal value V_{DC}^{nom}. If the variations of V_{DC} are expected to be large, then they can be compensated for by measuring V_{DC} and multiplying the output p of the controller with the correction factor V_{DC}^{nom}/V_{DC}.

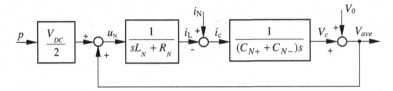

Figure 12.3 Block diagram of the uncontrolled neutral leg

12.2 Controller Design

The control objective for the plant shown in Figure 12.3 is to maintain a stable and balanced neutral point, i.e. to make V_{ave} as small as possible while maintaining the stability of the system. The disturbances are the neutral current i_N and the equivalent external disturbance V_0 defined in (12.5). Since the current i_c contains ripples (at the switching frequency), it must be sensed through a low-pass filter $F(s)$, as shown in Figure 12.4. The remaining ripples after this filter can be interpreted as a measurement noise n and the measured signal $V_i = Fi_c + n$ is available for feedback. Another signal available for feedback is the average voltage V_{ave}. In other words, the measured signal is $y = \begin{bmatrix} V_{ave} & V_i \end{bmatrix}^T$.

From the H^∞ control point of view, the simplest formulation of the problem would be to minimise the H^∞-norm of the transfer function from $\begin{bmatrix} i_N & V_0 & n \end{bmatrix}^T$ to the average voltage V_{ave}. Such a formulation would be unrealistic, since it would ignore the fact that the disturbances are expected in a certain frequency range and that the signal u_N should not be too large (because for obvious physical reasons, $V_- < u_N < V_+$). Thus, as is usual in applications of the H^∞ control theory, the weighting functions W_v and W_u are introduced to V_{ave} and u_N,

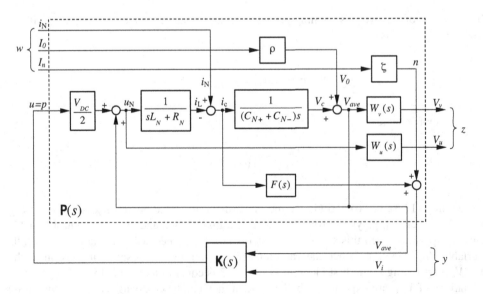

Figure 12.4 Formulation of the H^∞ control problem

respectively. Another two weighting factors I_0 and I_n, which are proportional to V_0 and n via the factors ρ and ζ, are also introduced to adjust the relative importance of the three disturbances i_N, V_0 and n in the H^∞-norm minimisation process. The H^∞ control problem is then formulated to minimise the H^∞-norm of the transfer function from $w = \begin{bmatrix} i_N & I_0 & I_n \end{bmatrix}^T$ to $z = \begin{bmatrix} V_v & V_u \end{bmatrix}^T$, denoted $\mathbf{T}_{zw} = \mathcal{F}_l(\mathbf{P}, \mathbf{K})$, as shown in Figure 12.4. The closed-loop system can be represented as

$$\begin{bmatrix} z \\ y \end{bmatrix} = \mathbf{P} \begin{bmatrix} w \\ u \end{bmatrix}, \quad u = \mathbf{K}\, y,$$

where $\mathbf{P}$ is the generalised plant and $\mathbf{K}$ is the controller to be designed. A nearly optimal $\mathbf{K}$ can then be obtained with the standard H^∞ control algorithm; see (Zhou and Doyle 1998; Green and Limebeer 1995) for details.

12.2.1 State-space Realisation of $\mathbf{P}$

If the state variables of the original plant are chosen to be the inductor current i_L and the voltage $V_c = V_{ave} - V_0$, i.e. $x = \begin{bmatrix} i_L \\ V_c \end{bmatrix}$, and if the control input u is p, then the following state equations can be obtained from Figure 12.4:

$$\dot{x} = Ax + B_1 w + B_2 u \tag{12.8}$$

with

$$A = \begin{bmatrix} -\dfrac{R_N}{L_N} & \dfrac{1}{L_N} \\ -\dfrac{1}{C_{N1} + C_{N2}} & 0 \end{bmatrix}, \quad B_1 = \begin{bmatrix} 0 & \dfrac{\rho}{L_N} & 0 \\ \dfrac{1}{C_{N1} + C_{N2}} & 0 & 0 \end{bmatrix}, \quad B_2 = \begin{bmatrix} \dfrac{V_{DC}}{2L_N} \\ 0 \end{bmatrix}.$$

The output equations are

$$V_{ave} = C_a x + D_{1a} w + D_{2a} u, \tag{12.9}$$

$$u_N = C_{1b} x + D_{11b} w + D_{12b} u,$$

$$i_c = C_{2b} x + D_{21b} w + D_{22b} u,$$

with

$$C_a = \begin{bmatrix} 0 & 1 \end{bmatrix}, \quad D_{1a} = \begin{bmatrix} 0 & \rho & 0 \end{bmatrix}, \quad D_{2a} = \begin{bmatrix} 0 \end{bmatrix},$$

$$C_{1b} = \begin{bmatrix} 0 & 1 \end{bmatrix}, \quad D_{11b} = \begin{bmatrix} 0 & \rho & 0 \end{bmatrix}, \quad D_{12b} = \begin{bmatrix} \dfrac{V_{DC}}{2} \end{bmatrix},$$

$$C_{2b} = \begin{bmatrix} -1 & 0 \end{bmatrix}, \quad D_{21b} = \begin{bmatrix} 1 & 0 & 0 \end{bmatrix}, \quad D_{22b} = \begin{bmatrix} 0 \end{bmatrix}.$$

Assume that the state-space realisation of the weighting function W_v is

$$W_v = \left[\begin{array}{c|c} A_v & B_v \\ \hline C_v & D_v \end{array}\right].$$

Then

$$\begin{aligned}
V_v &= \left[\begin{array}{c|c} A_v & B_v \\ \hline C_v & D_v \end{array}\right] V_{ave} \\
&= \left[\begin{array}{c|c} A_v & B_v \\ \hline C_v & D_v \end{array}\right] \left[\begin{array}{c|cc} A & B_1 & B_2 \\ \hline C_a & D_{1a} & D_{2a} \end{array}\right] \left[\begin{array}{c} w \\ u \end{array}\right] \\
&= \left[\begin{array}{cc|cc} A & 0 & B_1 & B_2 \\ B_v C_a & A_v & B_v D_{1a} & B_v D_{2a} \\ \hline D_v C_a & C_v & D_v D_{1a} & D_v D_{2a} \end{array}\right] \left[\begin{array}{c} w \\ u \end{array}\right].
\end{aligned} \quad (12.10)$$

Similarly, assume that the state-space realisation of the weighting function W_u is

$$W_u = \left[\begin{array}{c|c} A_u & B_u \\ \hline C_u & D_u \end{array}\right],$$

then

$$\begin{aligned}
V_u &= \left[\begin{array}{c|c} A_u & B_u \\ \hline C_u & D_u \end{array}\right] u_N \\
&= \left[\begin{array}{cc|cc} A & 0 & B_1 & B_2 \\ B_u C_{1b} & A_u & B_u D_{11b} & B_u D_{12b} \\ \hline D_u C_{1b} & C_u & D_u D_{11b} & D_u D_{12b} \end{array}\right] \left[\begin{array}{c} w \\ u \end{array}\right].
\end{aligned} \quad (12.11)$$

Moreover, assume that the implementation of the low-pass filter $F(s)$ is

$$F(s) = \left[\begin{array}{c|c} A_F & B_F \\ \hline C_F & D_F \end{array}\right],$$

then

$$\begin{aligned}
V_i &= \left[\begin{array}{c|c} A_F & B_F \\ \hline C_F & D_F \end{array}\right] i_c + \zeta I_n \\
&= \left[\begin{array}{c|c} A_F & B_F \\ \hline C_F & D_F \end{array}\right] \left[\begin{array}{c|cc} A & B_1 & B_2 \\ \hline C_{2b} & D_{21b} & D_{22b} \end{array}\right] \left[\begin{array}{c} w \\ u \end{array}\right] + \zeta I_n \\
&= \left[\begin{array}{cc|cc} A & 0 & B_1 & B_2 \\ B_F C_{2b} & A_F & B_F D_{21b} & B_F D_{22b} \\ \hline D_F C_{2b} & C_F & D_F D_{21b} + \begin{bmatrix} 0 & 0 & \zeta \end{bmatrix} & D_F D_{22b} \end{array}\right] \left[\begin{array}{c} w \\ u \end{array}\right].
\end{aligned} \quad (12.12)$$

Combining (12.8), (12.9), (12.10), (12.11) and (12.12), the state-space realisation of the generalised plant **P**, with inputs w, u and outputs z, y, is obtained as

$$\mathbf{P} = \left[\begin{array}{cccc|cc} A & 0 & 0 & 0 & B_1 & B_2 \\ B_v C_a & A_v & 0 & 0 & B_v D_{1a} & B_v D_{2a} \\ B_u C_{1b} & 0 & A_u & 0 & B_u D_{11b} & B_u D_{12b} \\ B_F C_{2b} & 0 & 0 & A_F & B_F D_{21b} & B_F D_{22b} \\ \hline D_v C_a & C_v & 0 & 0 & D_v D_{1a} & D_v D_{2a} \\ D_u C_{1b} & 0 & C_u & 0 & D_u D_{11b} & D_u D_{12b} \\ \hline C_a & 0 & 0 & 0 & D_{1a} & D_{2a} \\ D_F C_{2b} & 0 & 0 & C_F & D_F D_{21b} + \begin{bmatrix} 0 & 0 & \zeta \end{bmatrix} & D_F D_{22b} \end{array}\right].$$

12.2.2 State-space Realisation of the Closed-loop Transfer Function

Denote by $\tilde{\mathbf{P}}$ the transfer function from $\begin{bmatrix} w \\ u \end{bmatrix}$ to $\begin{bmatrix} z' \\ y \end{bmatrix}$. Then

$$\tilde{\mathbf{P}} = \left[\begin{array}{cc|cc} A & 0 & B_1 & B_2 \\ B_F C_{2b} & A_F & B_F D_{21b} & B_F D_{22b} \\ C_a & 0 & D_{1a} & D_{2a} \\ C_{1b} & 0 & D_{11b} & D_{12b} \\ \hline C_a & 0 & D_{1a} & D_{2a} \\ D_F C_{2b} & C_F & D_F D_{21b} + \begin{bmatrix} 0 & 0 & \zeta \end{bmatrix} & D_F D_{22b} \end{array}\right] \doteq \left[\begin{array}{c|cc} \tilde{A} & \tilde{B}_1 & \tilde{B}_2 \\ \hline \tilde{C}_1 & \tilde{D}_{11} & \tilde{D}_{12} \\ \tilde{C}_2 & \tilde{D}_{21} & \tilde{D}_{22} \end{array}\right].$$

Assume that the controller is realised as

$$\mathbf{K} = \left[\begin{array}{c|c} A_K & B_K \\ \hline C_K & D_K \end{array}\right],$$

where, usually, $D_K = 0$ if **K** is obtained from the standard H^∞ algorithm. However, D_K may be non-zero after controller reduction. According to the star-product formula (Zhou and Doyle 1998), the transfer function from w to z' is

$$\mathbf{T}_{z'w} = \mathcal{F}_l(\tilde{\mathbf{P}}, \mathbf{K}) = \left[\begin{array}{cc|c} \tilde{A} + \tilde{B}_2 \tilde{R} D_K \tilde{C}_2 & \tilde{B}_2 R C_K & \tilde{B}_1 + \tilde{B}_2 \tilde{R} D_K \tilde{D}_{21} \\ B_K R \tilde{C}_2 & A_K + B_K R \tilde{D}_{22} C_K & B_K R \tilde{D}_{21} \\ \hline \tilde{C}_1 + \tilde{D}_{12} D_K R \tilde{C}_2 & \tilde{D}_{12} R C_K & \tilde{D}_{11} + \tilde{D}_{12} D_K R \tilde{D}_{21} \end{array}\right],$$

where

$$\tilde{R} = (I - D_K \tilde{D}_{22})^{-1}, \qquad R = (I - \tilde{D}_{22} D_K)^{-1}.$$

Hence, the closed-loop transfer function from $w' = \begin{bmatrix} i_N \\ V_0 \\ n \end{bmatrix} = \begin{bmatrix} 1 & 0 & 0 \\ 0 & \rho & 0 \\ 0 & 0 & \zeta \end{bmatrix} w$ to z', which is the relevant closed-loop transfer function in terms of the original variables, is

$$\mathbf{T}_{z'w'} = \mathbf{T}_{z'w} \begin{bmatrix} 1 & 0 & 0 \\ 0 & \dfrac{1}{\rho} & 0 \\ 0 & 0 & \dfrac{1}{\zeta} \end{bmatrix}.$$

12.3 Selection of Weighting Functions

It is not easy to choose suitable weighting functions for a specific H^∞ control problem. These functions have to reflect the relative weight of different signals, their frequency characteristics, and at the same time they must be chosen so that the problem is solvable. It is difficult to find specific guidelines in the literature although some general guidelines can be found in, e.g. (Green and Limebeer 1995; Zhou and Doyle 1998).

The weighting function W_v has to be large for the range of significant disturbance frequencies (50 Hz and a few multiples of it) and it has to decay for high frequencies, which cannot be controlled. Anyway, no significant high-frequency variations of V_{ave} is expected because of the two capacitors. The weighting function $W_v(s)$ can then be chosen as

$$W_v(s) = g \frac{s + \omega_h}{s + \omega_l} = \left[\begin{array}{c|c} -\omega_l & g \\ \omega_h - \omega_l & g \end{array} \right],$$

where g is a tuning parameter to move the Bode plot of W_v up or down.

The signal u_N is closely related (almost proportional) to $u = p$. In order to prevent u from becoming too large, especially at high frequencies, W_u is chosen small at low frequencies and large at high frequencies. Moreover, the normal H^∞ algorithm requires that the matrix $\begin{bmatrix} D_v D_{2a} \\ D_u D_{12b} \end{bmatrix}$ has full column rank. Since $D_{2a} = 0$, D_u cannot be zero. Hence, W_u has to be non-strictly proper. It can be chosen as

$$W_u(s) = k \frac{s + \omega_l}{s + \omega_h} = \left[\begin{array}{c|c} -\omega_h & k \\ \omega_l - \omega_h & k \end{array} \right],$$

where k is a tuning parameter to move the Bode plot of W_u up or down. In this application, the fundamental frequency is 50 Hz and the high frequency components under consideration can be up to the 31st harmonics. Therefore, ω_h is chosen to be 10,000 rad/s. The Bode plots of the weighting functions $W_u(s)$ and $W_v(s)$ are shown in Figure 12.5 for $k = 0.01$ and $g = 10$.

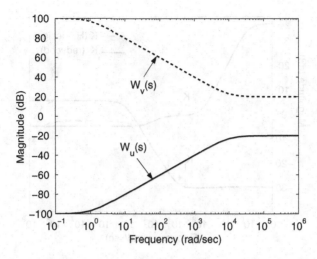

Figure 12.5 Bode plots of the weighting functions $W_u(s)$ and $W_v(s)$

12.4 Design Example

The parameters of the neutral leg are: $L_N = 2.5$ mH, $R_N = 0.2\,\Omega$, $C_{N1} = C_{N2} = 6600\,\mu$F and $V_{DC} = 850$ V. The switching frequency is $f_s = 10$ kHz and $F(s) = \frac{1000}{s+1000}$. The tuning parameters are chosen to be: $k = 0.01$, $g = 10$, $\rho = 0.01$, $\zeta = 0.01$, $\omega_l = 1$ rad/s and $\omega_h = 10000$ rad/s. Using the μ-analysis toolbox from MATLAB®, the H^∞ controller $\mathbf{K} = \begin{bmatrix} \mathbf{K}_v & \mathbf{K}_i \end{bmatrix}$ is obtained as

$$\mathbf{K}_v(s) = \frac{-0.0023529(s - 2.407 \times 10^{11})(s + 8.87 \times 10^4)(s + 10^4)(s + 80.03)(s + 74.28)}{(s + 1.556 \times 10^8)(s + 1.096 \times 10^5)(s + 8327)(s + 76.57)(s + 1)},$$

$$\mathbf{K}_i(s) = \frac{8.63 \times 10^8(s + 10^4)(s + 1000)(s + 80)}{(s + 1.556 \times 10^8)(s + 1.096 \times 10^5)(s + 8327)(s + 76.57)}.$$

This controller is somewhat unrealistic because the sampling frequency is usually limited. In order to make the controller implementable, any zeros or poles which correspond to a corner frequency much higher than $\omega_h = 10^4$ rad/s are substituted by a proportional gain, i.e. ignoring the s. Then the controller is reduced to

$$\mathbf{K}_r(s) = \begin{bmatrix} 2.9457 \frac{(s + 10^4)(s + 80.03)(s + 74.28)}{(s + 8327)(s + 76.57)(s + 1)} & 5.5475 \frac{(s + 10^4)(s + 1000)(s + 80)}{(s + 1.096 \times 10^5)(s + 8327)(s + 76.57)} \end{bmatrix}.$$

The Bode plots of these controllers are shown in Figure 12.6. It can be seen that the reduced-order controllers are very close to the original controller, in particular, at low frequencies. The Bode plots of the corresponding closed-loop transfer functions using the original H^∞ controller $\mathbf{K}$ and the reduced controller $\mathbf{K}_r$ are shown in Figure 12.7. The curves are very close to each other.

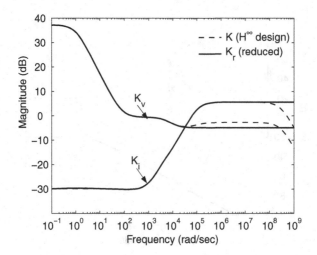

Figure 12.6 Bode plots of the controllers **K** and **K**$_r$

12.5 Simulation Results

A single-phase converter leg was used to create a current i_N flowing back through the neutral line. The phase leg had a load composed of a resistor $R = 5\ \Omega$ in series with an inductor $L = 5$ mH and it generated a sinusoidal neutral current of about 68 A peak at 50 Hz. The reduced controller **K**$_r$ was used in the simulation. A variable voltage component of $100 \sin 200\pi t$ V and a resistor of $0.5\ \Omega$ were put in series with the DC source of 850 V to make it more realistic. The simulation results are shown in Figure 12.8.

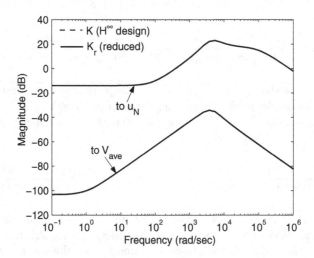

Figure 12.7 Bode plots of the closed-loop transfer functions from i_N to V_{ave} and to u_N with the controllers **K** and **K**$_r$

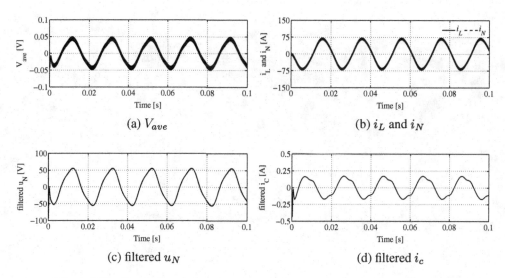

Figure 12.8 Simulation results when the neutral current is about 68 A peak at 50 Hz

The signals u_N and i_c shown were filtered by the hold filter $H(s) = \frac{1-e^{-sh}}{sh}$ with $h = 10^{-4}$ s, which is a notch filter with the notching frequencies at the multiples of the switching frequency $f_s = 10$ kHz, so that the effect of the ripples at the switching frequency was eliminated. The peak value of V_{ave} is less than 0.1 V; the inductor current is almost the same as the neutral current and the current i_c is very small. The neutral leg provides a voltage u_N of about 60 V peak. The neutral point is balanced with respect to the DC terminals in spite of a varying DC source and a large neutral current. Note that the filtered voltage is quite different from the one shown in Figure 11.10 because of the different values of L.

12.6 Summary

Based on (Zhong et al., 2002a, 2006), a linear model for the independently-controlled neutral leg that consists of a conventional neutral leg and a split DC link is developed at first. The conventional neutral leg provides a path for the neutral current and charges/discharges the capacitors to maintain the neutral point while the capacitors provide a way to independently control the conventional neutral leg. Hence, the role of the split DC capacitors has been changed from maintaining the neutral point to controlling the neutral point. Actually, as can be seen from the model, only one capacitor is necessary. This results in an asymmetric topology but does not change the control design. After introducing an artificial measurement noise signal n and selecting appropriate weighting functions, the control problem of the neutral leg is formulated as a standard H^∞ control problem and a controller is derived.

13

Parallel PI Voltage-H^∞ Current Control of a Neutral Leg

As explained in Chapter 11, in order to maintain a stable and balanced neutral point, the current flowing through the split DC capacitors should be close to zero. This fact is not fully reflected in the controllers designed in the previous two chapters. Here, this is taken into account and a parallel voltage-current control strategy is designed to improve the performance of the independently-controlled neutral leg. The current controller is designed by using the H^∞ control theory to minimise the current i_C flowing through the split DC capacitors. Moreover, an extra voltage controller is added to bring the voltage shift caused by the mismatch of capacitors and/or the non-linearity of the switches back to normal. Since the voltage controller mainly deals with DC signals and the current controller mainly deals with the harmonic current components, these two controllers are decoupled in the frequency domain and can be arranged in a parallel control structure instead of a cascaded control structure. This improves the stability of the system. The control strategy leads to very small deviations of the neutral point from the mid-point of the DC source, in spite of the possibly very large neutral current. Experimental results are presented to demonstrate the excellent performance of the control strategy.

13.1 Description of the Neutral Leg

Again, the neutral leg under consideration is the independently-controlled neutral leg shown in Figure 13.1, which is the combination of a split DC link and a conventional neutral leg (Zhong *et al.* 2005a, 2006).

The total DC-link voltage is the difference between the two capacitor voltages measured with respect to the neutral point, i.e.

$$V_{DC} = V_+ - V_-.$$

Control of Power Inverters in Renewable Energy and Smart Grid Integration, First Edition.
Qing-Chang Zhong and Tomas Hornik.
© 2013 John Wiley & Sons, Ltd. Published 2013 by John Wiley & Sons, Ltd.

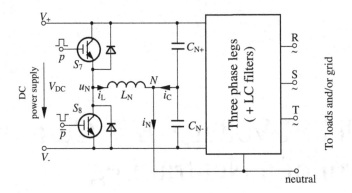

Figure 13.1 Independently-controlled neutral leg under consideration

Since the control objective is to maintain the point N as a neutral point, the average voltage

$$V_{ave} = \frac{V_+ + V_-}{2} \qquad (13.1)$$

is expected to be as close to 0 as possible all the time. As a result, i_L is expected to be almost equal to i_N. This means the majority of the neutral current should flow through the inductor L_N but not through the capacitors. A small current i_C would lead to a small variation of the neutral point N. The capacitors do not have to be chosen to be very large, like those needed in the conventional split DC link.

The block diagram of the independently-controlled neutral leg is shown in Figure 13.2, where p is the average of the control signal over one switching period normalised with respect to half of the DC-link voltage. This is very similar to Figure 12.3, but in a different form to highlight that the control objective is to make i_C as small as possible while maintaining the stability of the system. The duty cycle for the upper switch is

$$d = \frac{p+1}{2}.$$

The resistance of the inductor is taken into account and the capacitors C_{N+} and C_{N-} could be different.

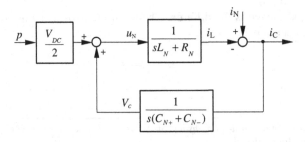

Figure 13.2 Block diagram of the independently-controlled neutral leg

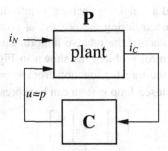

Figure 13.3 Block diagram of the H^∞ current control scheme

13.2 Design of an H^∞ Current Controller

13.2.1 Controller Description

The H^∞ current control scheme is shown in Figure 13.3, where **P** is the transfer function of the independently-controlled neutral leg and **C** is the transfer function of the controller. The controller **C** is designed by solving a weighted sensitivity H^∞ problem (Zhou and Doyle 1998), as formulated later, to assure the stability of the entire system. The neutral current i_N is regarded as an external disturbance and the control signal is $u = p$.

13.2.2 Formulation as a Standard H^∞ Problem

A measurement noise n is introduced when measuring the current i_C, as shown in Figure 13.4. The components of the current i_C are expected to be within a certain frequency range so a weighting function W is introduced to reflect this. A weighting factor μ is introduced to

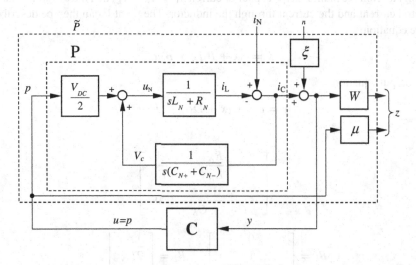

Figure 13.4 Formulation of the H^∞ problem to control the current flowing through the split DC capacitors

avoid a large control action and a weighting factor ξ is introduced so that the impact of the measurement noise n can be tuned during the process of controller design. The weighting factors ξ and μ can be used to adjust the relative importance of the disturbances i_N and n in the H^∞ norm. The H^∞ control problem, as shown in Figure 13.4, is then formulated to minimise the H^∞ norm of the transfer function from $w = [n \quad i_N]^T$ to $z = [z_1 \quad z_2]^T$, denoted $T_{zw} = \mathcal{F}_l(\tilde{P}, \mathbf{C})$. The closed-loop system can be represented as

$$\begin{bmatrix} z \\ y \end{bmatrix} = \tilde{P} \begin{bmatrix} w \\ u \end{bmatrix},$$
$$u = \mathbf{C}y,$$

where $\tilde{P}$ is the generalised plant that includes the original control plant and weighting functions, $\mathbf{C}$ is the controller to be designed and y is the measured current i_C that contains the effect of the measurement noise n, which is an imaginary variable introduced to facilitate the design of the controller so that the impact of the measurement noise n in the current i_C is reflected and minimised. When the controller is implemented, as shown in Figure 13.3, the feedback signal to the controller $\mathbf{C}$ is the measured current i_C, which contains the effect of the measurement noise, so there is no need to determine or calculate the value of the measurement noise n.

It is worth noting that the controlled variables z in the previous chapter are the neutral-point voltage shift and the neutral line inductor voltage but the controlled variables here are chosen as the current flowing through the split DC capacitors and the duty cycle for the switches. Moreover, a two-input-one-output controller is adopted in Chapter 12 but a single-input-single-output controller is adopted here, with the add-on of a voltage controller to be discussed later.

13.2.3 State-space Realisation of the Plant P

The inductor current i_L and the voltage V_c are chosen as state variables $x = \begin{bmatrix} i_L & V_c \end{bmatrix}^T$. The output signal from the plant **P** is the capacitor current $i_C = i_N - i_L$, i.e., the difference between the neutral current and the current through the inductor. The plant **P** can then be described by the state equation

$$\dot{x} = Ax + B_1 i_N + B_2 u \qquad (13.2)$$

and the output equation

$$y = e = C_1 x + D_1 i_N + D_2 u \qquad (13.3)$$

with

$$A = \begin{bmatrix} -\dfrac{R_N}{L_N} & \dfrac{1}{L_N} \\ -\dfrac{1}{C_{N+} + C_{N-}} & 0 \end{bmatrix},$$

$$B_1 = \begin{bmatrix} 0 \\ \dfrac{1}{C_{N+} + C_{N-}} \end{bmatrix}, \quad B_2 = \begin{bmatrix} \dfrac{V_{DC}}{2L_N} \\ 0 \end{bmatrix},$$

$$C_1 = \begin{bmatrix} -1 & 0 \end{bmatrix}, \quad D_1 = \begin{bmatrix} 1 \end{bmatrix}, \quad D_2 = \begin{bmatrix} 0 \end{bmatrix}.$$

The corresponding plant transfer function is then

$$\mathbf{P} = \begin{bmatrix} D_1 & D_2 \end{bmatrix} + C_1(sI - A)^{-1} \begin{bmatrix} B_1 & B_2 \end{bmatrix},$$

or, in short,

$$\mathbf{P} = \left[\begin{array}{c|cc} A & B_1 & B_2 \\ \hline C_1 & D_1 & D_2 \end{array} \right]. \tag{13.4}$$

13.2.4 State-space Realisation of the Generalised Plant $\tilde{P}$

Since there are harmonics in the neutral current, the weighting function W can be chosen as a resonant filter

$$W = \left[\begin{array}{c|c} A_w & B_w \\ \hline C_w & 0 \end{array} \right],$$

with a high gain at the vicinity of the line frequency while providing small gains at all the other frequencies. It can be designed to cover the fundamental frequency only or to cover the fundamental frequency and some harmonic frequencies, e.g. the 3rd and 5th harmonics, as well. From Figure 13.4, the following equations can be obtained:

$$y = i_C + \xi n = \xi n + \left[\begin{array}{c|cc} A & B_1 & B_2 \\ \hline C_1 & D_1 & D_2 \end{array} \right] \begin{bmatrix} i_N \\ u \end{bmatrix}$$

$$= \left[\begin{array}{cc|cc} A & 0 & B_1 & B_2 \\ \hline C_1 & \xi & D_1 & D_2 \end{array} \right] \begin{bmatrix} n \\ i_N \\ u \end{bmatrix}, \tag{13.5}$$

$$z_1 = W(i_C + \xi n)$$

$$= \left[\begin{array}{c|c} A_w & B_w \\ \hline C_w & 0 \end{array} \right] \left[\begin{array}{cc|cc} A & 0 & B_1 & B_2 \\ \hline C_1 & \xi & D_1 & D_2 \end{array} \right] \begin{bmatrix} n \\ i_N \\ u \end{bmatrix}$$

$$= \left[\begin{array}{cc|ccc} A & 0 & 0 & B_1 & B_2 \\ B_w C_1 & A_w & B_w \xi & B_w D_1 & B_w D_2 \\ \hline 0 & C_w & 0 & 0 & 0 \end{array} \right] \begin{bmatrix} n \\ i_N \\ u \end{bmatrix}, \tag{13.6}$$

$$z_2 = \mu u. \tag{13.7}$$

Combining equations (13.5), (13.6) and (13.7), the realisation of the generalised plant is obtained as

$$\tilde{P} = \left[\begin{array}{cc|ccc} A & 0 & 0 & B_1 & B_2 \\ B_w C_1 & A_w & B_w \xi & B_w D_1 & B_w D_2 \\ 0 & C_w & 0 & 0 & 0 \\ 0 & 0 & 0 & 0 & \mu \\ \hline C_1 & 0 & \xi & D_1 & D_2 \end{array} \right]. \tag{13.8}$$

Table 13.1 Parameters of the neutral leg

Parameters	Value	Parameters	Value
L_N	2.35 mH	C_{N+}, C_{N-}	1000 μF
R_N	0.54 Ω	f_s	5 kHz

The controller can then be obtained by using the standard H^∞ control algorithms to guarantee the stability of the system.

13.2.5 Design Example

The parameters of the neutral leg circuit are given in Table 13.1. The switching frequency is $f_s = 5$ kHz. The weighting parameters are chosen to be $\xi = 0.5$ and $\mu = 0.5$. This is to reflect the case that the neutral current i_N weighs more than the noise n and the output z_1 weighs more than the control signal u.

Controllers with different numbers of harmonics compensators (i.e., with different weighting functions W) are designed in this section. Controller $\mathbf{C}_1$ is designed according to the weighting function $W(s) = \dfrac{K_1 2\zeta \omega s}{s^2 + 2\zeta \omega s + \omega^2}$ that has a high gain at the fundamental grid frequency ω only. Controller $\mathbf{C}_3$ is designed according to $W(s) = \dfrac{K_1 2\zeta \omega s}{s^2 + 2\zeta \omega s + \omega^2} + \dfrac{K_3 6\zeta \omega s}{s^2 + 6\zeta \omega s + 9\omega^2}$, which has an extra high gain at the 3rd harmonic frequency to further improve the performance of the controller. Similarly, Controllers $\mathbf{C}_5$ and $\mathbf{C}_7$ are designed with weighting functions having an added high gain at the 5th and 7th harmonic frequencies, respectively. The gains in the weighting functions are chosen as $K_1 = 10$, $K_3 = 1.666$, $K_5 = 0.5$ and $K_7 = 0.5$ and the damping factor is chosen as $\zeta = 0.005$, according to the guidelines reported in (Castilla et al. 2009).

Using the MATLAB® *hinfsyn* algorithm, the H^∞ controllers $\mathbf{C}_1(s)$, $\mathbf{C}_3(s)$, $\mathbf{C}_5(s)$ and $\mathbf{C}_7(s)$ are obtained as

$$\mathbf{C}_1(s) = \frac{75.7417(s - 55.09)(s^2 + 296s + 1.393 \times 10^5)}{(s + 1147)(s + 82.99)(s^2 + 3.14s + 9.87 \times 10^4)}, \tag{13.9}$$

$$\mathbf{C}_3(s) = \frac{100.0978(s - 87.02)(s^2 + 295.9s + 1.442 \times 10^5)(s^2 - 149.6s + 6.693 \times 10^5)}{(s + 1323)(s + 71.9)(s^2 + 3.14s + 9.87 \times 10^4)(s^2 + 9.42s + 8.883 \times 10^5)}, \tag{13.10}$$

$$\mathbf{C}_5(s) = \frac{107.6544(s - 97.26)(s^2 + 295.3s + 1.454 \times 10^5)(s^2 - 168.3s + 6.659 \times 10^5)}{(s + 1374)(s + 69.27)(s^2 + 3.14s + 9.87 \times 10^4)(s^2 + 9.42s + 8.883 \times 10^5)}$$
$$\times \frac{(s^2 - 106.1s + 2.284 \times 10^6)}{(s^2 + 15.7s + 2.467 \times 10^6)}, \tag{13.11}$$

$$\mathbf{C}_7(s) = \frac{112.5476(s - 104)(s^2 + 294.8s + 1.461 \times 10^5)(s^2 - 178.5s + 6.666 \times 10^5)}{(s + 1405)(s + 67.71)(s^2 + 3.14s + 9.87 \times 10^4)(s^2 + 9.42s + 8.883 \times 10^5)}$$
$$\times \frac{(s^2 - 122.1s + 2.287 \times 10^6)(s^2 - 133s + 4.579 \times 10^6)}{(s^2 + 15.7s + 2.467 \times 10^6)(s^2 + 21.98s + 4.836 \times 10^6)}. \tag{13.12}$$

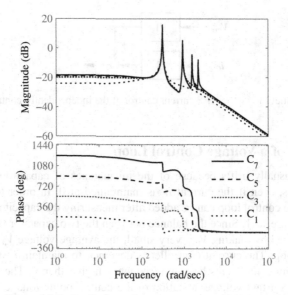

Figure 13.5 Bode plots of the current controllers designed

The Bode plots of these controllers are shown in Figure 13.5. It can be seen that high gains appear at the harmonic frequencies included in the corresponding weighting function W, which helps reduce the harmonic components in i_C. These controllers can be discretised for digital implementation. With a sampling frequency of 5 kHz, the discretised controllers can be obtained, e.g. by using the MATLAB® $c2d$ (ZOH) algorithm, as

$$\mathbf{C}_1(z) = \frac{0.013755(z - 1.011)(z^2 - 1.937z + 0.9425)}{(z - 0.7951)(z - 0.9835)(z^2 - 1.995z + 0.9994)}, \tag{13.13}$$

$$\mathbf{C}_3(z) = \frac{0.017515(z - 1.018)(z^2 - 1.937z + 0.9426)(z^2 - 2.003z + 1.03)}{(z - 0.7674)(z - 0.9857)(z^2 - 1.995z + 0.9994)(z^2 - 1.963z + 0.9981)}, \tag{13.14}$$

$$\mathbf{C}_5(z) = \frac{0.018432(z - 1.02)(z^2 - 1.937z + 0.9427)(z^2 - 2.007z + 1.034)}{(z - 0.7598)(z - 0.9862)(z^2 - 1.995z + 0.9994)(z^2 - 1.963z + 0.9981)}$$

$$\times \frac{(z^2 - 1.93z + 1.021)}{(z^2 - 1.899z + 0.9969)}, \tag{13.15}$$

$$\mathbf{C}_7(z) = \frac{0.018805(z - 1.021)(z^2 - 1.937z + 0.9428)(z^2 - 2.009z + 1.036)}{(z - 0.755)(z - 0.9865)(z^2 - 1.995z + 0.9994)(z^2 - 1.963z + 0.9981)}$$

$$\times \frac{(z^2 - 1.933z + 1.025)(z^2 - 1.844z + 1.027)}{(z^2 - 1.899z + 0.9969)(z^2 - 1.806z + 0.9956)}. \tag{13.16}$$

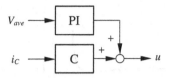

Figure 13.6 Parallel PI voltage–H^∞ current control of the independently-controlled neutral leg

13.3 Addition of a Voltage Control Loop

The neutral point usually drifts because of the mismatches of capacitors and/or the non-linearity of switches, even if the current i_C is maintained at 0. In order to avoid this from happening, a voltage control loop can be added after measuring the capacitor voltages V_+ and V_-, as shown in Figure 13.1. Since the balance between the two capacitor voltages is desired, which can be achieved by making V_{ave} very small, the average voltage V_{ave} is used to form a voltage control loop. The current controller is designed to maintain i_C to be nearly zero, that is, to remove any components having a frequency higher than 0. The voltage controller is just to bring the constant voltage deviation of the neutral-point voltage back to zero, that is, to work with DC components. Hence, the functions of the two controllers are decoupled in the frequency domain. As a result, the voltage controller and the current controller can be put together to form a parallel voltage–current control structure, as shown in Figure 13.6. The voltage controller can be chosen as a PI controller, which has a high DC gain. The output of the voltage controller enters the current control loop, taking the role of the measurement noise n in Figure 13.4. Hence, adding the voltage controller does not affect the stability of the current loop. As long as the voltage control loop is stable, which can be easily achieved by tuning the PI controller, the whole system is stable. It is worth noting that the derivation of the analytic stability condition could be difficult but the stability of the system can easily be achieved according to the above analysis. This control strategy not only leads to a stable neutral point (by the current controller) but also leads to a balanced neutral point with respect to the DC-link terminals (by the voltage controller).

13.4 Experimental Results

The system consists of an inverter board, a three-phase LC filter, a board consisting of voltage and current sensors, a dSPACE DS1104 R&D controller board with ControlDesk software, and MATLAB®/Simulink® SimPowerSystems™ software. The inverter board consists of two independent three-phase inverters and has the capability to generate PWM voltages from a constant 42 V DC voltage source. The first inverter was used to generate a three-phase output voltage and one leg of the second inverter was used to control the neutral point. By changing the load of the first inverter, different neutral currents can be generated. The parameters of the system are the same as those given in Table 13.1. The neutral current i_N and the inductor current i_L were measured and the current i_C was calculated; the DC-link voltage and the voltage across the capacitor C_{N-} were measured to calculate the average voltage V_{ave}.

The controllers were implemented to evaluate their steady-state and transient performance. The steady-state performance was tested under three scenarios when balanced resistive, unbalanced resistive and non-linear loads were connected to the three-phase inverter to generate nearly zero, clean and distorted currents, respectively. The transient performance was tested when the resistive load connected to the three-phase inverter was changed.

13.4.1 Steady-state Performance

13.4.1.1 When the Neutral Current is Clean

The three-phase inverter was connected to a resistive local load in the open-loop mode to generate i_N and the controllers $\mathbf{C}_1$ and $\mathbf{C}_3$ were tested when the RMS value of the neutral current i_N was $I_N = 0$ A, $I_N = 1$ A and $I_N = 2$ A, respectively. The results are shown in Figure 13.7 when $I_N = 0$ A ($R_A = R_B = R_C = 6$ Ω), in Figure 13.8 when $I_N = 1$ A ($R_A = R_C = 6$ Ω and $R_B = 12$ Ω) and in Figure 13.9 when $I_N = 2$ A ($R_A = R_C = 6$ Ω and $R_B = \infty$), respectively. The controllers maintained a small voltage shift in all cases. The majority of the neutral current was forced to flow through the inductor and i_C remained nearly 0. There was no visible fundamental component in V_{ave} when $\mathbf{C}_1$ and $\mathbf{C}_3$ were used. Although there was visible 3rd harmonics in V_{ave} when $\mathbf{C}_1$ was used, no visible 3rd harmonics appeared in V_{ave} when $\mathbf{C}_3$ was used and the variation of V_{ave} was very small. The neutral point shift remained almost unchanged when the neutral current was increased.

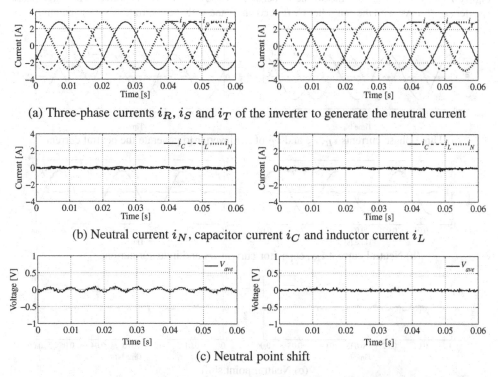

(a) Three-phase currents i_R, i_S and i_T of the inverter to generate the neutral current

(b) Neutral current i_N, capacitor current i_C and inductor current i_L

(c) Neutral point shift

Figure 13.7 Steady-state responses when a balanced resistive load was connected to the three-phase inverter ($I_N = 0$ A): with the controller $\mathbf{C}_1$ (left column) and with the controller $\mathbf{C}_3$ (right column)

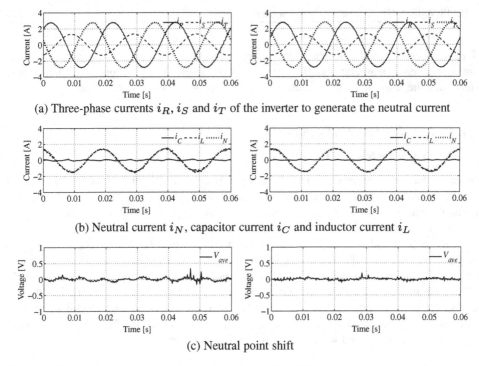

(a) Three-phase currents i_R, i_S and i_T of the inverter to generate the neutral current

(b) Neutral current i_N, capacitor current i_C and inductor current i_L

(c) Neutral point shift

Figure 13.8 Steady-state responses when a balanced resistive load was connected to the three-phase inverter ($I_N = 1$ A): with the controller C_1 (left column) and with the controller C_3 (right column)

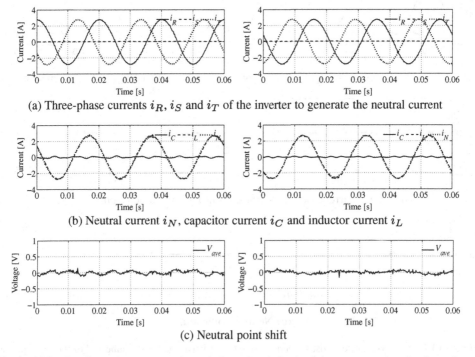

(a) Three-phase currents i_R, i_S and i_T of the inverter to generate the neutral current

(b) Neutral current i_N, capacitor current i_C and inductor current i_L

(c) Neutral point shift

Figure 13.9 Steady-state responses when a balanced resistive load was connected to the three-phase inverter ($I_N = 2$ A): with the controller C_1 (left column) and with the controller C_3 (right column)

13.4.1.2 When the Neutral Current is Distorted

In this experiment, a single-phase uncontrolled rectifier loaded with an LC filter $L = 150\,\mu H$, $C = 1000\,\mu F$ and a resistor $R = 20\,\Omega$ was connected to Phase S of the three-phase inverter. Experiments were carried out with the controllers $\mathbf{C}_1$, $\mathbf{C}_3$, $\mathbf{C}_5$ and $\mathbf{C}_7$ designed above. The results are shown in Figure 13.10. In all cases, there was no visible fundamental component in V_{ave}. When controller $\mathbf{C}_1$ was adopted, there was visible 3rd harmonics in V_{ave}. This was improved when controller $\mathbf{C}_3$ was adopted. However, after the 3rd harmonics was removed, there was visible 5th harmonics in V_{ave}. When controller $\mathbf{C}_5$ was adopted, the 5th harmonics was removed from V_{ave}; when controller $\mathbf{C}_7$ was adopted, the performance was further improved. This clearly shows that when the number of harmonics compensated for is increased,

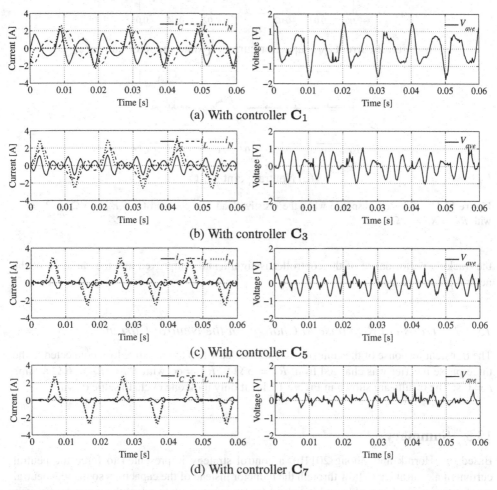

Figure 13.10 Steady-state responses when an unbalanced non-linear load was connected to the three-phase inverter: the neutral current i_N, the capacitor current i_C and the inductor current i_L (left column) and the neutral point shift V_{ave} (right column)

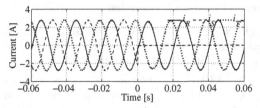

(a) Three-phase currents i_R, i_S and i_T of the inverter to generate the neutral current

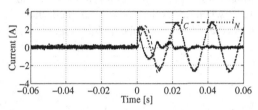

(b) Neutral current i_N, capacitor current i_C and inductor current i_L

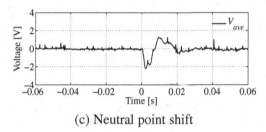

(c) Neutral point shift

Figure 13.11 Transient response when the resistive load was changed from $R_B = 6\,\Omega$ to $R_B = \infty$ with $R_A = R_C = 6\,\Omega$

the tracking performance of the controller is significantly improved and the variation of the neutral point is reduced.

13.4.2 Transient Response to Changes in the Neutral Current

The transient response of the controller $\mathbf{C}_7$ was tested when the resistive load connected to the three-phase inverter was changed from $R_B = 6\,\Omega$ to $R_B = \infty$ with $R_A = R_C = 6\,\Omega$ at time $t = 0$ s. The results are shown in Figure 13.11. It only took one cycle to settle down.

13.5 Summary

Based on (Hornik and Zhong 2011b), a control strategy is presented to force the neutral current of a neutral leg to flow through the inductor instead of the capacitors so that the neutral point is maintained as stable and balanced for an inverter, even when the neutral current is large. The strategy takes the form of a voltage controller and a current controller connected in parallel connection because their functions are decoupled in the frequency domain. The current controller is designed according to the H^∞ control theory and the weighting function is chosen so that the harmonic components flowing through the capacitors are eliminated. Intensive experimental results have shown that, for an increased neutral current i_N, the parallel voltage-H^∞ current controller is effective and is able to maintain a stable neutral point even when the neutral current contains significant harmonic components. The improvement of performance due to the individual treatment of harmonics is clearly demonstrated.

14

Applications in Single-phase to Three-phase Conversion

As an application of the neutral line provision, the conversion of a single-phase power supply to independent three phases is discussed in this chapter so that balanced or unbalanced, linear or non-linear, three-phase loads can be operated from a single-phase supply. It can be used in places, e.g., rural areas, where only a single-phase power supply is available. The converter consists of four legs: (i) one rectifier leg to generate a DC-link voltage; (ii) two phase legs to generate two independent phases to form balanced three-phase voltages together with the single-phase power supply; and (iii) one neutral leg to generate a neutral point, which is common to the single-phase supply and the two phases generated. Decoupled control strategies are developed to make sure that (i) the current drawn from the single-phase supply is sinusoidal and in phase with the supply voltage; (ii) the generated phase voltages contain low voltage harmonics even when the load is non-linear; and (iii) the neutral point is maintained as stable. Simulation results are provided.

14.1 Introduction

In some remote areas, it is quite normal to only have a single-phase power supply even though three-phase distribution systems are very common (Cipriano dos Santos *et al.* 2011). But some electric appliances, e.g., air conditioners and motors above a certain power level, require three-phase voltages. Hence, a device that converts a single-phase supply to balanced three-phase voltages is often needed. This converter is expected to have the capability of powering single-phase and/or three-phase, balanced or unbalanced, linear or non-linear loads and drawing a clean current that is in phase with the supply voltage.

Normally, this conversion process involves an uncontrolled rectifier bridge cascaded with a three-phase full-bridge inverter via a DC link as shown in Figure 14.1(a), which are widely adopted in motor drives (Bose 2001). This topology causes high current harmonics and low input power factor to the supply. Moreover, the power can only flow one way from the supply

---

Control of Power Inverters in Renewable Energy and Smart Grid Integration, First Edition.
Qing-Chang Zhong and Tomas Hornik.
© 2013 John Wiley & Sons, Ltd. Published 2013 by John Wiley & Sons, Ltd.

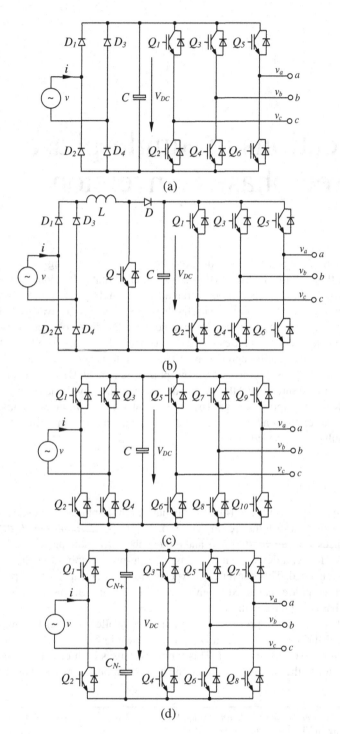

Figure 14.1 Conventional single-phase to three-phase converters

to the motor. The current harmonics and low input power factor can be improved by adding a boost converter between the rectifier and the inverter (Singh and Singh 2010), as shown in Figure 14.1(b). Figure 14.1(c) shows another popular topology consisting of a controlled single-phase full-bridge rectifier with a three-phase full-bridge inverter (Bose 2001; Kwak and Toliyat 2005). Although it is able to provide a clean grid current that is in phase with the grid voltage as well as high quality output voltage, it requires ten power semiconductor switches to process the full power. For economic and reliable operation, it is better to reduce the number of switches. One of the rectifier legs can be replaced with a split DC link to reduce the number of switches (Lee and Kim 2007; Singh and Singh 2010), as shown in Figure 14.1(d). However, it has some drawbacks, such as high output voltage distortion and large DC-link capacitors (Blaabjerg *et al.* 1997; Lee and Kim 2007). For these conventional topologies, the full load power needs to be processed by the converter, which increases the cost of the converter.

The topology shown in Figure 14.2 only processes partial load power because one phase is powered directly by the supply (Machado *et al.* 2006). It takes advantage of the supply voltage as the line-to-line voltage and generates another line to form a balanced three-phase voltages. The three-phase converter in this topology acts as an active power filter to provide the reactive and harmonic currents required by the load so that the supply current is clean and in phase with the supply voltage. Note that the generated line voltage depends on the load power. Another drawback is that no neutral line is available for unbalanced loads. The topology shown in Figure 14.3 is another one with a single-phase to three-phase universal active power filter (Cipriano dos Santos *et al.* 2011), where a controlled full-bridge rectifier with four switches and a transformer are added to the topology shown in Figure 14.2. The generated phase voltages are balanced and have the same voltage level as the supply voltage but a transformer is needed and no neutral line is available for the operation of unbalanced three-phase loads. Moreover, for these two topologies, there may be significant amount of harmonics in the output voltage as well since no strategies are designed to obtain high power quality under high power non-linear loads.

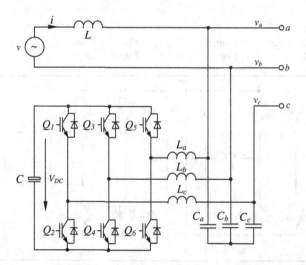

Figure 14.2 Line-interactive single-phase to three-phase converter. *Source:* Machado *et al.* 2006

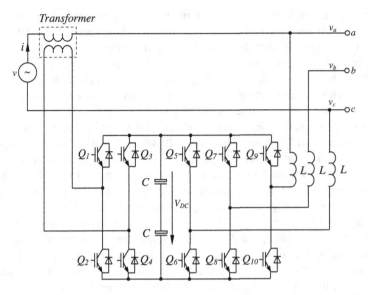

Figure 14.3 Single-phase to three-phase converter with a universal active power filter. *Source:* Cipriano dos Santos *et al.* 2011

14.2 The Topology under Consideration

According to the analysis in Chapters 10–13, an independently-controlled neutral leg is able to maintain a stable and balanced neutral point that is the mid-point of the DC link. This can be applied to form a single-phase to three-phase converter with three independent phases. Moreover, this neutral line can be shared with the single-phase supply to reduce the number of power semiconductor switches needed. The resulting topology is shown in Figure 14.4.

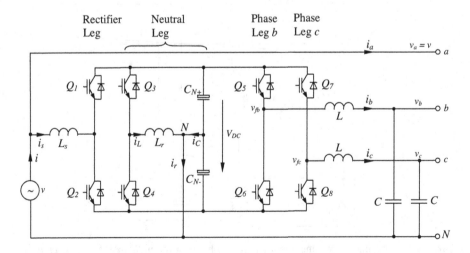

Figure 14.4 The single-phase to three-phase converter under consideration

It consists of an independently-controlled neutral leg, a rectifier leg and two phase legs. The neutral leg, taking the topology in (Zhong *et al.* 2006), consists of two switches, one inductor and two split DC-link capacitors. It is controlled to maintain the neutral point close to the mid-point of the DC link, which is common to the single-phase supply and to the two phases generated. The provision of the neutral point makes the three-phase voltages independent so that unbalanced loads can be operated and the operation of one phase does not affect others. The rectifier leg, together with the neutral leg, form a rectifier and the two phase legs, together with the neutral leg, form two single-phase inverters to generate two phase voltages that are $\pm \frac{2\pi}{3}$ displaced from the single-phase supply. The rectifier leg is to provide a compensation current so that the current drawn from the supply is clean and in phase with the voltage, to pass the power for the other two phases and to maintain a stable DC-link voltage. It operates as a boost converter so that the DC-link voltage can be maintained at a value needed by the two phase legs without a transformer. The two phase legs generate two independent phase voltages which, together with the single-phase supply, form independent balanced three phase voltages. An LC filter is connected to each phase leg to filter out the harmonics caused by the switching. It is worth noting that the provided three-phase voltages have the same level as the single-phase supply voltage and, hence, there is no need to use transformers for loads rated at the supply voltage level, which is normally the case.

With this topology, the load of Phase a is fed directly by the single-phase supply and the loads of Phases b and c are supplied by the converter. Hence, the converter does not process the full power, which reduces the cost and improves the efficiency. If balanced loads are connected, only two-thirds of the power is provided by the converter. If no loads are connected to Phases b and c, then the converter operates as an active power filter and a reactive power compensator for Phase a so that the current drawn from the grid is clean and in phase with the supply voltage. Another important property of the topology is that the rectifier leg, the neutral leg and the phase legs can be controlled independently, which offers considerable freedom in designing the controllers. In principle, any control strategies developed for reactive power compensation and active power filters can be applied to the rectifier leg. Any control strategies developed for neutral legs can be applied to the neutral leg and any control strategies developed for inverters can be applied to the phase legs.

14.3 Basic Analysis

Assume that the single-phase supply voltage v is

$$\dot{V} = V \angle 0,$$

where V is its RMS value, and that the generated two phase voltages together with the supply voltage form balanced three-phase voltages to supply three-phase linear loads with the load angle (lagging) of θ, θ_b and θ_c, respectively. Moreover, assume that the power loss in the converter is negligible. Then the (real) power supplied is equal to the (real) power consumed by the load, i.e.,

$$VI = VI_a \cos\theta + V_b I_b \cos\theta_b + V_c I_c \cos\theta_c,$$
$$= V I_a \cos\theta + VI_b \cos\theta_b + VI_c \cos\theta_c$$

where I_a, I_b and I_c are the RMS values of the three-phase load currents. Hence, the RMS value of the current drawn from the supply is

$$I = I_a \cos\theta + I_b \cos\theta_b + I_c \cos\theta_c.$$

If the loads are balanced, i.e. when $\theta = \theta_b = \theta_c$ and $I_a = I_b = I_c$, then the supply current is

$$I = 3I_a \cos\theta.$$

According to Kirchhoff's current law, there are

$$i_s = i - i_a, \tag{14.1}$$

and

$$i_r = i - i_a - i_b - i_c = i_s - i_b - i_c. \tag{14.2}$$

As a result, the phasor diagram of the system can be obtained as shown in Figure 14.5, where the cases with linear balanced and unbalanced three-phase loads are illustrated. The current I_s is

$$I_s = \sqrt{I^2 + I_a^2 - 2I_a I \cos\theta}.$$

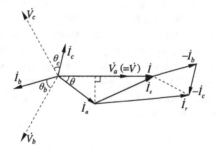

(a) With linear unbalanced three-phase loads

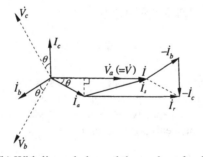

(b) With linear balanced three-phase loads

Figure 14.5   Phasor diagrams of the system

When the loads are linear and balanced, there are

$$I_s = \sqrt{I_a^2 + (3I_a \cos\theta)^2 - 6I_a^2 \cos^2\theta} = \sqrt{1 + 3\cos^2\theta}\, I_a,$$

and $i_r = i$, which means

$$I_r = I = 3I_a \cos\theta.$$

Hence,

$$I_s \leq 2I_a \quad \text{and} \quad I_r \leq 3I_a.$$

These can be applied to determine the current ratings of the switches of the rectifier leg and the neutral leg.

14.4 Controller Design

As mentioned before, the controller of the four legs can be decoupled and designed independently. The main function of the controller for the neutral leg is to make the voltage of the common point of the split capacitors with respect to the mid-point of the DC link stable, and close to zero and to provide a return current path for the rectifier leg and the phase legs. As a result, the common point of the split capacitors can be used as the neutral point N that is common to the supply and the two phases generated. This is vital to the operation of the system. The main function of the controller for the rectifier leg is to draw a clean sinusoidal current that is in phase with the supply voltage and to maintain a constant DC link, which automatically transfers power to the two phases. This is quite similar to the function of a combined active power filter and reactive power compensator. The main function of the controller for the phase legs is to convert the DC link voltage into two phase voltages with low harmonics to form balanced three phase voltages together with the single-phase supply.

14.4.1 Synchronisation Unit

In order to make sure that the current drawn from the supply is in phase with the supply voltage and also that the generated phase voltages are able to form balanced three-phase voltages together with the supply voltage, a synchronisation unit is needed to provide the phase information of the supply. This can be obtained in many ways, e.g. with phase-locked loops (PLL) (da Silva et al. 2010; Rodriguez et al. 2007b) or sinusoidal tracking algorithms (STA) (Ziarani and Konrad 2004). In this chapter, a STA is adopted to form a synchronisation unit, as shown in Figure 14.6(a). Once the phase information of the supply ωt is obtained, the phase information of the generated voltages can be obtained via adding $\pm\frac{2\pi}{3}$ to it. The unit also provides the RMS value of the supply voltage to be used as the reference RMS value for the generated phase voltages.

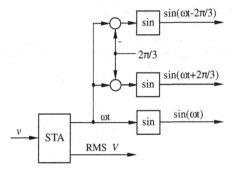

(a) Synchronisation unit

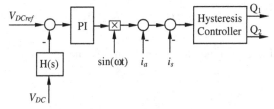

(b) Controller for the rectifier leg

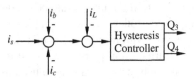

(c) Controller for the neutral leg

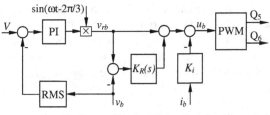

(d) Controller for Phase Leg b

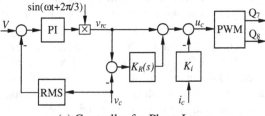

(e) Controller for Phase Leg c

Figure 14.6 Control strategy for the single-phase to three-phase converter

14.4.2 Control of the Rectifier Leg

As mentioned before, an important function of the rectifier leg is to inject the right amount of current to maintain a stable DC-link voltage. This can be achieved by introducing a PI controller, as shown in Figure 14.6(b). Because there are ripples in the DC-link voltage at the doubled supply frequency, a low-pass filter, e.g. the hold filter

$$H(s) = \frac{1 - e^{-Ts/2}}{Ts/2},$$

where T is the fundamental period of the supply, can be adopted to remove the ripples of V_{DC} for feedback. The output of the PI controller actually plays the role of the reference peak amplitude of the current drawn from the supply. The reference current drawn from the supply is then obtained by multiplying it with the phase signal $\sin \omega t$ of the supply voltage provided by the synchronisation unit shown in Figure 14.6(a). This makes sure that no reactive power is drawn from the supply. The reference current of i_s for the rectifier leg is then obtained by subtracting i_a from the reference current drawn from the supply. As a result, all the harmonic current components in i_a are automatically diverted into the reference current of i_s and no extra effort is needed to suppress the current harmonics in the supply current. What is left is to design a current controller so that the current i_s tracks the reference current.

Many current controllers, e.g., hysteresis controllers (Tilli and Tonielli 1998) that have a variable switching frequency and repetitive controllers (Hornik and Zhong 2011a) that have a fixed switching frequency, can be applied to track the reference current of i_s. Here, a simple hysteresis controller, which is easily implemented, is adopted. For this purpose, the current i_s is measured for feedback. Since both the supply voltage and the supply current $i = i_a + i_s$ are available, a power meter can be easily embedded into the system for metering.

It is worth noting that a suitable band for the hysteresis controller should be selected. A small band offers good tracking performance but leads to high switching frequencies while a large band offers low switching frequencies but leads to a large tracking error.

14.4.3 Control of the Neutral Leg

The neutral leg included in Figure 14.4, which consists of switches Q_3 and Q_4, one inductor L_r and two split DC-link capacitors, was studied in detail in Chapters 10–13. It is the combination of a split DC link and a neutral leg with the advantage of being controlled separately (Zhong et al. 2006). Assume the voltages across the capacitors C_{N+} and C_{N-} with respect to the neutral point N are V_+ and V_-, respectively. Then the DC-link voltage is

$$V_{DC} = V_+ - V_- \qquad (14.3)$$

and the voltage of the neutral point N with respect to the mid-point of the DC link, which is the reference point common to the supply voltage and the phase voltages generated, is

$$V_{ave} = V_+ - \frac{V_{DC}}{2} = \frac{V_+ + V_-}{2}. \qquad (14.4)$$

The main task of the neutral-leg controller is to maintain the voltage V_{ave} close to zero via controlling Q_3 and Q_4. At the same time, it provides a return current path for the rectifier leg and the phase legs. If the switches are controlled so that $i_L \approx i_r$ and $i_C \approx 0$, then almost no current flows through the split DC-link capacitors. As a result, the voltage V_{ave} is stable and close to 0. According to (14.2), the current i_r is $i_s - i_b - i_c$. If this is applied as the reference current for the inductor L_r, then the current i_C can be controlled to be nearly 0 by operating the switches Q_3 and Q_4. Hence, the control of the neutral leg is also a current-tracking problem and many control strategies can be applied. Here, the inductor current i_L is measured for feedback and a hysteresis controller is applied so that it is able to track the reference inductor current $i_s - i_b - i_c$, as shown in Figure 14.6(c).

14.4.4 Control of the Phase Legs

Since a neutral point that is close to the mid-point of the DC link is available, each phase leg could take the form of a half bridge connected with an LC filter, as shown in Figure 14.4. This topology has been widely studied, e.g. in (Liang et al. 2009; Srikanthan and Mishra 2010). One important aspect in this application is that the phase of the voltage generated should be synchronised with the supply voltage with Phase b lagging the supply voltage by $\frac{2\pi}{3}$ rad and Phase c leading the supply voltage by $\frac{2\pi}{3}$ rad. This requires the phase shift at the fundamental frequency caused by the LC filter to be small. The second important aspect is that the output voltage should contain low voltage harmonics even when the load is non-linear. There are many control strategies available for this; see e.g. (Weiss et al. 2004) for repetitive control-based strategy. The third important aspect is that the RMS value of the generated voltage should be the same as that of the single-phase supply voltage.

14.4.4.1 Generation of a Clean Voltage with the Right Phase

In order to address the first and second aspects, the simple and effective strategy in (Zhong et al. 2011) that bypasses the harmonic currents can be adopted; see Chapter 8. As shown in Figure 14.6(d) or Figure 14.6(e), it consists of a current feedback loop to force the output impedance of the phase to be resistive and a voltage loop to track the reference phase voltage v_{rb} or v_{rc}, respectively. The voltage loop is able to reduce the output impedance at harmonic frequencies, which is able to reduce the harmonic components of the output voltage.

Take Phase b as an example. The following equations hold for the voltage loop and the current loop:

$$u_b = v_{rb} - K_i i_b + K_R(s)(v_{rb} - v_b)$$

and

$$u_{fb} = sLi_b + v_b.$$

Since the switches are operated so that the average of u_{fb} during a switching period is the same as u_b, there is

$$v_{rb} - K_i i_b + K_R(s)(v_{rb} - v_b) = sLi_b + v_b.$$

That is,

$$v_b = v_{rb} - Z_o(s)i_b,$$

where the output impedance $Z_o(s)$ is

$$Z_o(s) = \frac{sL + K_i}{1 + K_R(s)}.$$

When there is no load, the output voltage v_b is the same as the reference voltage

$$v_{rb} = \sqrt{2}V \sin(\omega t - \frac{2\pi}{3})$$

and, similarly, the output voltage v_c is the same as the reference voltage

$$v_{rc} = \sqrt{2}V \sin(\omega t + \frac{2\pi}{3}).$$

There is not any phase shift or voltage drop.

If there is a load, then the voltage changes slightly because of the load effect. The smaller the output impedance, the smaller the voltage change. In particular, this is true for the harmonic voltage components. In order to improve the THD of the output voltage, the output impedance at harmonic frequencies should be small. This can be done by selecting appropriate $K_R(s)$. If the real part of $K_R(s)$ is positive, then the THD of the phase voltage can be reduced. The block $K_R(s)$ can be chosen to have high gains to obtain a small output impedance at harmonic frequencies. There are many ways to design K_R. One of them is to use the resonant harmonic compensator (Castilla *et al.* 2009; Shen *et al.* 2010)

$$K_R(s) = \sum_{h=3,5,\cdots} \frac{2\xi h\omega s}{s^2 + 2\xi h\omega s + (hw)^2} \times K_h$$

of which the gain at frequency hw is K_h with zero phases; see e.g. the Bode plot shown in Figure 14.7 for $1 + K_R(s)$ with $\xi = 0.01$ that is to be used in simulations later. It is almost 1 everywhere apart from around the harmonic frequencies. This means the output impedance can be tuned to be different values at different harmonic frequencies to improve the THD. For most cases, the coefficients K_h should be between 1 and 20 with large values for low-order harmonics and small values for high-order harmonics.

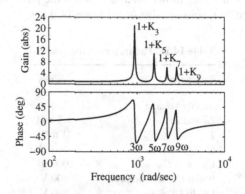

Figure 14.7 Bode plot of $1 + K_R(s)$ used in simulations

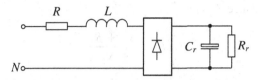

Figure 14.8 Model of non-linear loads

14.4.4.2 Regulation of the RMS Voltage

In order to address the third aspect, another loop is added to regulate the RMS value of the phase voltages generated, as shown in Figure 14.6(d) or Figure 14.6(e). The controller takes the form of a simple PI controller, with the RMS value of the phase voltage generated as feedback. This makes sure that all three-phase voltages have the same RMS value even when loaded. The output of the PI controller plays the role of the peak amplitude of the reference phase voltage, which is multiplied with the phase information $\sin(\omega t - \frac{2\pi}{3})$ or $\sin(\omega t + \frac{2\pi}{3})$ provided by the synchronisation unit shown in Figure 14.6(a) to form the reference phase voltage v_{rb} or v_{rc}.

14.5 Simulation Results

Simulations were carried out in MATLAB®/Simulink® to verify the topology and the control strategy. The solver used was ode23tb with a maximum step size of 1 μs. Different simulations with linear balanced loads and non-linear unbalanced loads were carried out. The model of non-linear loads is shown in Figure 14.8 and the other parameters of the system are given in Table 14.1. The band of the hysteresis controllers is 0.5 A and the K_h are 20, 10, 5 and 5, respectively, for the 3rd, 5th, 7th and 9th harmonics. $K_i = 4$, the parameters of the PI controller for the DC-link voltage are $K_P = 0.8$, $K_I = 10$ and the parameters of the PI controller for the RMS phase voltages are $K_P = 16$, $K_I = 248$.

14.5.1 With Three-phase Linear Balanced Loads

In this simulation, the three-phase loads were linear and balanced, consisting of a resistor of 5 Ω in series with an inductor of 10 mH for each phase. The simulation was started at 0 s with the rectifier leg and the neutral leg turned on at 0.04 s and the phase legs turned on at 0.12 s, respectively. The simulation results are shown in Figure 14.9.

Table 14.1 Parameters of the system

Parameters	Values
Supply voltage	220 V
L_s and L_r	1.5 mH
L	1 mH
C	20 μF
C_{N+} and C_{N-}	4000 μF
Initial DC-link voltage	600 V
DC-link voltage	800 V

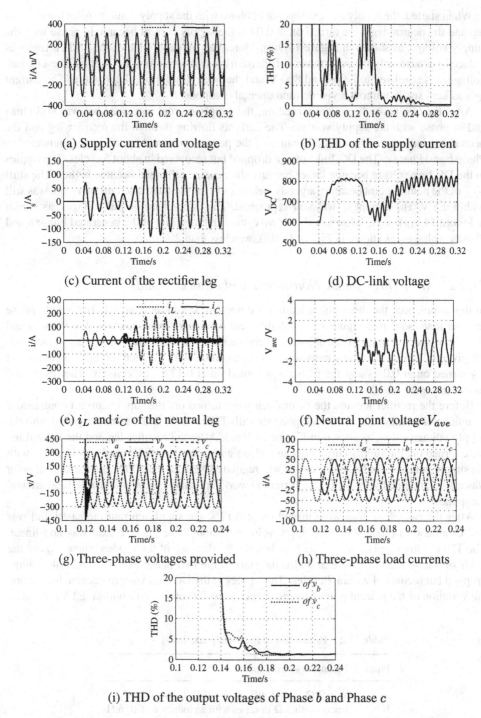

Figure 14.9 Simulation results for the case with linear balanced three-phase loads

When started, the supply current was not in phase with the supply voltage. After the rectifier leg and the neutral leg were turned on at 0.04 s, the supply current became in phase with the supply voltage, as shown in Figure 14.9(a). Moreover, the amplitude of the grid current was reduced because only the real power (of Phase a) was drawn from the supply. The DC-link voltage was regulated to be around 800 V and the neutral point was kept stable. The current i_C was kept small, without a visible fundamental component.

After the two phase legs were turned on, the supply current increased but was still clean and in phase with the supply voltage. The currents flowing through the rectifier leg and the neutral-leg inductor all increased because of the power consumed by the loads connected to Phase b and Phase c. The DC-link voltage dropped but recovered in about 5 cycles. The ripples in the DC link voltage became larger because the power exchanged became higher. The shift V_{ave} of the neutral point became larger as well because of the increased current i_L but was still within ± 3 V. The current i_C was still kept small. The three-phase output voltages, as shown in Figure 14.9(g), settled down in about two cycles and the THD of the generated Phase b and Phase c voltages, as shown in Figure 14.9(i), was less than 2%.

14.5.2 With Three-phase Non-linear Unbalanced Loads

In this simulation, the three-phase loads were non-linear and unbalanced. Two single-phase rectifier loads shown in Figure 14.8 with different parameters were connected to Phase a and Phase c, respectively, and a linear load was connected to Phase b. The parameters of the loads are given in Table 14.2. The simulation was started at 0 s with the rectifier leg and the neutral leg turned on at 0.04 s and the phase legs turned on at 0.12 s, respectively. The results are shown in Figure 14.10.

Before the rectifier leg and the neutral leg were turned on, the supply current contained a significant amount of harmonic components with THD >40% and was not in phase with the supply voltage either, as shown in Figure 14.10(a). After the rectifier leg and the neutral leg were turned on, the supply current became clean with a very low THD and was in phase with the supply voltage. The DC-link voltage was regulated well around 800 V and the neutral point was also maintained very well. The current i_C was kept small, without a visible fundamental component.

After the two phase legs were turned on at 0.12 s, the supply current increased and was still clean and in phase with the supply voltage, although the Phase c load was non-linear. The THD of the supply current was kept below 1%. Because of the sudden connection of the loads on Phase b and Phase c and also the start of the two phase legs, the DC-link voltage dropped but recovered within 7 cycles. The ripples in the DC-link voltage became higher and the variation of the neutral point became bigger as well (but was still within ± 2 V), because

Table 14.2 Parameters of the three-phase loads

Phase	Parameters of the load
a	$R = 3\,\Omega$, $L = 1$ mH, $R_r = 10\,\Omega$, $C_r = 800\,\mu$F
b	linear, with $5\,\Omega$ in series with an inductor of 10 mH
c	$R = 10\,\Omega$, $L = 5$ mH, $R_r = 15\,\Omega$, $C_r = 1000\,\mu$F

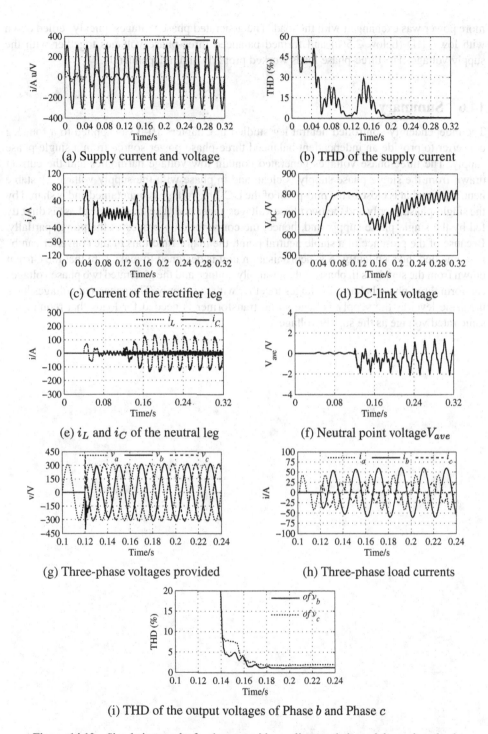

(i) THD of the output voltages of Phase b and Phase c

Figure 14.10 Simulation results for the case with non-linear unbalanced three-phase loads

more power was exchanged with the load. The generated phase voltages quickly settled down with low THD (below 2.5%) and formed balanced three-phase voltages together with the supply voltage. The three-phase loads worked properly and independently.

14.6 Summary

The independently-controlled neutral leg studied in Chapters 10–13 is applied to a four-leg converter to provide an independent balanced three-phase power source from a single-phase supply. The three-phase voltages generated contain low voltage harmonics and the current drawn from the single-phase supply is clean and in phase with the supply voltage. A stable neutral point that is close to the mid-point of the DC-link voltage is maintained and is shared by the single-phase supply and the two phase voltages generated. The load of one phase is directly fed by the single-phase supply and, hence, the converter only processes the power partially. Because of the presence of a stable neutral point, the major functions of the converter can be implemented independently. A synchronisation unit is designed to make sure that the current drawn from the supply is in phase with the supply voltage and the generated two phase voltages can form three-phase balanced voltages together with the supply. The generated voltages have the same level as the supply voltage so no transformer is needed for loads that require the same rated voltage as the supply voltage.

Part III
Power Flow Control

15

Current Proportional–Integral Control

Proportional–integral (PI) controllers are so far the most widely applied controllers in industry. In this chapter, a PI controller is applied to grid-connected inverters to track a reference current so that a desired current can be injected into the grid. This is done in the synchronously rotating reference (dq) frame. Its equivalent in the natural (abc) frame is also discussed. Experimental results are provided to demonstrate the performance of the PI control scheme.

15.1 Control Structure

15.1.1 In the Synchronously Rotating Reference (dq) Frame

The transfer function of a PI controller (Aström and Hägglund 1988, 1995, 2006; Tan et al. 1999; Visioli 2006; Yu 1999) is given by

$$C_{PI}(s) = K_p + \frac{K_i}{s}, \quad (15.1)$$

where K_p and K_i are the proportional and integral gains, respectively. As discussed in Chapter 2, three-phase signals in the natural frame can be transformed into DC signals in the synchronously rotating reference (dq) frame. Hence, for inverters, PI controllers are normally designed in the synchronously rotating reference (dq) frame (Teodorescu and Blaabjerg 2004; (Timbus et al. 2006c, 2009).

The block diagram of a current-controlled VSI in the synchronously rotating reference (dq) frame is shown in Figure 15.1. A PLL is adopted to provide the phase information of the grid voltage, which is needed to transform the grid currents into their dq components I_d and I_q in the dq frame and then transform the voltage control signal back to the abc frame. Two PI controllers are adopted to regulate I_d and I_q according to the current references I_d^* and I_q^*,

---

Control of Power Inverters in Renewable Energy and Smart Grid Integration, First Edition.
Qing-Chang Zhong and Tomas Hornik.
© 2013 John Wiley & Sons, Ltd. Published 2013 by John Wiley & Sons, Ltd.

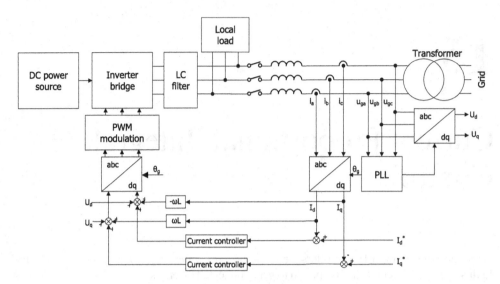

Figure 15.1 Block diagram of a current-controlled VSI in the synchronously rotating reference (dq) frame

which determines the real power and reactive power exchanged with the grid, respectively. The inverter in Figure 15.1 is assumed to be powered by a constant DC power source and, hence, no controller is needed to regulate the DC-link voltage. Otherwise, a controller can be introduced to regulate the DC-link voltage and to generate I_d^* accordingly.

A particular feature of the control structure shown in Figure 15.1 is that the grid voltage is feed-forwarded, after being transformed into the dq-components U_d and U_q, and added to the output of the current controllers after the $abc \rightarrow dq$ transformation. This allows the voltage of the inverter to be the same as the grid voltage when the circuit breaker is not turned on. It also improves the dynamics of the controller during grid voltage fluctuations, leading to a fast response for the control system (Timbus et al. 2009). In order to improve the performance of the PI controller in this structure, cross-coupling terms ($\pm\omega L$) are used (Blaabjerg et al. 2006; Timbus et al. 2006c).

The PI controller implemented in the dq frame is traditionally considered to be a good solution for regulating sinusoidal currents in balanced three-phase systems. However, it is not good at correcting unbalanced disturbance currents unless the positive- and negative-sequence currents are dealt with separately. Moreover, the compensation capability of low-order harmonics is poor. The PI controller in the synchronously rotating reference frame is more complex than the PR controller in the stationary frame to be discussed in Chapter 16 because it requires two coordinate transformations with the knowledge of the phase information of the grid voltage.

The scheme has the particular advantage of independent control of the real and reactive components of the grid currents, which can be directly translated into real and reactive power (Twining and Holmes 2003). As a result, it is possible to set the real power and reactive power sent to the grid directly.

15.1.2 Equivalent Structure in the Natural (abc) Frame

PI controllers are normally associated with the dq frame but it is possible to convert the controller in the dq frame to the abc frame. According to (Zmood *et al.* 1999, 2001), the PI controller implemented in the dq frame is equivalent to the proportional-resonant (PR) controller implemented in the $\alpha\beta$ frame at the fundamental frequency. Note that this is different from the harmonic compensator discussed in Chapter 16, which is tuned at harmonic frequencies. Using the transformation techniques presented in (Zmood *et al.* 1999, 2001), the equivalent PI controller in the abc frame is derived in (Twining and Holmes 2003) as

$$C_{PI_{abc}}(s) = \begin{bmatrix} K_p + K_i \dfrac{s}{s^2+\omega^2} & -\dfrac{K_p}{2} - \dfrac{K_i s + \sqrt{3}K_i\omega}{2(s^2+\omega^2)} & -\dfrac{K_p}{2} - \dfrac{K_i s - \sqrt{3}K_i\omega}{2(s^2+\omega^2)} \\ -\dfrac{K_p}{2} - \dfrac{K_i s - \sqrt{3}K_i\omega}{2(s^2+\omega^2)} & K_p + K_i \dfrac{s}{s^2+\omega^2} & -\dfrac{K_p}{2} - \dfrac{K_i s + \sqrt{3}K_i\omega}{2(s^2+\omega^2)} \\ -\dfrac{K_p}{2} - \dfrac{K_i s + \sqrt{3}K_i\omega}{2(s^2+\omega^2)} & -\dfrac{K_p}{2} - \dfrac{K_i s - \sqrt{3}K_i\omega}{2(s^2+\omega^2)} & K_p + K_i \dfrac{s}{s^2+\omega^2} \end{bmatrix}.$$

(15.2)

It consists of cross-coupling terms, in addition to the PR controllers on the diagonal. The equivalent control structure in the abc frame is shown in Figure 15.2. The cross-coupling terms indicate that the three-phases are not independent. This could be a problem when the system is not balanced.

The phase-lead low-pass filter (3.1) in Chapter 3 is again adopted to feed-forward the grid voltage, which is added to the output of the current controller. This forces the output voltage of the inverter to be the same as the grid voltage when the current references are set to be zero, which facilitates the grid connection of the inverter. It also improves the dynamic performance of the inverter when there are fluctuations in the grid voltage.

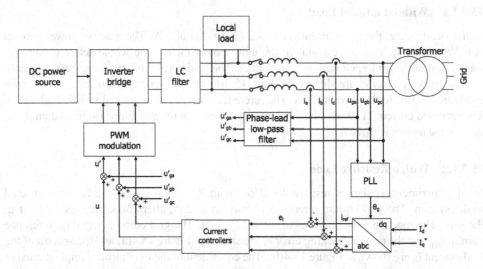

Figure 15.2 Equivalent control strategy in the natural frame (abc) for the one shown in Figure 15.1

15.2 Controller Implementation

According to (Timbus *et al.* 2006c; Zmood *et al.* 1999), the controller gains remain unchanged when the reference frame is changed from one to another. The proportional gain K_p and the integral gain K_i can be determined by trial-and-error. The ones to be derived in Chapter 16 can also be used. As can be seen from (15.2), the controller in the *abc* frame is quite complex due to the cross-coupling terms but there are only three different dynamic terms. In order to facilitate the implementation of this controller, these three terms can be discretised separately for implementation.

15.3 Experimental Results

The control scheme was tested on an experimental set-up, which consists of an inverter board, a three-phase LC filter, a three-phase grid interface inductor, a board consisting of voltage and current sensors, a step-up Wye-Wye transformer (12 V/230 V/50 Hz), a dSPACE DS1104 R&D controller board with ControlDesk software, and MATLAB® Simulink®/SimPower software package. The inverter board consists of two independent three-phase inverters and has the capability to generate PWM voltages from a constant 42 V DC voltage source. One inverter was used to generate a stable neutral line for the three-phase inverter. The generated three-phase voltage was connected to the grid via a controlled circuit breaker and a step-up transformer. The PWM switching frequency was 12 kHz and the sampling frequency was 5 kHz. A Yokogawa power analyser WT1600 was used to measure the THD.

The PI control scheme implemented in the *dq* frame and its equivalent in the *abc* frame were evaluated in the grid-connected mode under five different scenarios to test the steady-state responses without a local load and with a resistive, non-linear and unbalanced resistive local load connected to the system, and the transient response without a local load.

15.3.1 Steady-state Performance

15.3.1.1 Without a Local Load

In the steady state, the grid current reference I_d^* was set at 3 A. The reactive power was set at 0 Var ($I_q^* = 0$). This corresponds to the unity power factor. The whole active power was transferred to the grid via a step-up transformer. The grid current output i_a, current reference i_{ref} and tracking error e_i are shown in Figure 15.3(a) and the spectra of current i_a are shown in Figure 15.3(b) for both controllers. The reference currents were tracked well in both cases. The recorded current THD was slightly better in the case of the *abc* frame (4.38%) than in the case of the *dq* frame (4.76%).

15.3.1.2 With a Resistive Load

In this experiment, a balanced resistive local load with $R_A = R_B = R_C = 12\,\Omega$ was connected to the system. The grid current reference I_d^* was set at 2 A, after connecting the inverter to the grid. The reactive power was set at 0 Var ($I_q^* = 0$). The grid current output i_a, reference current i_{ref} and the current tracking error e_i are shown in Figure 15.4(a) and the spectra of the grid current i_a are shown in Figure 15.4(b). The equivalent of the PI controller implemented in the *abc* frame demonstrated better tracking performance and lower THD than the PI controller itself. The 7th harmonics in the case of the *dq* frame is remarkably increased compared to that

Current Proportional–Integral Control

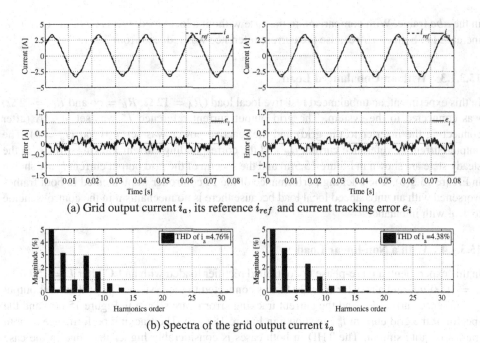

(a) Grid output current i_a, its reference i_{ref} and current tracking error e_i

(b) Spectra of the grid output current i_a

Figure 15.3 Experimental results for a PI controller without a local load implemented in the dq frame (left column) and its equivalent in the abc frame (right column)

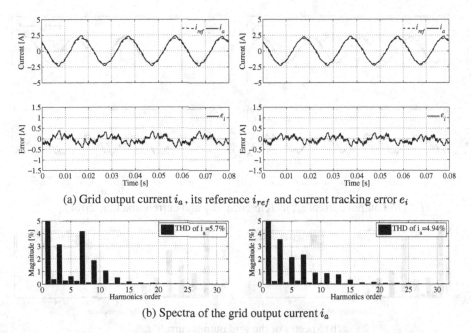

(a) Grid output current i_a, its reference i_{ref} and current tracking error e_i

(b) Spectra of the grid output current i_a

Figure 15.4 Experimental results for a PI controller with a balanced resistive local load implemented in the dq frame (left column) and its equivalent in the abc frame (right column)

in the *abc* frame. With comparison to the case without a local load, the THD of the current i_a increased in both cases because the current reference was reduced from 3 A to 2 A.

15.3.1.3 With an Unbalanced Load

In this experiment, an unbalanced resistive local load ($R_A = 12\,\Omega$, $R_B = \infty$ and $R_C = 12\,\Omega$) was connected to the system. The grid output current reference I_d^* was set at 2 A (after connecting the inverter to the grid) and the reactive power was set at 0 Var ($I_q^* = 0$). The output current i_a, reference current i_{ref} and the corresponding current tracking error e_i in the steady state are shown in Figure 15.5(a) and the spectra of the grid output current i_a are shown in Figure 15.5(b). The tracking performance of the PI controller implemented in both frames worsened with an unbalanced local load because there is no mechanism in the control scheme to deal with imbalance.

15.3.1.4 With a Non-linear Load

In this experiment, a three-phase uncontrolled rectifier loaded with an LC filter $L = 150\,\mu H$, $C = 1000\,\mu F$ and a resistor $R = 20\,\Omega$ was connected to the system. The grid current output i_a, reference current i_{ref} and the current tracking error e_i are shown in Figure 15.6(a) and the spectra of the grid current i_a are shown in Figure 15.6 (b). The tracking performance in both frames is quite similar. The THD in both cases is considerably higher than that in the case with resistive loads although the THD in the dq frame is slightly better than that in the *abc* frame. This is because there is no mechanism in the control scheme to deal with harmonics.

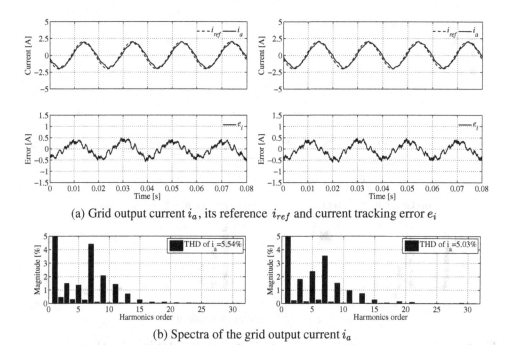

(a) Grid output current i_a, its reference i_{ref} and current tracking error e_i

(b) Spectra of the grid output current i_a

Figure 15.5 Experimental results for a PI controller with an unbalanced resistive local load implemented in the dq frame (left column) and its equivalent in the *abc* frame (right column)

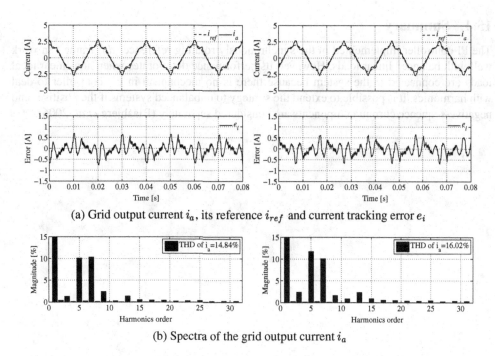

(a) Grid output current i_a, its reference i_{ref} and current tracking error e_i

(b) Spectra of the grid output current i_a

Figure 15.6 Experimental results for a PI controller with a balanced non-linear local load implemented in the dq frame (left column) and its equivalent in the abc frame (right column)

15.3.2 Transient Performance

In this experiment, a step change in the current I_d^* reference from 2 A to 3 A was applied (while keeping $I_q^* = 0$) in the grid-connected mode without a local load. The grid current output i_a, current reference i_{ref} and tracking error e_i are shown in Figure 15.7. The PI controllers demonstrated fast dynamics, following closely the reference signal i_{ref}.

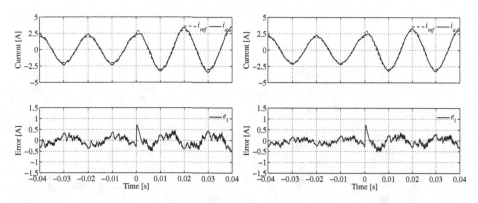

Figure 15.7 Transient response of the PI controller in the grid-connected mode without a local load implemented in the dq frame (left column) and its equivalent in the abc frame (right column)

15.4 Summary

The PI controller implemented in the synchronously rotating reference (dq) frame can work well with balanced linear loads. However, its performance deteriorates when non-linear local loads are connected to the system because there is no mechanism in the controller to deal with harmonics. It is possible to extend the strategy to unbalanced systems if the positive- and negative-sequence current components are considered separately (Blaabjerg *et al.* 2006).

16

Current Proportional-Resonant Control

The proportional-resonant (PR) controller is one of the most popular controllers used for grid-connected inverters to regulate the current injected into the grid. In this chapter, the PR current controller is designed and implemented for three-phase inverters, in the stationary reference frame and in the natural reference frame.

16.1 Proportional-resonant Controller

For inverters, the controller deals with sinusoidal signals, which makes it difficult to design the controller with the correct gain that is able to regulate the performance at the fundamental frequency and also to reject harmonic disturbances. PI controllers, having a pole (with an infinite gain) at the zero frequency, are not able to eliminate the steady-state error at the fundamental frequency (Blaabjerg et al. 2006) unless it is adopted in the *dq* frame, as done in Chapter 15. Alternatively, PR controllers can be used.

A PR controller is the combination of a proportional term and a resonant term given by

$$C_{PR}(s) = K_p + K_i \frac{s}{s^2 + \omega^2} \quad (16.1)$$

where ω is the resonant frequency. Such a controller has a high gain around the resonant frequency and, thus, is capable of eliminating the steady-state error when tracking or rejecting a sinusoidal signal (Blaabjerg et al. 2006; Sera et al. 2005; Timbus et al. 2006b), according to the internal model principle (Francis and Wonham 1975). As a result, PR controllers are widely used in inverter control. In order to improve the performance of handling harmonics, a harmonic compensator given by

$$C_{HC}(s) = \sum_{h=3,5,7,\ldots} K_{ih} \frac{s}{s^2 + (\omega h)^2} \quad (16.2)$$

Control of Power Inverters in Renewable Energy and Smart Grid Integration, First Edition.
Qing-Chang Zhong and Tomas Hornik.
© 2013 John Wiley & Sons, Ltd. Published 2013 by John Wiley & Sons, Ltd.

where h is the harmonic order, can be easily added to a PR controller (Timbus et al. 2006b). It is worth noting that, in order to maintain good performance of the controller, the resonant frequency should be maintained close to the system frequency. If the system frequency varies significantly, adaptive mechanisms, e.g. the one reported in (Timbus et al. 2006a), can be adopted to adjust the resonant frequency according to the system frequency.

16.2 Control Structure

16.2.1 In the Stationary Reference ($\alpha\beta$) Frame

A PR controller is usually adopted in the stationary reference ($\alpha\beta$) frame for inverter control (Sera et al. 2005; Teodorescu et al. 2006; Timbus et al. 2006a; Zmood et al. 1999, 2001), but it can easily be implemented in the abc frame as well.

The block diagram of a current-controlled VSI in the stationary reference frame with the PR controller is shown in Figure 16.1. The three-phase currents are transformed into α-, β-components i_α and i_β using the $abc \rightarrow \alpha\beta$ transformation. The α-component is the same as the Phase a component and the β-component is the combination of Phase b and Phase c components (when the currents are balanced). Hence, only two current channels i_α and i_β need to be controlled, with two separate PR controllers. A PLL is adopted to generate the phase information of the grid voltage, which is needed by the $dq \rightarrow \alpha\beta$ transformation to convert the d-, q-current references I_d^* and I_q^* into the reference currents i_α^* and i_β^*. The PR controllers, often equipped with the harmonic compensator in (16.2), forces i_α and i_β to track the reference currents i_α^* and i_β^*. The output of the PR controllers is then converted back to the abc frame before being converted into PWM signals to drive the switches. The real power and reactive power exchanged with the grid are directly determined by I_d^* and I_q^* because the voltage is determined by the grid.

The grid voltages are feed-forwarded via a phase-lead low-pass filter to improve the performance, as shown in Chapter 3, etc. The filter, chosen as (3.1), is introduced to compensate

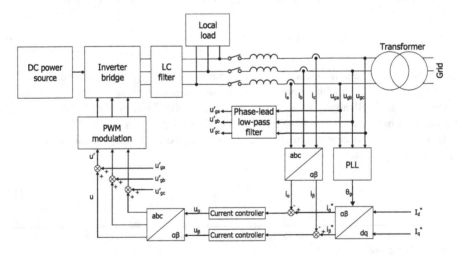

Figure 16.1 Block diagram of a current-controlled VSI in the stationary reference frame ($\alpha\beta$)

the phase shift and gain attenuation caused by the computational delay, PWM modulation, the inverter bridge and the LC filter. It also attenuates the harmonics in the feed-forwarded grid voltages and improves the dynamics during grid voltage fluctuations (Timbus *et al.* 2009).

16.2.2 Equivalent Controller in the abc Frame

As mentioned before, the PR controllers can easily be implemented in the abc frame. Actually, it is straightforward to do so. Instead of using the $abc \rightarrow \alpha\beta$ transformation, each phase can be equipped with a PR controller, with the current reference generated from I_d^* and I_q^* with the $dq \rightarrow abc$ transformation. According to (Timbus *et al.* 2009), the controller matrix in the abc frame is given as

$$C_{PR_{abc}}(s) = \begin{bmatrix} K_p + K_i \frac{s}{s^2+\omega^2} & 0 & 0 \\ 0 & K_p + K_i \frac{s}{s^2+\omega^2} & 0 \\ 0 & 0 & K_p + K_i \frac{s}{s^2+\omega^2} \end{bmatrix}. \quad (16.3)$$

Because of this, the PR controller can be designed for each phase and then applied to the stationary reference frame.

16.3 Controller Design

16.3.1 Model of the Plant

The model of the control plant can be derived from the single-phase diagram of the inverter shown in Figure 16.2, which mainly consists of the LCL filter of the inverter. The inverter is operated so that the average of u_f over a switching period is the same as u and, hence, the effect of the PWM conversion and the switching is negligible.

According to Kirchhoff's current law (KCL), there is

$$i_c = i_1 - i_2. \quad (16.4)$$

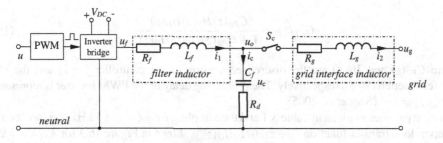

Figure 16.2 Single-phase representation of an inverter

According to Kirchhoff's voltage law (KVL), there are

$$u_f = i_1(R_f + sL_f) + i_c \left(\frac{1}{sC_f} + R_d\right), \quad (16.5)$$

and

$$u_g = -i_2(R_g + sL_g) + i_c \left(\frac{1}{sC_f} + R_d\right), \quad (16.6)$$

where s is the Laplace operator. Re-writing (16.4)–(16.6) in terms of impedances, then

$$\begin{bmatrix} u_f \\ u_g \end{bmatrix} = \begin{bmatrix} Z_{11} & -Z_{12} \\ Z_{21} & -Z_{22} \end{bmatrix} \begin{bmatrix} i_1 \\ i_2 \end{bmatrix} \quad (16.7)$$

with

$$Z_{11} = R_f + sL_f + R_d + \tfrac{1}{sC_f}, \quad Z_{12} = R_d + \tfrac{1}{sC_f},$$

$$Z_{21} = R_d + \tfrac{1}{sC_f}, \quad Z_{22} = R_g + sL_g + R_d + \tfrac{1}{sC_f}.$$

The transfer function from the inverter voltage u_f to the grid output current i_2 is

$$H_P(s) = \frac{i_2}{u_f} = \frac{Z_{21}}{Z_{11}Z_{22} - Z_{12}Z_{21}}$$

$$= \frac{sC_f R_d + 1}{s^3 C_f L_f L_g + s^2 C_f (L_f R_g + R_f L_g + L_g R_d + L_f R_d) + sC_f (R_f R_g + R_d R_g + R_f R_d + L_f + L_g) + R_f + R_g}$$

(16.8)

16.3.2 Design Example

For a PR controller, there are two parameters K_p and K_i, in addition to the known system frequency ω. The parameter K_i is the gain for the resonant term so it can be chosen to be large. What is to be determined is the gain K_p. Here, this is done with the root-locus approach.

The closed-loop system transfer function in the discrete-time domain is given by

$$G_{CL}(z) = \frac{C_{PR}(z)H_P(z)H_D(z)}{1 + C_{PR}(z)H_P(z)H_D(z)}, \quad (16.9)$$

where $C_{PR}(z)$ and $H_P(z)$ are the discretised version of the controller (16.1) and the plant transfer function (16.8), respectively. The processing delay of the PWM inverter is represented by $H_D(z) = z^{-1}$ (Sera et al. 2005).

For the parameters given in Table 3.1 and the sampling frequency of 5 kHz, the root locus of the open-loop transfer function $C_{PR}(z)H_P(z)H_D(z)$ is shown in Figure 16.3 for $K_i/K_p = 200$. When the proportional gain K_p is chosen as $K_p = 1.12$, the damping ratio of the system is

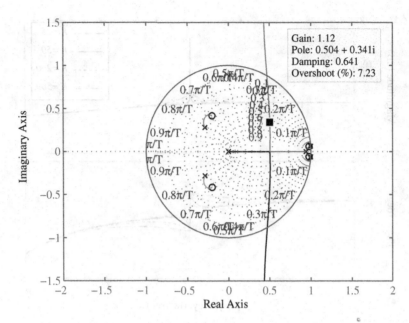

Figure 16.3 Root locus for the controller design

$\xi = 0.641$. Figure 16.4 shows the Bode plots of the open-loop system for different integral gain K_i with $K_p = 1.12$, from which the integral gain is chosen as $K_i = 200$. Since the controller has a very high gain at the resonant frequency, it has a very good ability to reduce the steady-state error. The resulting controller in the discrete-time domain is

$$C_{PR}(z) = \frac{1.12z^2 - 2.196z + 1.08}{z^2 - 1.996z + 1}. \tag{16.10}$$

16.4 Experimental Results

The PR controllers implemented in both the $\alpha\beta$ frame and the abc frame were evaluated with the test rig described in Chapter 15 in the grid-connected mode under five different scenarios to test the steady-state responses without a local load and with a resistive, non-linear and unbalanced resistive local load connected to the system, and the transient response without a local load.

16.4.1 Steady-state Performance

16.4.1.1 Without a Local Load

In the steady state, the grid current reference I_d^* was set at 3 A. The reactive power was set at 0 Var ($I_q^* = 0$). This corresponds to the unity power factor. All the active power was transferred to the grid via the step-up transformer. The grid current output i_a and its spectra, the current reference i_{ref} and the tracking error e_i are shown in the left column of Figure 16.5 for the

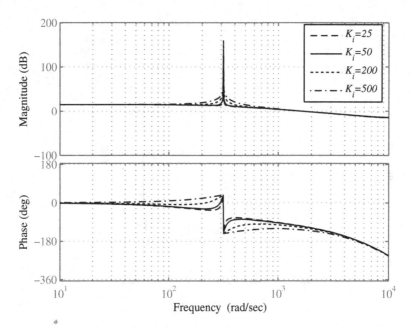

Figure 16.4 Bode plots of the open-loop system for different K_i with $K_p = 1.12$

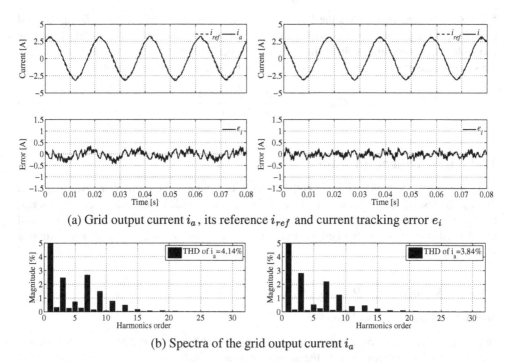

(a) Grid output current i_a, its reference i_{ref} and current tracking error e_i

(b) Spectra of the grid output current i_a

Figure 16.5 Experimental results for a PR controller without a local load implemented in the $\alpha\beta$ frame (left column) and in the abc frame (right column)

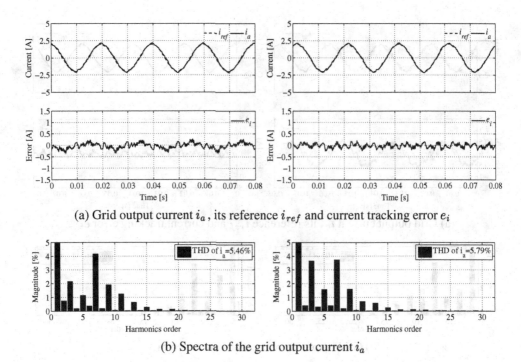

(a) Grid output current i_a, its reference i_{ref} and current tracking error e_i

(b) Spectra of the grid output current i_a

Figure 16.6 Experimental results for a PR controller with a resistive local load implemented in the $\alpha\beta$ frame (left column) and in the abc frame (right column)

implementation in the $\alpha\beta$ frame and in the right column of Figure 16.5 for the implementation in the abc frame. The recorded current THD, shown in Figure 16.5(b), was 3.84% (right) in the abc frame compared to 4.14% in the $\alpha\beta$ frame (left). Both the tracking performance and the recorded current THD in the abc frame were better than that in the $\alpha\beta$ frame.

16.4.1.2 With a Resistive Load

In this experiment a balanced resistive local load with $R_A = R_B = R_C = 12\,\Omega$ was connected to the system. The grid current reference I_d^* was set at 2 A, after connecting the inverter to the grid. The reactive power was set at 0 Var ($I_q^* = 0$), which corresponds to the unity power factor. The grid current output i_a and its spectra, the reference current i_{ref} and the current tracking error e_i are shown in the left column of Figure 16.6 for the implementation in the $\alpha\beta$ frame and in the right column of Figure 16.6 for the implementation in the abc frame. The PR controller implemented in the abc frame again demonstrated better tracking performance, although with a higher THD due to the increased 3rd harmonics.

16.4.1.3 With an Unbalanced Load

In this experiment, an unbalanced resistive local load with $R_A = 12\,\Omega$, $R_B = \infty$ and $R_C = 12\,\Omega$ was connected to the system. The grid output current reference I_d^* was set at 2 A (after connecting the inverter to the grid) and the reactive power was set at 0 Var ($I_q^* = 0$). The output

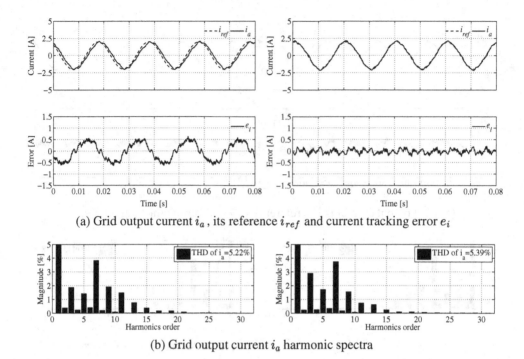

(a) Grid output current i_a, its reference i_{ref} and current tracking error e_i

(b) Grid output current i_a harmonic spectra

Figure 16.7 Experimental results for a PR controller with an unbalanced resistive local load implemented in the $\alpha\beta$ frame (left column) and in the abc frame (right column)

current i_a and its spectra, the reference current i_{ref} and the corresponding current tracking error e_i in the steady state are shown in the left column of Figure 16.7 for the implementation in the $\alpha\beta$ frame and in the right column of Figure 16.7 for the implementation in the abc frame. The tracking performance of the PR controller implemented in the $\alpha\beta$ frame was remarkably worse than the one implemented in the abc frame, although the recorded THD in the $\alpha\beta$ frame was slightly better than that in the abc frame.

16.4.1.4 With a Non-linear Load

In this experiment, a three-phase uncontrolled rectifier loaded with an LC filter $L = 150\ \mu H$, $C = 1000\ \mu F$ and a resistor $R = 20\ \Omega$ was connected to the system. The grid current output i_a and its spectra, the reference current i_{ref} and the current tracking error e_i are shown in the left column of Figure 16.8 for the implementation in the $\alpha\beta$ frame and in the right column of Figure 16.8 for the implementation in the abc frame. While the tracking performance in both frames were quite similar, the current THD in the abc frame was slightly better than that in the $\alpha\beta$ frame.

16.4.2 Transient Performance

In this experiment, a step change in the current I_d^* reference from 2 A to 3 A was applied (while keeping $I_q^* = 0$). The grid current output i_a, the current reference i_{ref} and the tracking

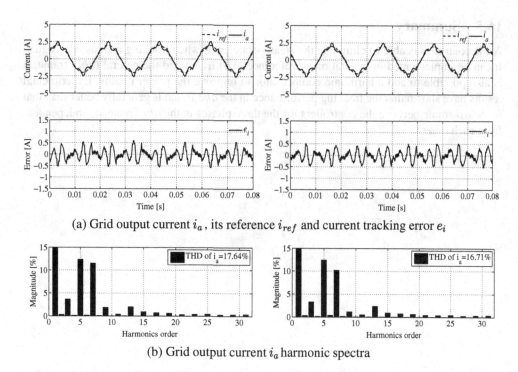

(a) Grid output current i_a, its reference i_{ref} and current tracking error e_i

(b) Grid output current i_a harmonic spectra

Figure 16.8 Experimental results for a PR controller with a non-linear local load implemented in the $\alpha\beta$ frame (left column) and in the abc frame (right column)

error e_i are shown in the left column of Figure 16.9 for the implementation in the $\alpha\beta$ frame and in the right column of Figure 16.9 for the implementation in the abc frame. Both controllers demonstrated very fast dynamics, following the reference signal i_{ref} closely. The implementation in the abc frame demonstrated better performance than the implementation in the $\alpha\beta$ frame.

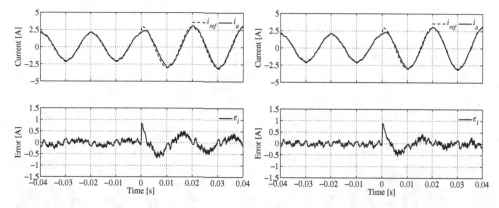

Figure 16.9 Transient response of the PR controller in the grid-connected mode without a local load implemented in the $\alpha\beta$ frame (left column) and in the abc frame (right column)

16.5 Summary

A PR controller is able to eliminate the steady-state error when tracking a sinusoidal signal. It can be implemented in the stationary reference ($\alpha\beta$) frame and the natural (abc) frame. It is also possible to add a harmonic compensator to improve the power quality. Experimental results have shown that the tracking performance in the abc frame is generally better than that in the $\alpha\beta$ frame because the controllers for the three phases in the abc frame are independent from each other.

17

Current Deadbeat Predictive Control

As shown in the previous two chapters, PI controllers in the dq frame and PR controllers in the $\alpha\beta$ or abc frame can be adopted to track the reference set for the grid current so that the power exchanged with the grid is regulated. They are able to track pure sinusoidal reference currents without steady-state errors under balanced conditions. In this chapter, a deadbeat (DB) predictive controller is discussed so that the grid currents can track the grid current references within two sampling steps so that the dynamic performance is improved. The controller is designed and implemented in the natural (abc) reference frame.

17.1 Control Structure

The block diagram of a current-controlled VSI in the natural (abc) frame with the DB predictive controller is shown in Figure 17.1. Each phase current is independently controlled via a DB predictive controller, which takes the reference current, the grid current and the grid voltage as inputs. The three-phase current references are generated from the dq current references I_d^* and I_q^* with a $dq \to abc$ transformation.

17.2 Controller Design

The basic idea of a DB controller is for the controlled variable to reach its desired value within fixed periods of sampling. In inverter control, this is widely employed, with the capability of predicting the reference current and the grid voltage, to track sinusoidal currents (Buso et al. 2001; Habetler 1993; Holmes and Martin 1996; Kawabata et al. 1990, 1991; Kojabadi et al. 2006; Lindgren 1996; Lindgren and Svensson 1998; Mohamed and El-Saadany 2007, 2008b; Wu and Lehn 2006; Zeng and Chang 2008). A DB predictive controller is able to provide a fast dynamic response but it is sensitive to system uncertainties (Timbus et al. 2009; Zeng and Chang 2005).

Control of Power Inverters in Renewable Energy and Smart Grid Integration, First Edition.
Qing-Chang Zhong and Tomas Hornik.
© 2013 John Wiley & Sons, Ltd. Published 2013 by John Wiley & Sons, Ltd.

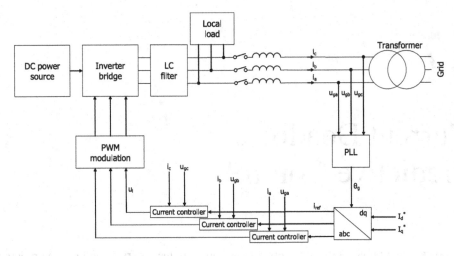

Figure 17.1 Block diagram of a current-controlled VSI in the natural frame equipped with DB predictive current controllers

According to (Buso *et al.* 2001; Timbus *et al.* 2009), the model of an inverter can be simplified to only an inductive components as shown in Figure 17.2 for the purpose of designing a DB controller. In this case, the grid current i can be expressed as

$$\frac{di(t)}{dt} = -\frac{R_f + R_g}{L_f + L_g} i(t) + \frac{1}{L_f + L_g} (u_i(t) - u_g(t)). \tag{17.1}$$

With the sampling period of T_s, this can be discretised (Franklin *et al.* 1990; Kuo 1992; Middleton and Goodwin 1990) as

$$i((k+1)T_s) = ai(kT_s) + b\left(u_i(kT_s) - u_g(kT_s)\right), \tag{17.2}$$

where

$$a = e^{-\frac{R_f + R_g}{L_f + L_g} T_s}, \tag{17.3}$$

$$b = \frac{1}{L_f + L_g} \int_0^{T_s} e^{-\frac{R_f + R_g}{L_f + L_g} \eta} d\eta = \frac{1}{R_f + R_g} \left(1 - e^{-\frac{R_f + R_g}{L_f + L_g} T_s}\right), \tag{17.4}$$

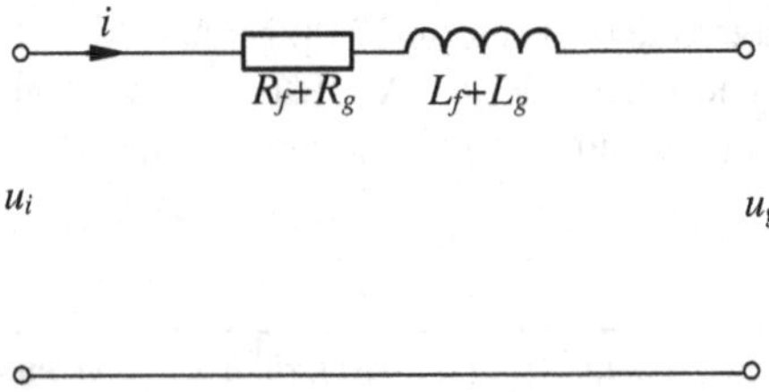

Figure 17.2 Single-phase representation for the design of DB controllers

or in the simplified form as

$$i(k+1) = ai(k) + bu_i(k) - bu_g(k). \tag{17.5}$$

Due to the one-step computation delay, the current i can be controlled to reach its reference in a minimum of two steps.

By incrementing (17.5) with one step, then

$$i(k+2) = ai(k+1) + bu_i(k+1) - bu_g(k+1). \tag{17.6}$$

Substituting (17.5) into (17.6), then the output current i at step $k+2$ is obtained as

$$i(k+2) = a^2 i(k) + abu_i(k) - abu_g(k) + bu_i(k+1) - bu_g(k+1). \tag{17.7}$$

By rearranging (17.7), the required inverter command voltage is

$$u_i(k+1) = \frac{1}{b}i(k+2) - \frac{a^2}{b}i(k) - au_i(k) + au_g(k) + u_g(k+1), \tag{17.8}$$

which requires the knowledge of the output current $i(k+2)$ and the grid voltage $u_g(k+1)$. Here, the output current $i(k+2)$ can be predicted (expected) to be the reference current $i_{ref}(k+2)$ and estimated via the linear extrapolation (Holmes and Martin 1996) as

$$i_{ref}(k+2) = 4i_{ref}(k-1) - 3i_{ref}(k-2). \tag{17.9}$$

Similar to (17.9), the grid voltage can be estimated as

$$u_g(k+1) = 2u_g(k) - u_g(k-1). \tag{17.10}$$

Substituting (17.9) and (17.10) into (17.8), the control signal $u_i(k+1)$ can then be obtained as

$$\begin{aligned} u_i(k+1) = & \tfrac{1}{b}(4i_{ref}(k-1) - 3i_{ref}(k-2)) - \tfrac{a^2}{b}i(k) \\ & -au_i(k) + (2+a)u_g(k) - u_g(k-1). \end{aligned} \tag{17.11}$$

Hence, the inputs of the DB controller are i_{ref}, i and u_g.

17.3 Experimental Results

The control strategy was evaluated in the grid-connected mode under the same five different scenarios tested in Chapters 15 and 16.

17.3.1 Steady-state Performance

17.3.1.1 Without a Local Load

In the steady state, the current reference I_d^* was set at 3 A. The reactive power was set at 0 Var ($I_q^* = 0$). This corresponds to the unity power factor. Since there is no local load included in the experiment, all generated active power was injected into the grid via a step-up transformer. The output current i_a, reference current i_{ref} and the corresponding current tracking error e_i in the steady state are shown in the left column of Figure 17.3. The DB predictive controller demonstrated very good tracking performance.

The current controllers discussed in Chapters 3 and 15–17 were compared and their corresponding values of THD of the grid current are listed in Table 17.1. It can be seen that the current H^∞ repetitive controller considerably outperformed the other controllers.

17.3.1.2 With a Resistive Load

In this experiment, a balanced resistive local load with $R_A = R_B = R_C = 12\,\Omega$ was connected to the system. The grid output current reference I_d^* was set at 2 A (after connecting the inverter to the grid) and the reactive power was set at 0 Var ($I_q^* = 0$). The output current i_a, reference current i_{ref} and the corresponding current tracking error e_i in the steady state are shown in the right column of Figure 17.3. Again, the tracking performance of the controller was satisfactory and remained more or less unchanged with comparison to the previous experiment. The current THD increased because the total amount of the current injected into the grid was reduced.

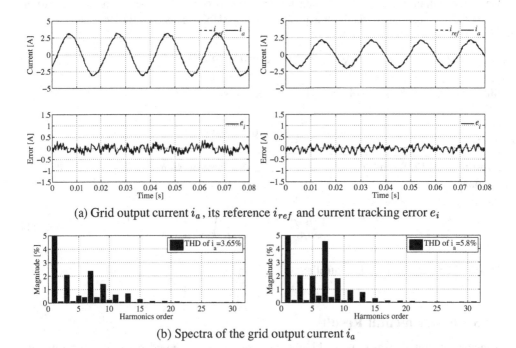

(a) Grid output current i_a, its reference i_{ref} and current tracking error e_i

(b) Spectra of the grid output current i_a

Figure 17.3 Experimental results for a DB controller without a local load (left column) and with a resistive local load (right column)

Table 17.1 Comparison of current controllers in the grid-connected mode without a local load

Controller	Current i_a THD [%]	Grid voltage v_{ga} THD [%]
H^∞ repetitive	1.03	1.57
PR	3.84	1.45
PI	4.38	1.52
DB	3.65	1.61

The current controllers discussed in Chapters 3 and 15–17 were compared and their corresponding values of THD of the grid current are listed in Table 17.2. It can be seen that the current H^∞ repetitive controller considerably outperformed the other controllers.

17.3.1.3 With an Unbalanced Load

In this experiment, an unbalanced resistive local load with $R_A = 12\ \Omega$, $R_B = \infty$ and $R_C = 12\ \Omega$ was connected to the system. The grid output current reference I_d^* was set at 2 A (after connecting the inverter to the grid) and the reactive power was set at 0 Var ($I_q^* = 0$). The output current i_a, reference current i_{ref} and the corresponding current tracking error e_i in the steady state are shown in the left column of Figure 17.4. The tracking performance of the controller remained satisfactory.

The current controllers discussed in Chapters 3 and 15–17 were compared and their corresponding values of THD of the grid current are listed in Table 17.3. It can be seen that the current H^∞ repetitive controller considerably outperformed the other controllers.

17.3.1.4 With a Non-linear Load

In this experiment, a three-phase uncontrolled rectifier loaded with an LC filter $L = 150\ \mu H$, $C = 1000\ \mu F$ and a resistor $R = 20\ \Omega$ was connected to the system. The grid output current reference I_d^* was set at 2 A (after connecting the inverter to the grid) and the reactive power was set at 0 Var ($I_q^* = 0$). The output current i_a, reference current i_{ref} and the corresponding current tracking error e_i in the steady state are shown in the right column of Figure 17.4. The tracking performance worsened and the current THD increased considerably.

The current controllers discussed in Chapters 3 and 15–17 were compared and their corresponding values of THD of the grid current are listed in Table 17.4. The THD of the grid current obtained by the current H^∞ repetitive controller is below one-third of those obtained from the other current controllers.

Table 17.2 Comparison of current controllers in the grid-connected mode with a resistive local load

Controller	Current i_a THD [%]	Grid Voltage v_{ga} THD [%]
H^∞ Repetitive	1.53	1.63
PR	5.79	1.41
PI	4.93	1.41
DB	5.8	1.59

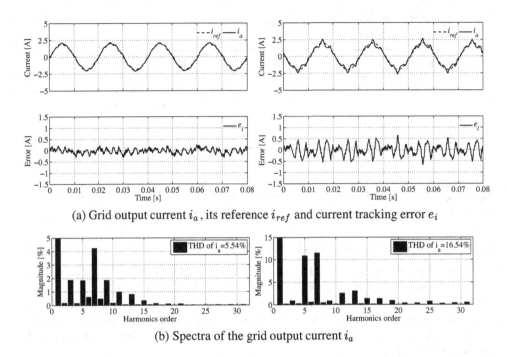

(a) Grid output current i_a, its reference i_{ref} and current tracking error e_i

(b) Spectra of the grid output current i_a

Figure 17.4 Experimental results for a DB controller with an unbalanced local load (left column) and with a non-linear local load (right column).

Table 17.3 Comparison of current controllers in the grid-connected mode with an unbalanced local load

Controller	Current i_a THD [%]	Grid Voltage v_{ga} THD [%]
H^∞ Repetitive	1.55	1.59
PR	5.39	1.45
PI	5.03	1.48
DB	5.54	1.52

Table 17.4 Comparison of current controllers in the grid-connected mode with a non-linear local load

Controller	Current i_a THD [%]	Grid voltage v_{ga} THD [%]
H^∞ Repetitive	4.91	1.60
PR	16.71	1.47
PI	16.02	1.41
DB	16.54	1.49

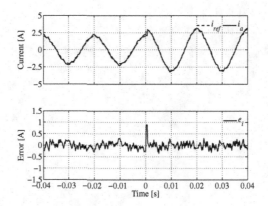

Figure 17.5 Transient responses of the DB predictive controller without a local load

17.3.2 *Transient Performance*

A step change in the current reference I_d^* from 2 A to 3 A was applied (while keeping $I_q^* = 0$). As there is no local load connected to the system in this experiment, all the generated active power was injected into the grid via a step-up transformer. The transient response of the output current i_a, reference current i_{ref} and the corresponding current tracking error e_i are shown in Figure 17.5. The DB controller demonstrated very good dynamic performance and its transient response is indeed very fast.

17.4 Summary

A DB predictive controller is designed to control the grid current of inverters connected to the grid. The controller demonstrated very good performance in tracking sinusoidal reference currents when there was no non-linear local load. When there was a non-linear local load, the THD of the current increased considerably. Also, because of the fast dynamics, the sensitivity to parameter variations is high.

The current controllers presented in Chapters 3 and 15–17 are compared. The current H^∞ repetitive controller clearly outperformed the other controllers in terms of power quality.

18

Synchronverters: Grid-friendly Inverters that Mimic Synchronous Generators

As discussed in Chapters 3 and 15–17, the current paradigm in the control of inverters associated with renewable energy sources is to extract the maximum power from the power source and inject it all into the power grid using current-controlled inverters; see for example (Busquets-Monge *et al.* 2008b; Carrasco *et al.* 2006b; Ekanayake *et al.* 2003). This is a good policy as long as such power sources constitute a negligible part of the grid power capacity. Indeed, any random power fluctuations of the renewable power generators in this case can be compensated for by the controllers associated with large conventional generators because some of these generators are dedicated to take care of the overall power balance, system stability and fault ride-through. When the penetration of renewable power generators reaches a certain level, such "irresponsible" behaviour will become untenable. In responding to the daily increasing share of electricity generated from distributed generation and renewable energy sources, it is important for these sources to feed power to the grid in the form of voltage sources instead of current sources, in a way similar to the conventional power generators (Brabandere *et al.* 2007; Loix *et al.* 2007; Piagi and Lasseter 2006; Sao and Lehn 2005). This is particularly true when the grid is weak or when an inverter or a microgrid works in the stand-alone mode.

In this chapter, the idea of operating an inverter to mimic a synchronous generator is developed after establishing a model for synchronous generators to cover all dynamics without any assumptions on the signals. This means that the well-established theory/algorithms used to control synchronous generators can still be used in power systems when a significant proportion of the generating capacity is inverter-based. Such inverters are called synchronverters (Zhong and Weiss 2011). The implementation and operation of synchronverters are described in detail. The real and reactive power delivered by synchronverters connected in parallel can

---

Control of Power Inverters in Renewable Energy and Smart Grid Integration, First Edition.
Qing-Chang Zhong and Tomas Hornik.
© 2013 John Wiley & Sons, Ltd. Published 2013 by John Wiley & Sons, Ltd.

be automatically shared by using the well-known frequency and voltage droop mechanism. Synchronverters can also easily be operated in the stand-alone mode and hence they provide an ideal solution for microgrids and smart grids. Extensive simulation and experimental results are presented.

18.1 Mathematical Model of Synchronous Generators

The model of a synchronous generator can be found in many sources, e.g., (Fitzgerald *et al.* 2003; Grainger and Stevenson 1994; Kundur 1994; Walker 1981). Most of the references make various assumptions, e.g. steady state and/or balanced sinusoidal voltages/currents, to simplify the analysis. Here, a model that is a passive dynamic system without any assumptions on the signals is established, from the perspective of system analysis and controller design, for a round rotor machine (without damper windings), with p pairs of poles per phase (and p pairs of poles on the rotor) and with no saturation effects in the iron core.

Since a synchronous generator is an electrical machine, it has an electrical part and a mechanical part.

18.1.1 Electrical Part

The details of the geometry of the windings can be found in (Fitzgerald *et al.* 2003; Walker 1981). The field and the three identical stator windings are distributed in slots around the periphery of the uniform air gap. The stator windings can be regarded as concentrated coils having self-inductance L and mutual inductance $-M$ ($M > 0$ with a typical value $\frac{1}{2}L$, the negative sign is due to the phase angle of $\frac{2\pi}{3}$ rad), as shown in Figure 18.1. The field (or rotor) windings can be regarded as a concentrated coil having self-inductance L_f. The mutual inductance between the field coil and each of the three stator coils varies with the (electrical) rotor angle θ as follows:

$$M_{af} = M_f \cos(\theta),$$

$$M_{bf} = M_f \cos\left(\theta - \frac{2\pi}{3}\right),$$

$$M_{cf} = M_f \cos\left(\theta - \frac{4\pi}{3}\right).$$

The flux linkages of the windings are

$$\Phi_a = L i_a - M i_b - M i_c + M_{af} i_f,$$

$$\Phi_b = -M i_a + L i_b - M i_c + M_{bf} i_f,$$

$$\Phi_c = -M i_a - M i_b + L i_c + M_{cf} i_f,$$

$$\Phi_f = M_{af} i_a + M_{bf} i_b + M_{cf} i_c + L_f i_f,$$

Synchronverters: Grid-friendly Inverters that Mimic Synchronous Generators

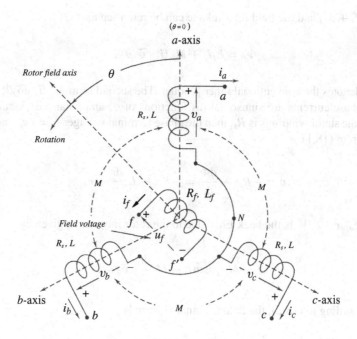

Figure 18.1 Structure of an idealised three-phase round-rotor synchronous generator with $p = 1$, modified from Grainger and Stevenson 1994, Figure 3.4

where i_a, i_b and i_c are the stator phase currents and i_f is the rotor excitation current. Denote

$$\Phi = \begin{bmatrix} \Phi_a \\ \Phi_b \\ \Phi_c \end{bmatrix}, \quad i = \begin{bmatrix} i_a \\ i_b \\ i_c \end{bmatrix}$$

and

$$\widetilde{\cos}\,\theta = \begin{bmatrix} \cos\theta \\ \cos\left(\theta - \frac{2\pi}{3}\right) \\ \cos\left(\theta - \frac{4\pi}{3}\right) \end{bmatrix}, \quad \widetilde{\sin}\,\theta = \begin{bmatrix} \sin\theta \\ \sin\left(\theta - \frac{2\pi}{3}\right) \\ \sin\left(\theta - \frac{4\pi}{3}\right) \end{bmatrix}.$$

Assume for the moment that the neutral line is not connected, then

$$i_a + i_b + i_c = 0.$$

It follows that the stator flux linkages can be rewritten as

$$\Phi = L_s i + M_f i_f \widetilde{\cos}\,\theta, \tag{18.1}$$

where $L_s = L + M$, and the field flux linkage can be rewritten as

$$\Phi_f = L_f i_f + M_f \langle i, \widetilde{\cos} \theta \rangle, \tag{18.2}$$

where $\langle \cdot, \cdot \rangle$ denotes the conventional inner product. The second term $M_f \langle i, \widetilde{\cos}\theta \rangle$ is constant if the three phase currents are sinusoidal (as functions of θ) and balanced. Assume that the resistance of the stator windings is R_s, then the phase terminal voltages $v = [\, v_a \; v_b \; v_c \,]^T$ can be obtained from (18.1) as

$$v = -R_s i - \frac{d\Phi}{dt} = -R_s i - L_s \frac{di}{dt} + e, \tag{18.3}$$

where $e = [\, e_a \; e_b \; e_c \,]^T$ is the back EMF due to the rotor movement given by

$$e = M_f i_f \dot{\theta} \widetilde{\sin} \theta - M_f \frac{di_f}{dt} \widetilde{\cos} \theta. \tag{18.4}$$

Similarly, according to (18.2), the field terminal voltage is

$$v_f = -R_f i_f - \frac{d\Phi_f}{dt}, \tag{18.5}$$

where R_f is the resistance of the rotor windings. However, this is not used here because the field current i_f, instead of v_f, is used as an adjustable constant input. This completes modelling the electrical part of the machine.

18.1.2 Mechanical Part

The mechanical part of the machine is governed by

$$J\ddot{\theta} = T_m - T_e - D_p \dot{\theta}, \tag{18.6}$$

where J is the moment of inertia of all parts rotating with the rotor, T_m is the mechanical torque, T_e is the electromagnetic toque and D_p is a damping factor. T_e can be found from the energy E stored in the machine magnetic field, i.e.,

$$\begin{aligned}
E &= \frac{1}{2} \langle i, \Phi \rangle + \frac{1}{2} i_f \Phi_f \\
&= \frac{1}{2} \langle i, L_s i + M_f i_f \widetilde{\cos} \theta \rangle + \frac{1}{2} i_f (L_f i_f + M_f \langle i, \widetilde{\cos} \theta \rangle) \\
&= \frac{1}{2} \langle i, L_s i \rangle + M_f i_f \langle i, \widetilde{\cos} \theta \rangle + \frac{1}{2} L_f i_f^2.
\end{aligned}$$

From simple energy considerations (see, e.g., (Ellison 1965; Fitzgerald *et al.* 2003)), there is

$$T_e = \left.\frac{\partial E}{\partial \theta_m}\right|_{\Phi,\, \Phi_f \text{ constant}}.$$

It is not difficult to verify (using the formula for the derivative of the inverse of a matrix function) that this is equivalent to

$$T_e = -\left.\frac{\partial E}{\partial \theta_m}\right|_{i,\, i_f \text{ constant}}.$$

Since the mechanical rotor angle θ_m satisfies $\theta = p\theta_m$,

$$T_e = -p \left.\frac{\partial E}{\partial \theta}\right|_{i,\, i_f \text{ constant}}$$

$$= -pM_f i_f \left\langle i,\, \frac{\partial}{\partial \theta}\widetilde{\cos}\,\theta \right\rangle$$

$$= pM_f i_f \left\langle i,\, \widetilde{\sin}\,\theta \right\rangle. \qquad (18.7)$$

Note that if $i = i_0 \widetilde{\sin}\,\varphi$ (as would be the case in sinusoidal steady state), then

$$T_e = pM_f i_f i_0 \left\langle \widetilde{\sin}\,\varphi,\, \widetilde{\sin}\,\theta \right\rangle = \frac{3}{2} pM_f i_f i_0 \cos(\theta - \varphi).$$

Note also that if i_f is constant (as is usually the case), then (18.7) with (18.4) yields

$$T_e \dot{\theta}_m = \langle i,\, e \rangle.$$

18.1.3 Presence of a Neutral Line

The above analysis is based on the assumption that the neutral line is not connected. If the neutral line is connected, then

$$i_a + i_b + i_c = i_N,$$

where i_N is the current flowing through the neutral line. Then, the formula for the stator flux linkages (18.1) becomes

$$\Phi = L_s i + M_f i_f \widetilde{\cos}\,\theta - \begin{bmatrix} 1 \\ 1 \\ 1 \end{bmatrix} M i_N$$

and the phase terminal voltages (18.3) become

$$v = -R_s i - L_s \frac{di}{dt} + \begin{bmatrix} 1 \\ 1 \\ 1 \end{bmatrix} M \frac{di_N}{dt} + e,$$

where e is given by (18.4). The other formulae are not affected.

The presence of a neutral line makes the system model somewhat more complicated. However, in a synchronverter to be designed in the next section, M is a design parameter that can be chosen to be 0. The physical meaning of this is that there is no magnetic coupling between the stator windings, which does not happen in a physical synchronous generator but can easily be implemented in a synchronverter. When a neutral line is needed, it is advantageous to take $M = 0$ and then to provide a neutral line with the strategies discussed in Part II. The choice of M and L individually is irrelevant; what matters is only $L_s = L + M$. In the sequel, the model of a synchronous generator consisting of (18.3), (18.4), (18.6) and (18.7) will be used to operate an inverter as a synchronverter.

18.2 Implementation of a Synchronverter

In this section, the details of how to implement an inverter as a synchronverter are described. A synchronverter consists of a power part and an electronic part. The power part is a simple inverter used to convert DC power into three-phase AC as shown in Figure 18.2. The electronic part is an electronic controller that runs a program in a processor to control the switches shown in Figure 18.2. The core of the electronic part is the mathematical model of a synchronous generator, shown in Figure 18.3. These two parts interact via the signals e and i, in addition to v and v_g that are used for controlling the synchronverter.

18.2.1 Power Part

This part consists of three phase legs and a three-phase LC filter, which is used to suppress the switching noise. If the inverter is to be connected to the grid, then three more inductors and a

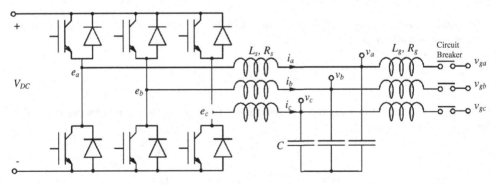

Figure 18.2 Power part of a synchronverter: a basic inverter

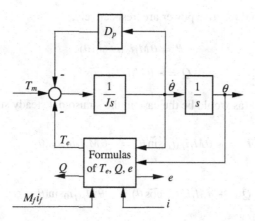

Figure 18.3 Electronic part of a synchronverter (without control)

circuit breaker can be adopted to interface with the grid. If a neutral line is needed, then the strategies discussed in (Zhong et al. 2005a, 2006) and Part II to provide a neutral line without affecting the control of the three-phase inverter can be used.

It is advantageous to assume that the field (rotor) winding of the synchronverter is fed by an adjustable DC current source i_f instead of a voltage source v_f. In this case, the terminal voltage v_f varies, but this is irrelevant. As long as i_f is constant, the generated voltage from (18.4) is

$$e = \dot{\theta} M_f i_f \widetilde{\sin} \theta. \tag{18.8}$$

The terminal voltages $v = [\, v_a \ v_b \ v_c \,]^T$ given in (18.3) should be the capacitor voltages, as shown in Figure 18.2. The inductance L_s and resistance R_s of the inductor can be chosen to represent the stator impedance of a synchronous generator. The switches in the inverter are operated so that the average values of e_a, e_b and e_c over a switching period should be equal to e given in (18.8), which can be achieved by the usual PWM technique. Also shown in Figure 18.2 are three interfacing inductors L_g (with series resistance R_g) and a circuit breaker to facilitate the synchronisation/connection with the grid.

18.2.2 Electronic Part

Define the generated real power P and reactive power Q (as seen from the inverter legs) as

$$P = \langle i, e \rangle \quad \text{and} \quad Q = \langle i, e_q \rangle,$$

where e_q has the same amplitude as e but with a phase delayed from that of e by $\frac{\pi}{2}$, i.e.,

$$e_q = \dot{\theta} M_f i_f \widetilde{\sin} \left(\theta - \frac{\pi}{2} \right) = -\dot{\theta} M_f i_f \widetilde{\cos} \theta.$$

Then, the real power and reactive power are, respectively,

$$P = \dot{\theta} M_f i_f \langle i, \widetilde{\sin}\theta \rangle,$$
$$Q = -\dot{\theta} M_f i_f \langle i, \widetilde{\cos}\theta \rangle. \tag{18.9}$$

Note that if $i = i_0 \widetilde{\sin} \varphi$ (as would be the case in the sinusoidal steady state), then

$$P = \dot{\theta} M_f i_f \langle i, \widetilde{\sin}\theta \rangle = \frac{3}{2} \dot{\theta} M_f i_f i_0 \cos(\theta - \varphi),$$

$$Q = -\dot{\theta} M_f i_f \langle i, \widetilde{\cos}\theta \rangle = \frac{3}{2} \dot{\theta} M_f i_f i_0 \sin(\theta - \varphi).$$

These coincide with the conventional definitions for real power and reactive power, usually expressed in the dq coordinates. When the voltage and current are in phase, i.e. when $\theta - \varphi = 0$, the product of the RMS values of the voltage and current gives the real power P. When the voltage and current are $\frac{\pi}{2}$ rad out of phase, this product gives the reactive power Q. Positive Q corresponds to an inductive load. The above formulae for P and Q are used when regulating the real and reactive power of an SG.

Equation (18.6) can be written as

$$\ddot{\theta} = \frac{1}{J}(T_m - T_e - D_p \dot{\theta}),$$

where the input is the mechanical torque T_m, while the electromagnetic torque T_e depends on i and θ, according to (18.7). This equation, together with (18.7), (18.8) and (18.9), are implemented as the core of the electronic part of a synchronverter shown in Figure 18.3. Thus, the state variables of the synchronverter are i (which are actual currents), θ and $\dot{\theta}$ (which are a virtual angle and a virtual angular speed). The control inputs of the synchronverter are T_m and $M_f i_f$. In order to operate the synchronverter in a useful way, a controller should be added to generate the signals T_m and $M_f i_f$ so that the system stability is maintained and the desired values of real and reactive power are followed. The significance of Q will be discussed in the next section.

18.3 Operation of a Synchronverter

18.3.1 Regulation of Real Power and Frequency Droop Control

For synchronous generators, the rotor speed is maintained by the prime mover and it is known that the damping factor D_p is due to mechanical friction. An important mechanism for SGs to share the load evenly is to vary the real power it delivers according to the grid frequency, a property called "frequency droop". When the real power demand increases, the speed of the SGs drops due to increased T_e in (18.6). The speed regulation system of the prime mover then increases the mechanical power, e.g. by widening the throttle valve of an engine, so that a new power balance is achieved. This mechanism can be implemented in a synchronverter by comparing the virtual angular speed $\dot{\theta}$ with the angular frequency reference $\dot{\theta}_r$, e.g. the

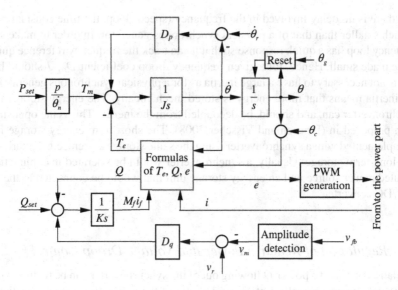

Figure 18.4 Electronic part of a synchronverter with the function of frequency and voltage control, and real and reactive power regulation

nominal angular speed $\dot{\theta}_n$, before feeding it into the damping block D_p; see the upper part of Figure 18.4. As a result, the damping factor D_p actually plays the role of the frequency droop coefficient, which is defined as the ratio of the required change of torque ΔT to the change of speed (frequency) $\Delta\dot{\theta}$. That is,

$$D_p = \frac{\Delta T}{\Delta\dot{\theta}} = \frac{\Delta T}{T_{mn}} \frac{\dot{\theta}_n}{\Delta\dot{\theta}} \frac{T_{mn}}{\dot{\theta}_n},$$

where T_{mn} is the nominal mechanical torque. Note that in some references, e.g. (Sao and Lehn 2005), D_p is defined as $\frac{\Delta\dot{\theta}}{\Delta T}$. For example, it can be set for the torque (power) to change 100% for a frequency change of 1%. The active torque T_m can be obtained from the set point of real power P_{set} by dividing it with the nominal mechanical speed $\frac{\dot{\theta}_n}{p}$. This completes the feedback loop for real power; see the upper part of Figure 18.4. Because of the built-in frequency droop mechanism, a synchronverter automatically shares the load with other inverters of the same type and with SGs connected on the same bus. The power regulation loop is very simple, because no mechanical devices are involved and no extra measurements are needed for real power regulation (all the variables are available internally).

The regulation mechanism of the real power (torque) shown in the upper part of Figure 18.4 has a cascaded control structure, of which the inner loop is the frequency (speed) loop and the outer loop is the real power (torque) loop. The time constant of the frequency loop is $\tau_f = \frac{J}{D_p}$. In other words, J can be chosen as

$$J = D_p \tau_f.$$

Because there is no delay involved in the frequency (speed) loop, the time constant τ_f can be made much smaller than that of a physical synchronous generator. In order to make sure that the frequency loop has a quick response so that it can track the frequency reference quickly, τ_f should be made small. Hence, for a given frequency droop coefficient D_p, J should be made small. It is not necessary to have a large inertia as for a physical synchronous generator, where a larger inertia means that more energy is stored mechanically. The energy storage function of a synchronverter can, and should, be decoupled from the inertia. This is the opposite of the approach proposed in (Driesen and Visscher 2008). The short-term energy storage function can be implemented with a synchronverter using the same storage system, e.g. batteries, that is used for long-term storage. Usually, a synchronverter would be operated in conjunction with a distributed power source and an energy storage unit that would be connected to the DC bus via a DC/DC converter.

18.3.2 Regulation of Reactive Power and Voltage Droop Control

The regulation of reactive power Q flowing out of the synchronverter can be realised similarly. Define the voltage droop coefficient D_q as the ratio of the required change of reactive power ΔQ to the change of voltage Δv, i.e.,

$$D_q = \frac{\Delta Q}{\Delta v} = \frac{\Delta Q}{Q_n} \frac{v_n}{\Delta v} \frac{Q_n}{v_n},$$

where Q_n is the nominal reactive power, which can be chosen as the nominal power, and v_n is the nominal amplitude of the terminal voltage v. Again, note that in much of the literature, e.g. (Sao and Lehn 2005), D_q is defined as $\frac{\Delta v}{\Delta Q}$. The regulation mechanism for the reactive power can be realised as shown in the lower part of Figure 18.4. The difference between the reference voltage v_r and the amplitude of the feedback voltage v_{fb} is amplified with the voltage droop coefficient D_q before adding to the difference between the set point Q_{set} and the reactive power Q, which is calculated according to (18.9). The resulting signal is then fed into an integrator with a gain $\frac{1}{K}$ to generate $M_f i_f$. Here, K is dual to the inertia J. Note that there is no need to measure reactive power Q because it is available internally.

Similarly, the regulation mechanism of the reactive power shown in the lower part of Figure 18.4 also has a cascaded control structure, if the effect of the LC filter is ignored or compensated, which means $v_{fb} \approx e$. The inner loop is the (amplitude) voltage loop and the outer loop is the reactive power loop. The time constant τ_v of the voltage loop is

$$\tau_v = \frac{K}{\dot{\theta} D_q} \approx \frac{K}{\dot{\theta}_n D_q}$$

as the variation of $\dot{\theta}$ is very small. Hence, K can be chosen as

$$K = \dot{\theta}_n D_q \tau_v.$$

The amplitude v_m of the terminal voltage v can be obtained from the RMS values but it can also be obtained as follows. Assume that $v_a = v_{am}\sin\theta_a$, $v_b = v_{bm}\sin\theta_b$ and $v_c = v_{cm}\sin\theta_c$, then

$$v_a v_b + v_b v_c + v_c v_a$$
$$= v_{am}v_{bm}\sin\theta_a\sin\theta_b + v_{bm}v_{cm}\sin\theta_b\sin\theta_c + v_{cm}v_{am}\sin\theta_c\sin\theta_a$$
$$= \frac{v_{am}v_{bm}}{2}\cos(\theta_a - \theta_b) + \frac{v_{bm}v_{cm}}{2}\cos(\theta_b - \theta_c) + \frac{v_{cm}v_{am}}{2}\cos(\theta_c - \theta_a)$$
$$- \frac{v_{am}v_{bm}}{2}\cos(\theta_a + \theta_b) - \frac{v_{bm}v_{cm}}{2}\cos(\theta_b + \theta_c) - \frac{v_{cm}v_{am}}{2}\cos(\theta_c + \theta_a).$$

When the terminal voltages are balanced, i.e. when $v_{am} = v_{bm} = v_{cm} = v_m$ and $\theta_b = \theta_a - \frac{2\pi}{3} = \theta_c + \frac{2\pi}{3}$, the last three terms in the above equality are balanced, having a doubled frequency. Hence,

$$v_a v_b + v_b v_c + v_c v_a = -\frac{3}{4}v_m^2,$$

and the amplitude v_m of the actual terminal voltage v can be obtained as

$$v_m = \frac{2}{\sqrt{3}}\sqrt{-(v_a v_b + v_b v_c + v_c v_a)}. \quad (18.10)$$

In real-time implementation, a low-pass filter is needed to filter out the ripples at the doubled frequency as the terminal voltages may be unbalanced. This also applies to T_e and Q.

18.4 Simulation Results

The idea described above was verified with simulations. The parameters of the inverter for carrying out the simulations are given in Table 18.1.

The inverter is connected to a three-phase 400 V 50 Hz grid via a circuit breaker and a step-up transformer. The frequency droop coefficient is chosen as $D_p = 0.2026$ so that the frequency drops 0.5% when the torque (power) increases 100%. The voltage droop coefficient is chosen as $D_q = 117.88$ so that the voltage drops 5% when the reactive power increases 100%. The time constant of the frequency loop is chosen as $\tau_f = 0.002$ s and that of the voltage loop is

Table 18.1 Parameters of the synchronverter for simulations and experiments

Parameters	Values	Parameters	Values
L_s	0.45 mH	L_g	0.45 mH
R_s	0.135 Ω	R_g	0.135 Ω
C	22 μF	Frequency	50 Hz
R (parallel to C)	1000 Ω	Voltage (line-line)	20.78 Vrms
Rated power	100 W	DC-link voltage	42 V

chosen as $\tau_v = 0.02$ s. The simulations were carried out in MATLAB® 7.4 with Simulink®. The solver used in the simulations is ode23tb with a relative tolerance 10^{-3} and a maximum step size of 0.2 ms. The synchronverter can feed pre-set real power and reactive power to the grid (called the set mode) and can automatically change the real power and reactive power fed to the grid according to the grid frequency and voltage (called the droop mode).

The simulation was started at $t = 0$ to allow the PLL and synchronverter to start up (in real applications, these two can be started separately). The dynamics in the first half second is omitted. The circuit breaker was turned on at $t = 1$ s; the real power $P_{set} = 80$ W was applied at $t = 2$ s and the reactive power $Q_{set} = 60$ Var was applied at $t = 3$ s. The droop mechanism was enabled at $t = 4$ s and then the grid voltage decreased by 5% at $t = 5$ s.

18.4.1 Under Different Grid Frequencies

The system responses when the grid frequency was 50 Hz are shown in the left column of Figure 18.5. The synchronverter tracked the grid frequency very well all the time. The voltage difference between v and v_g before any power was applied was very small and the synchronisation process was very quick. There was no problem when turning the circuit breaker on at $t = 1$ s; there was not much transient response after this event either. The synchronverter responded to both the real power command at $t = 2$ s and the reactive command at $t = 3$ s very quickly and settled down in less than 10 cycles without any error. The coupling effect between the real power and the reactive power was reasonably small. When the droop mechanism was enabled at $t = 4$ s, there was not much change to the real power output as the frequency was not changed but the reactive power dropped by about 53 Var, about 50% of the power rating, because the local terminal voltage v was about 2.5% higher than the nominal value. When the grid voltage dropped by 5% at $t = 4$ s, the local terminal voltage dropped to just below the nominal value. The reactive power output then increased to just above the set point 60 Var.

The same simulation was repeated but with a grid frequency of 49.95 Hz, that is 0.1% lower than the nominal value. The system responses are also shown in the left column of Figure 18.5 as well for comparison. The synchronverter followed the grid frequency very well. When the synchronverter worked at the set mode, i.e. before $t = 4$ s, the real power and reactive power all responded to the setpoint exactly, respectively. After the droop mechanism was enabled at $t = 4$ s, the synchronverter increased the real power output by 20 W, that is 20% of the rated power, corresponding to 0.1% drop of the frequency. This did not cause much extra change to the reactive power output, just slight adjustment corresponding to the slight change of the local voltage v.

18.4.2 Under Different Load Conditions

The same simulation with the nominal grid frequency was repeated but with the local load changed from $R = 1000$ Ω to $R = 4$ Ω at $t = 1.5$ s. The system responses are shown in the right column of Figure 18.5. The local voltage v dropped immediately, with some short big transients, after the load R was changed to 4Ω at $t = 1.5$ s and the voltage difference between v and v_g increased because of the load current drawn from the grid. There was some

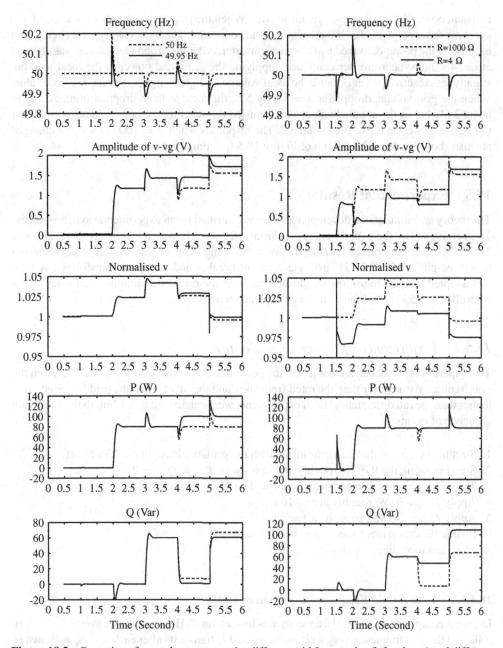

Figure 18.5 Operation of a synchronverter under different grid frequencies (left column) and different load conditions (right column)

transient power drawn from the synchronverter. When the power commands were applied, the synchronverter responded as if there was no load connected. The local voltage v recovered a bit because of the power delivered by the synchronverter. After the droop mechanism was enabled at $t = 4$ s, the synchronverter continued supplying the same real power to the local load but slightly less reactive power because the local voltage is slightly higher than the nominal value. When the grid voltage dropped at $t = 5$ s by 5%, the local voltage dropped immediately to about 2.5% below the nominal value, which caused the synchronverter to send an extra 50 Var of reactive power on top of the command. The responses when $R = 1000$ Ω was not changed are also shown in the right column of Figure 18.5 for comparison.

18.5 Experimental Results

The theory and simulations developed above were verified on an experimental synchronverter. The parameters of the experimental synchronverter were largely the same as those given in Table 18.1 and the control parameters were the same. The synchronverter was connected to a three-phase 400 V 50 Hz grid via a circuit breaker and a step-up transformer. A PLL was adopted to synchronise the synchronverter with the grid. The sampling frequency of the controller was 5 kHz and the switching frequency was 12 kHz.

18.5.1 Performance of Power Flow Control

Two experiments were carried out to test the performance of power flow control, one when the grid frequency was lower than the rated frequency and the other when the grid frequency was higher than the rated frequency. Both experiments were carried out according to the following sequence of events:

1. Start the system so that it synchronises with the grid, but keep all the IGBTs off.
2. Start operating the IGBTs, roughly at $t = 6$ s with $P_{set} = Q_{set} = 0$.
3. Turn the circuit breaker on, roughly at $t = 11$ s.
4. Apply $P_{set} = 70$ W, roughly at $t = 16$ s.
5. Apply $Q_{set} = 30$ Var, roughly at $t = 21$ s.
6. Enable the droop mechanism, roughly at $t = 26$ s.
7. Stop data recording, roughly at $t = 30$ s.

18.5.1.1 With a Grid Frequency Lower than 50 Hz

During this experiment, the grid frequency was lower than 50 Hz. The synchronverter frequency followed the grid frequency very well, with noticeable transients after each action, as shown in Figure 18.6(a). The local terminal voltage synchronised with the grid voltage very quickly once the inverter output was enabled, roughly at $t = 6$ s, as shown in Figures 18.6(b) and 18.6(c). The connection to the grid went very smoothly and there were no noticeable transients in the frequency or power, as shown in Figure 18.6. The power remained around 0, but with bigger spikes. After the real power demand was raised, it took less than 10 cycles to reach the setpoint, which is very fast, and the overshoot was very small; see Figure 18.6(d). This action

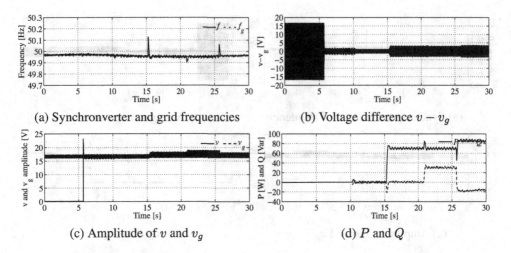

Figure 18.6 Experimental results: when the grid frequency was lower than 50 Hz

caused the frequency to respond with a big spike; the synchronverter initially "stored" some reactive power but then released it very quickly. After the reactive power demand was raised, it took less than 10 cycles to reach the setpoint with a small overshoot, as shown in Figure 18.6(d). The frequency dropped slightly and the real power increased a bit, but all returned to normal very quickly. Because the droop mechanism was not enabled, the real power and reactive power delivered by the synchronverter followed the reference values without any error. After the droop mechanism was enabled, roughly at $t = 26$ s, the synchronverter responded to the deviations of the grid frequency and the voltage from their nominal values. The real power delivered was increased because the grid frequency was lower than the nominal value, while the reactive power was decreased because the local terminal voltage was higher than the nominal value.

18.5.1.2 With a Grid Frequency Higher than 50 Hz

During this experiment, the grid frequency was higher than 50 Hz, as shown in Figure 18.7. There was not much difference from the previous experiment before the droop mechanism was enabled and the synchronverter responded well to the instructions. After the droop mechanism was enabled, roughly at $t = 26$ s, the synchronverter started responding to the grid frequency, which was higher than the nominal value, and dropped the real power output by 10 W to 60 W. The reactive power delivered was decreased because the local terminal voltage was higher than the nominal value.

18.5.2 *Loading Performance in the Stand-alone Mode*

Experiments were carried out in the stand-alone mode to test the loading performance of the sychronverter, under a balanced resistive load with $R_A = R_B = R_C = 12 \ \Omega$, a three-phase

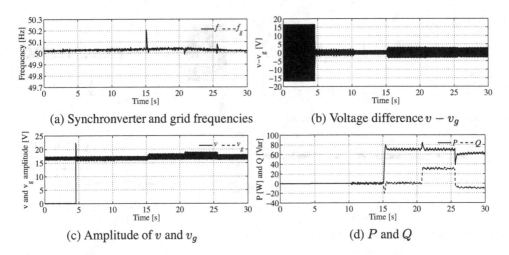

Figure 18.7 Experimental results: when the grid frequency was higher than 50 Hz

uncontrolled rectifier loaded with an LC filter $L = 150\ \mu\text{H}$, $C = 1000\ \mu\text{F}$ and a resistor $R = 20\ \Omega$, and an unbalanced resistive load with $R_A = R_C = 12\ \Omega$ and $R_B = \infty$.

18.5.2.1 With a Balanced Linear Load

The results are shown in Figure 18.8. The output voltage was very clean and was maintained well.

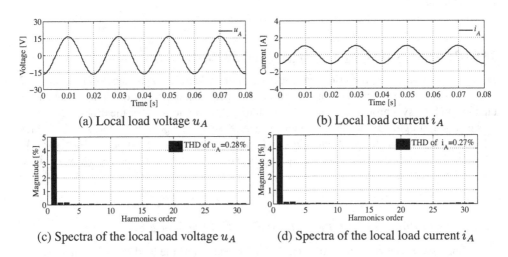

Figure 18.8 Loading performance in the stand-alone mode with a linear load

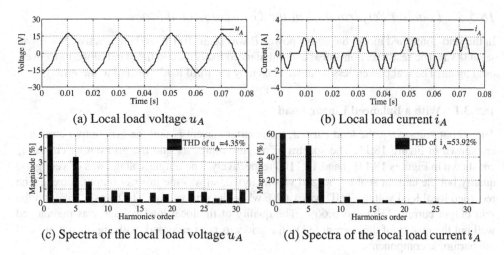

Figure 18.9 Loading performance in the stand-alone mode with a non-linear load

18.5.2.2 With a Non-linear Load

The results are shown in Figure 18.9. Although the load current was extremely distorted with a THD of 53.92%, the output voltage was clean with a THD below 5%.

18.5.2.3 With an Unbalanced Load

The results are shown in Figure 18.10. Although there was no load connected to Phase B, the other two phases worked well. The THD of the output voltage was 0.48%.

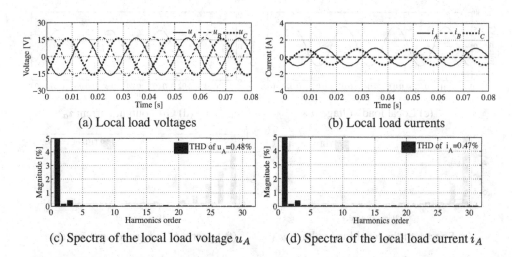

Figure 18.10 Loading performance in the stand-alone mode with an unbalanced load

18.5.3 Loading Performance in the Grid-connected Mode

In the grid-connected mode, the active power P was set to 100 W and the reactive power was set at 0 Var. The resistive, non-linear and unbalanced loads used in the stand-alone mode were used again. The real power, less what was taken by the local load, was fed to the grid.

18.5.3.1 With a Balanced Linear Load

The output voltage u_A, the load current, the inverter current i_A and the grid output current i_a are shown in Figure 18.11. The spectra of output voltage u_A and the grid output current i_a are shown in Figures 18.11(e) and 18.11(f) respectively. The output voltage was of very good quality but the current sent to the grid was distorted, so was the inverter output current. The recorded local load voltage THD was 1.62% while the grid voltage THD was 1.15% and the grid output current THD was 5.66%. The quality of the local load voltage was maintained well but the quality of the current fed to the grid was not very good, with significant 3rd and 5th harmonic components.

18.5.3.2 With a Non-linear Load

The output voltage u_A, the load current, the inverter current i_A and the grid output current i_a are shown in Figure 18.12. The spectra of output voltage u_A and the grid output

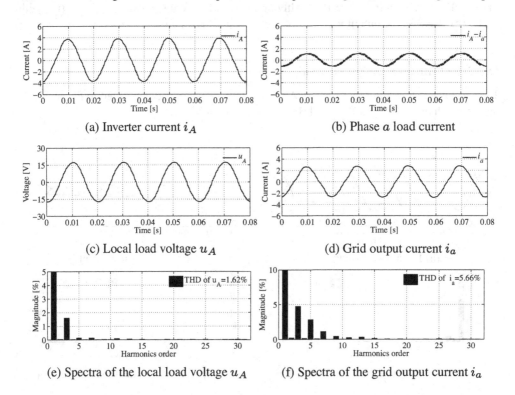

(a) Inverter current i_A

(b) Phase a load current

(c) Local load voltage u_A

(d) Grid output current i_a

(e) Spectra of the local load voltage u_A

(f) Spectra of the grid output current i_a

Figure 18.11 Loading performance in the grid-connected mode with a resistive local load

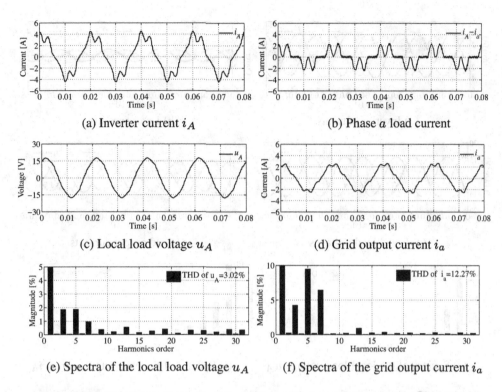

Figure 18.12 Loading performance in the grid-connected mode with a non-linear local load

current i_a are shown in Figures 18.12(e) and 18.12(f) respectively. The quality of the local load voltage remained very good in spite of the heavily non-linear local load. The current sent to the grid was distorted. The recorded local load voltage THD was 3.02% while the grid voltage THD was 1.08% and the grid output current THD was 12.27%. Again, the local load voltage was controlled very well but the current exchanged with the grid was not. There were significant 5th and 7th harmonic components.

18.5.3.3 With an Unbalanced Load

The local load voltages, the local load currents, the inverter output currents and the grid currents are shown in Figure 18.13. Although the local load currents were not balanced, the output voltages were balanced. The inverter output currents were nearly balanced, which made the grid currents unbalanced. The spectra of output voltage u_A and the grid output current i_a are shown in Figures 18.13(e) and 18.13(f) respectively. The output voltage was of very good quality but the current sent to the grid was distorted, so was the inverter output current. The recorded local load voltage THD was 1.78% while the grid voltage THD was 1.15% and the grid output current THD was 5.54%. The quality of the local load voltage and the quality of the current sent to the grid were similar to results with balanced resistive load.

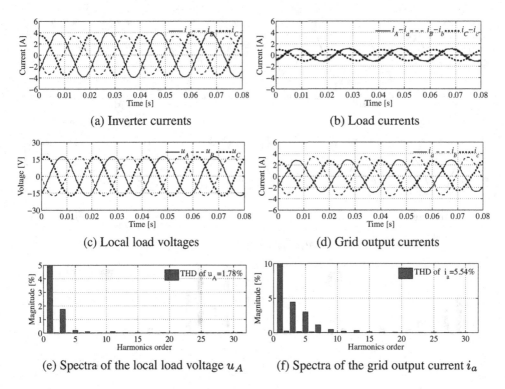

Figure 18.13 Loading performance in the grid-connected mode with an unbalanced local load

18.6 Summary

Based on (Zhong and Weiss 2009, 2011), the idea of operating an inverter as a synchronous generator is developed in this chapter after establishing a model for synchronous generators to cover all the dynamics without any assumptions about the signals. The implementation and operation of such an inverter, including power regulation and load sharing, are developed and described in detail. The mathematical model developed here can be used to investigate the stability of power systems dominated by parallel-operated inverters in distributed generation. Both simulation and experimental results are provided.

19

Parallel Operation of Inverters

Inverters may have to be operated in parallel directly or indirectly. How to share the load among the inverters operated in parallel is a challenging problem. The synchronverter discussed in Chapter 18 is able to share the load accurately. In this chapter, the inherent limitations of the conventional droop control scheme are revealed at first. It is shown that, when the conventional droop control scheme is adopted, parallel-operated inverters should have the same per-unit output impedance in order for them to share the load accurately in proportion to their power ratings. The droop controllers should also generate the same voltage set-point for the inverters. Both conditions are difficult to meet in practice, which results in errors in proportional load sharing. A robust droop controller is then presented to achieve accurate proportional load sharing without meeting these two requirements and to reduce the load voltage drop due to the load effect and the droop effect. The load voltage can be maintained within the desired range around the rated value. The strategy is robust against numerical errors, disturbances, noises, feeder impedance, parameter drifts and component mismatches, etc. The only sharing error comes from the error in measuring the load voltage. When there are errors in the voltage measured, a trade-off between the voltage drop and the sharing accuracy appears. It is also explained that, in order to avoid errors in power sharing, the global settings of the rated voltage and frequency should be accurate. The cases with R-, L- and C-inverters are discussed, with experimental results provided.

19.1 Introduction

There are several reasons why inverters are needed to operate in parallel. One obvious reason is because of the limited availability of high current power electronic devices. Another reason is that parallel-operated inverters are able to provide system redundancy and high reliability needed by critical customers. Moreover, the parallel operation of inverters also eases the difficulties in thermal management and design for high-power inverters.

Control of Power Inverters in Renewable Energy and Smart Grid Integration, First Edition.
Qing-Chang Zhong and Tomas Hornik.
© 2013 John Wiley & Sons, Ltd. Published 2013 by John Wiley & Sons, Ltd.

A natural problem for parallel-operated inverters is how to share the load among them. A key technique is to use droop control (Barklund et al. 2008; Brabandere et al. 2007; Guerrero et al. 2005, 2007, 2011; Majumder et al. 2010; Mohamed and El-Saadany 2008a; Tuladhar et al. 1997; Zhong and Weiss 2011), which is widely used in conventional power generation systems (Diaz et al. 2010). The advantage is that no external communication mechanism is needed among the inverters to achieve good sharing for linear and/or non-linear loads (Borup et al. 2001; Chandorkar et al. 1993; Coelho et al. 2002; Guerrero et al. 2004, 2006a; Tuladhar et al. 1997, 2000).

The equal sharing of linear and non-linear loads has been intensively investigated (Borup et al. 2001; Guerrero et al. 2004, 2005, 2006a, 2007). A voltage bandwidth droop control was used to share non-linear loads in (Tuladhar et al. 1997) and a small signal injection method was proposed to improve the reactive power sharing accuracy in (Tuladhar et al. 2000), which can also be extended to harmonic current sharing. Injecting a harmonic voltage according to the output harmonic current can be used to improve the total harmonic distortion (THD) of the voltage (Borup et al. 2001). In (Guerrero et al. 2004, 2005), it was pointed out that the output impedance of inverters could play an important role in power sharing and a droop controller for inverters with resistive output impedances was proposed for sharing linear and non-linear loads (Guerrero et al. 2006a, 2007).

Although significant progress has been made for the equal sharing of linear and non-linear loads, it was reported that the accuracy of reactive power sharing (for the $Q - E$ and $P - \omega$ droop) is not high (Guerrero et al. 2006b; Lee et al. 2010; Li and Kao 2009). Moreover, some approaches developed for equal sharing, e.g. the one proposed in (Marwali et al. 2004), cannot be directly applied to proportional sharing. Another issue is that the output voltage drops due to the increase of the load and also due to the droop control (Sao and Lehn 2005).

Adding an integral action to the droop controller is able to improve the accuracy of load sharing for grid-connected inverters (Dai et al. 2008a; Li et al. 2004; Marwali et al. 2004) but it does not work for inverters operated in the stand-alone mode and also there is an issue associated with the change of the operation mode. Adding a virtual inductor and estimating the effect of the line impedance is able to improve the situation by changing the droop coefficients (Li and Kao 2009). A Q–V dot droop control method was proposed in (Lee et al. 2010) to improve the accuracy of reactive power sharing following the idea of changing the droop coefficients in (Li and Kao 2009) but a mechanism to avoid the output voltage variation was necessary, which reduces the accuracy of the reactive power sharing, as can be seen from the results given there. These strategies are sensitive to numerical computational errors, parameter drifts and component mismatches.

Other trends to solve the power sharing problems for parallel-operated inverters are to make the output impedance as accurate as possible over a wide range of frequencies (see (Guerrero et al. 2008; Yao et al. 2011) and the references therein) and to introduce the secondary control to bring the deviated voltage and frequency back to the rated values (Guerrero et al. 2009, 2011). However, this inevitably needs communication among the inverters (even of low bandwidth) and the advantage of droop control is lessened. The secondary control also leads to slow responses and/or instability because of the delay introduced in the measurement and communication channel. The complexity of the system is also increased, as evidenced by the number of control loops/levels involved in such systems. This, again, increases the chance of instability.

In this chapter, it is proved that in order for parallel-connected inverters to share the load in proportion to their power ratings, the inverters should have the same per-unit impedance when the conventional droop controller is used (Zhong 2012c). It also requires that the RMS voltage set-points for the inverters be the same. Both are very strong conditions and this is the main reason why many different approaches have been proposed to improve the accuracy of power sharing. After a thorough analysis of the error in power sharing, a robust droop controller is presented to achieve accurate proportional load sharing among inverters that are operated in parallel. The strategy works for the stand-alone mode and, naturally, for the grid-connected mode as well. The accuracy of sharing does not depend on the output impedance of the inverters (as long as they are of the same type) nor on the RMS voltage set-point and, hence, it is robust to numerical computational errors, disturbances, noises, parameter drifts and component mismatches. Moreover, the robust droop controller is able to regulate the load voltage so that the voltage drop due to the load effect and the droop effect is reduced. The sensitivity to the errors in the global settings of the rated frequency and rated voltage is also analysed, which shows that these settings should be accurate.

The robust droop controller is discussed for R-inverters at first, followed by C-inverters and L-inverters, with experimental results.

19.2 Problem Description

Figure 19.1 shows two inverters connected in parallel. The line impedance is omitted, assuming that the output impedances of the inverters are designed to dominate the impedance from the inverter to the AC-bus. The reference voltages of the two inverters are, respectively,

$$v_{r1} = \sqrt{2}E_1 \sin(\omega_1 t + \delta_1),$$

$$v_{r2} = \sqrt{2}E_2 \sin(\omega_2 t + \delta_2).$$

The power ratings of the inverters are $S_1^* = E^* I_1^*$ and $S_2^* = E^* I_2^*$. They share the same load voltage

$$v_o = v_{r1} - R_{o1} i_1 = v_{r2} - R_{o2} i_2. \tag{19.1}$$

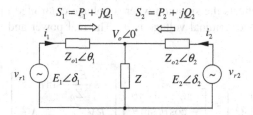

Figure 19.1 Two inverters connected in parallel

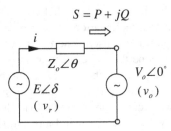

Figure 19.2 Power delivered to a voltage source through an impedance

Note that the voltage V_o drops when the load increases. This is called the load effect. In order for the inverters to share the real power and reactive power in proportion to their power ratings, appropriate control strategies should be deployed.

19.3 Power Delivered to a Voltage Source

Figure 19.2 illustrates a voltage source v_r delivering power to another voltage source $V_o \angle 0°$ through an impedance $Z_o \angle \theta$. Since the current flowing through the terminal is

$$i = \frac{E \angle \delta - V_o \angle 0°}{Z_o \angle \theta}$$

$$= \frac{E \cos \delta - V_o + jE \sin \delta}{Z_o \angle \theta},$$

the real power and reactive power delivered by the source to the terminal via the impedance can then be obtained as

$$P = \left(\frac{EV_o}{Z_o} \cos \delta - \frac{V_o^2}{Z_o}\right) \cos \theta + \frac{EV_o}{Z_o} \sin \delta \sin \theta,$$

$$Q = \left(\frac{EV_o}{Z_o} \cos \delta - \frac{V_o^2}{Z_o}\right) \sin \theta - \frac{EV_o}{Z_o} \sin \delta \cos \theta,$$

where δ is the phase difference between the supply and the terminal, often called the power angle. This actually reflects the situation when a voltage-controlled inverter is connected to an infinite bus where the terminal voltage is v_o. The real power and reactive power can be re-written as

$$\begin{bmatrix} P \\ Q \end{bmatrix} = \begin{bmatrix} \sin \theta & \cos \theta \\ -\cos \theta & \sin \theta \end{bmatrix} \begin{bmatrix} \dfrac{EV_o}{Z_o} \sin \delta \\ \dfrac{EV_o}{Z_o} \cos \delta - \dfrac{V_o^2}{Z_o} \end{bmatrix}.$$

Define

$$\begin{bmatrix} \tilde{P} \\ \tilde{Q} \end{bmatrix} = \begin{bmatrix} \sin\theta & -\cos\theta \\ \cos\theta & \sin\theta \end{bmatrix} \begin{bmatrix} P \\ Q \end{bmatrix}. \qquad (19.2)$$

Then

$$\begin{bmatrix} \tilde{P} \\ \tilde{Q} \end{bmatrix} = \begin{bmatrix} \dfrac{EV_o}{Z_o}\sin\delta \\ \dfrac{EV_o}{Z_o}\cos\delta - \dfrac{V_o^2}{Z_o} \end{bmatrix}.$$

Hence, for a small δ,

$$\tilde{P} = \frac{EV_o}{Z_o}\sin\delta \approx \frac{EV_o}{Z_o}\delta,$$

$$\tilde{Q} = \frac{EV_o}{Z_o}\cos\delta - \frac{V_o^2}{Z_o} \approx \frac{E - V_o}{Z_o}V_o,$$

which means $\tilde{P}$ and $\tilde{Q}$ can be controlled by controlling δ and E separately. This is the basis of the widely-used droop control strategies (Brabandere *et al.* 2007; Guerrero *et al.* 2005, 2006b, 2008; Yao *et al.* 2011). What is important is that a transformation involving the impedance angle θ should be applied to calculate the real power P and reactive power Q.

19.4 Conventional Droop Control

A voltage-controlled inverter can be modelled as an ideal voltage source v_r in series with its output impedance $Z_o \angle \theta$, as shown in Figure 19.2. For different types of output impedances, different droop control strategies can be obtained.

19.4.1 For R-inverters

When the output impedance is resistive, $\theta = 0°$. Then

$$P = \frac{EV_o}{Z_o}\cos\delta - \frac{V_o^2}{Z_o} \quad \text{and} \quad Q = -\frac{EV_o}{Z_o}\sin\delta.$$

When δ is small,

$$P \approx \frac{V_o}{Z_o}E - \frac{V_o^2}{Z_o} \quad \text{and} \quad Q \approx -\frac{EV_o}{Z_o}\delta,$$

and, roughly,

$$P \sim E \quad \text{and} \quad Q \sim -\delta,$$

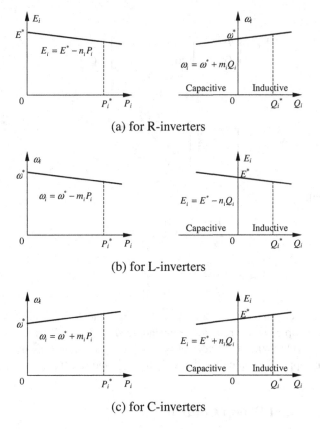

Figure 19.3 Droop control strategies

where $\sim$ means in proportion to. Hence, the conventional droop control strategy takes the form

$$E_i = E^* - n_i P_i,$$
$$\omega_i = \omega^* + m_i Q_i,$$

This strategy, consisting of the $Q - \omega$ and $P - E$ droop, is illustrated in Figure 19.3(a). In this case, $\tilde{P} = -Q$ and $\tilde{Q} = P$.

19.4.2 For L-inverters

When the output impedance is inductive, $\theta = 90°$. Then

$$P = \frac{EV_o}{Z_o} \sin\delta \quad \text{and} \quad Q = \frac{EV_o}{Z_o} \cos\delta - \frac{V_o^2}{Z_o}.$$

When δ is small,

$$P \approx \frac{EV_o}{Z_o}\delta \quad \text{and} \quad Q \approx \frac{V_o}{Z_o}E - \frac{V_o^2}{Z_o},$$

and, roughly,

$$P \sim \delta \quad \text{and} \quad Q \sim E.$$

As a result, the conventional droop control strategy for inverters with an inductive Z_o takes the form

$$E_i = E^* - n_i Q_i,$$
$$\omega_i = \omega^* - m_i P_i.$$

This strategy, consisting of the $Q - E$ and $P - \omega$ droop, is illustrated in Figure 19.3(b). In this case, $\tilde{P} = P$ and $\tilde{Q} = Q$.

19.4.3 For C-inverters

When the output impedance is capacitive, then $\theta = -90°$ and

$$P = -\frac{EV_o}{Z_o}\sin\delta \quad \text{and} \quad Q = -\frac{EV_o}{Z_o}\cos\delta + \frac{V_o^2}{Z_o}.$$

When δ is small,

$$P \approx -\frac{EV_o}{Z_o}\delta \quad \text{and} \quad Q \approx -\frac{V_o}{Z_o}E + \frac{V_o^2}{Z_o},$$

and, roughly,

$$P \sim -\delta \quad \text{and} \quad Q \sim -E.$$

Hence, the conventional droop control strategy for C-inverters takes the form

$$E_i = E^* + n_i Q_i,$$
$$\omega_i = \omega^* + m_i P_i,$$

This is illustrated in Figure 19.3(c). In this case, $\tilde{P} = -P$ and $\tilde{Q} = -Q$. Note that, in order to make sure that the $Q - E$ loop and the $P - \omega$ loop are of a negative feedback, respectively, so that the droop controller is able to regulate the frequency and the voltage, the signs before $n_i Q_i$ and $m_i P_i$ are all positive, which makes them boost terms.

19.4.4 Experimental Results with R-inverters

Experiments were carried out with a laboratory set-up, which consists of two single-phase inverters controlled by dSPACE kits and powered by separate 42 VDC power supplies. The values of the inductors and capacitors are 2.35 mH and 22 μF, respectively. The switching frequency is 7.5 kHz and the frequency of the system is 50 Hz. The nominal output voltage is 12 V RMS. K_i was chosen as 4 for both inverters and the droop coefficients are: $n_1 = 0.4$ and $n_2 = 0.8$; $m_1 = 0.1$ and $m_2 = 0.2$. Hence, it is expected that $P_1 = 2P_2$.

The first experiment was carried out for a linear load with a rheostat of about 9 Ω and the results are shown in the left column of Figure 19.4. The second experiment was carried out with a non-linear load, consisting of a rectifier loaded with an LC filter and the same rheostat of about 9 Ω, and the results are shown in the right column of Figure 19.4. In both experiments, the load was connected to Inverter 2 initially and Inverter 1 was synchronised with Inverter 2. Inverter 1 was then put in parallel operation with Inverter 2 before it was disconnected. In both cases, the two inverters were not able to share the load in the ratio of 2 : 1 as designed. The output voltage was significantly lower than the nominal voltage as well. An important feature is that both E_1 and E_2 were lower than the rated voltage 12 V because there is no mechanism to increase the voltage set-point in the conventional droop control scheme (note that what is dropped is the voltage set-point). The THD of the output voltage when the load was non-linear was high as well, although this is not inherent with the droop control.

19.5 Inherent Limitations of Conventional Droop Control

As shown before, the conventional droop control strategy failed to share the load in proportion to the power rating. The voltage also dropped significantly. These are due to the inherent limitations of the conventional droop control, which will be revealed in this section by taking R-inverters as an example. Although the analysis in the sequel is done for the case with two inverters connected in parallel, as shown in Figure 19.5, it can be applied to multiple inverters connected in parallel as well.

In this case, the active and reactive power of each inverter injected into the bus (Guerrero *et al.* 2004, 2005) are

$$P_i = \frac{E_i V_o \cos \delta_i - V_o^2}{R_{oi}}, \qquad (19.3)$$

$$Q_i = -\frac{E_i V_o}{R_{oi}} \sin \delta_i. \qquad (19.4)$$

The conventional droop controller

$$E_i = E^* - n_i P_i, \qquad (19.5)$$

$$\omega_i = \omega^* + m_i Q_i \qquad (19.6)$$

takes the form shown in Figure 19.6 to generate the amplitude and frequency of the voltage reference v_{ri} for Inverter i (Guerrero *et al.* 2006a, 2007), where ω^* is the rated frequency.

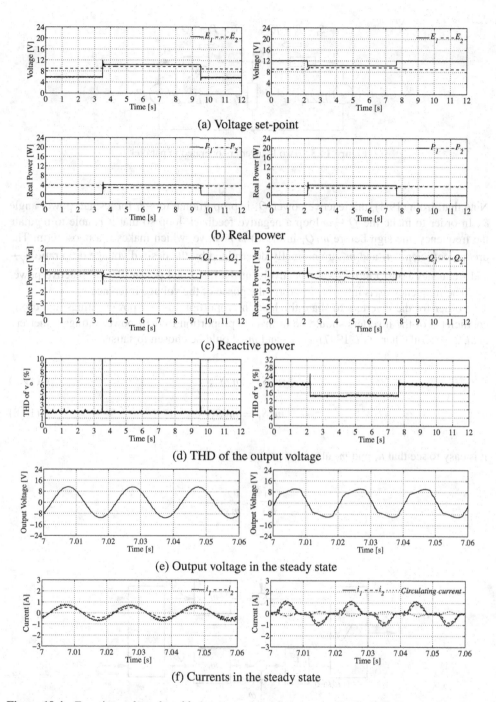

Figure 19.4 Experimental results with the conventional droop controller for R-inverters: with a linear load (left column) and with a non-linear load (right column)

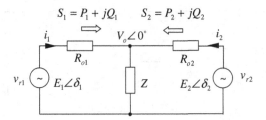

Figure 19.5 Two R-inverters operated in parallel

Note that, from (19.4), the reactive power Q_i is proportional to $-\delta_i$ for a small power angle δ_i. In order to make the $Q - \omega$ loop a negative feedback loop so that it is able to regulate the frequency, the sign before $m_i Q_i$ in (19.6) is positive, which makes it a boost term. The droop coefficients n_i and m_i are normally determined by the desired voltage drop ratio $\frac{n_i P_i^*}{E^*}$ and frequency boost ratio $\frac{m_i Q_i^*}{\omega^*}$, respectively, at the rated real power P^* and reactive power Q^*. The frequency ω_i is integrated to form the phase of the voltage reference v_{ri}.

In order for the inverters to share the load in proportion to their power ratings, the droop coefficients of the inverters should be in inverse proportion to their power ratings (Guerrero et al. 2008; Tuladhar et al. 1997), i.e. n_i and m_i should be chosen to satisfy

$$n_1 S_1^* = n_2 S_2^*, \tag{19.7}$$

$$m_1 S_1^* = m_2 S_2^*. \tag{19.8}$$

It is easy to see that n_i and m_i also satisfy

$$\frac{n_1}{m_1} = \frac{n_2}{m_2}.$$

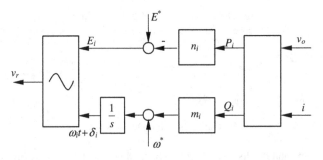

Figure 19.6 Conventional droop control scheme for R-inverters

19.5.1 Real Power Sharing

Substituting (19.5) into (19.3), the real power of the two inverters can be obtained as

$$P_i = \frac{E^* \cos \delta_i - V_o}{n_i \cos \delta_i + R_{oi}/V_o}. \tag{19.9}$$

Substituting (19.9) into (19.5), the voltage amplitude deviation of the two inverters is

$$\Delta E = E_2 - E_1 = \frac{E^* \cos \delta_1 - V_o}{\cos \delta_1 + \frac{R_{o1}}{n_1 V_o}} - \frac{E^* \cos \delta_2 - V_o}{\cos \delta_2 + \frac{R_{o2}}{n_2 V_o}}. \tag{19.10}$$

It is known from (Tuladhar *et al.* 1997) that the voltage deviation of the two units leads to considerable errors in load sharing. Indeed, in order for

$$n_1 P_1 = n_2 P_2 \quad \text{or} \quad \frac{P_1}{S_1^*} = \frac{P_2}{S_2^*}$$

to hold, the voltage deviation ΔE should be 0 according to (19.5). This is a very strict condition because there are always numerical computational errors, disturbances, parameter drifts and component mismatches. This condition is satisfied[1] if

$$\frac{n_1}{R_{o1}} = \frac{n_2}{R_{o2}} \tag{19.11}$$

and

$$\delta_1 = \delta_2. \tag{19.12}$$

In other words, n_i should be chosen to be proportional to its output impedance R_{oi}.

Taking (19.7) into account, in order to achieve accurate sharing of real power, the (resistive) output impedance should be designed to satisfy

$$R_{o1} S_1^* = R_{o2} S_2^*. \tag{19.13}$$

Since the per-unit output impedance of Inverter i is

$$\gamma_i = \frac{R_{oi}}{E^*/I_i^*} = \frac{R_{oi} S_i^*}{(E^*)^2},$$

the condition (19.13) is equivalent to

$$\gamma_1 = \gamma_2.$$

---

[1] Note that this set of conditions is sufficient but not necessary.

This simply means that the per-unit output impedances of all inverters operated in parallel should be the same in order to achieve accurate proportional real power sharing for the conventional droop control scheme. This is the basis for the virtual output impedance approach (Guerrero *et al.* 2006b) to work properly. If this is not met, then the voltage set-points E_i are not the same and errors appear in real power sharing.

According to (19.5), the real power deviation ΔP_i due to the voltage set-point deviation ΔE_i is

$$\Delta P_i = -\frac{1}{n_i} \Delta E_i.$$

For two inverters operated in parallel with real power consumption of P_1 and P_2, the relative sharing error is defined (Zhong 2012c) as

$$e_P\% = \left(\frac{P_1}{n_2} - \frac{P_2}{n_1}\right) \frac{n_1 + n_2}{P_1 + P_2}.$$

When $P_1 + P_2 = P_1^* + P_2^*$, the relative real power sharing error due to the voltage set-point deviation $\Delta E = E_2 - E_1 = \Delta E_2 - \Delta E_1$ is

$$e_P\% = \frac{P_1}{P_1^*} - \frac{P_2}{P_2^*} = \frac{\Delta P_1}{P_1^*} - \frac{\Delta P_2}{P_2^*} = \frac{E^*}{n_i P_i^*} \frac{\Delta E}{E^*},$$

where $\frac{E^*}{n_i P_i^*}$ is the inverse of the voltage drop ratio at the rated power for Inverter i. The smaller the droop coefficient (or the voltage drop ratio), the bigger the sharing error; the bigger the voltage set-point deviation ΔE, the bigger the sharing error. For example, for a voltage drop ratio of $\frac{n_i P_i^*}{E^*} = 10\%$ and a voltage set-point deviation of $\frac{\Delta E}{E^*} = 10\%$, which is very possible for the reasons mentioned before, the error in real power sharing is 100%. The accuracy is very low.

19.5.2 Reactive Power Sharing

When the system is in the steady state, the two inverters work under the same frequency, i.e. $\omega_1 = \omega_2$. It is well known that this guarantees the accuracy of reactive power sharing for R-inverters (or the accuracy of real power sharing for L-inverters); see e.g. (Li and Kao 2009). Indeed, from (19.6), there is

$$m_1 Q_1 = m_2 Q_2.$$

Since the coefficients m_i are chosen to satisfy (19.8), reactive power sharing proportional to their power ratings is (always) achieved, i.e.,

$$\frac{Q_1}{S_1^*} = \frac{Q_2}{S_2^*}.$$

Alternatively, according to (19.4), there is

$$m_1 \frac{E_1 V_o}{R_{o1}} \sin \delta_1 = m_2 \frac{E_2 V_o}{R_{o2}} \sin \delta_2. \quad (19.14)$$

If $\delta_1 = \delta_2$ and $E_1 = E_2$ then

$$\frac{m_1}{R_{o1}} = \frac{m_2}{R_{o2}}. \quad (19.15)$$

For inverters designed to have resistive output impedances, if the system is stable, then the following two sets of conditions are equivalent:

$$\mathbf{C}_1 : \begin{cases} E_1 = E_2 \\ \dfrac{n_1}{R_{o1}} = \dfrac{n_2}{R_{o2}} \end{cases} \iff \mathbf{C}_2 : \begin{cases} \delta_1 = \delta_2 \\ \dfrac{m_1}{R_{o1}} = \dfrac{m_2}{R_{o2}} \end{cases}.$$

If $\mathbf{C}_1$ holds, then proportional real power sharing is achieved according to (19.5). As a result, (19.12) holds, according to (19.10) and (19.14). Furthermore, reactive power sharing proportional to their ratings is achieved and (19.15) holds. Conversely, if $\mathbf{C}_2$ holds, then $E_1 = E_2$ according to (19.14). Furthermore, (19.11) holds according to (19.10).

A by-product from this is that n_i and m_i should be proportional. In other words, it is questionable for the conventional droop strategy to achieve different sharing ratios for real power and reactive power. It also indicates that if R-inverters achieve accurate proportional real power sharing under condition $\mathbf{C}_1$, then they also achieve proportional reactive power sharing. If they achieve proportional reactive power sharing under condition $\mathbf{C}_2$, then they also achieve proportional real power sharing. However, this is almost impossible in reality. It is difficult to maintain $E_1 = E_2$ or $\delta_1 = \delta_2$ because there are always numerical computational errors, disturbances and noises. It is also difficult to maintain $\gamma_1 = \gamma_2$ because of different feeder impedances, parameter drifts and component mismatches. The reality is that none of these conditions would be met although the reactive power sharing is accurate (note that conditions $\mathbf{C}_1$ and $\mathbf{C}_2$ are only sufficient but not necessary). A mechanism is needed to guarantee that accurate proportional load sharing can be achieved when these uncertain factors appear.

19.6 Robust Droop Control of R-inverters

19.6.1 Control Strategy

As a matter of fact, the voltage droop (19.5) can be re-written as

$$\Delta E_i = E_i - E^* = -n_i P_i,$$

and the voltage E_i can be implemented via integrating ΔE_i, that is,

$$E_i = \int_0^t \Delta E_i \mathrm{d}t.$$

This works for the grid-connected mode where ΔE_i is eventually 0 (so that the desired power is sent to the grid without error), as in (Dai *et al.* 2008a; Li *et al.* 2004; Marwali *et al.* 2004).

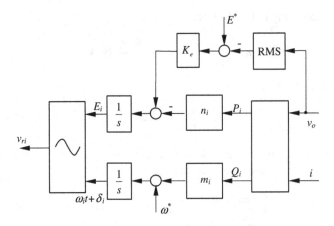

Figure 19.7 Robust droop controller for R-inverters

However, it does not work for the stand-alone mode because the actual power P_i is determined by the load and ΔE_i cannot be 0. This is why different controllers had to be used for the stand-alone mode and the grid-connected mode, respectively. When the operation mode changes, the controller needs to be changed as well. It would be advantageous if the change of controller could be avoided when the operation mode changes.

Another issue is that, according to (19.1), the load voltage v_o drops when the load increases. The voltage also drops due to the droop control, according to (19.5). The smaller the coefficient n_i, the smaller the voltage drop. However, the coefficient n_i needs to be big to obtain a fast response. In order to make sure that the voltage remains within a certain required range, the load voltage drop $E^* - V_o$ needs to be fed back in a certain way, according to the basic principles of control theory. It can be added to ΔE_i via an amplifier K_e. This actually results in an improved droop controller shown in Figure 19.7. This strategy is able to eliminate (at least considerably reduce) the impact of computational errors, noises and disturbances. As will be explained below, it is also able to maintain accurate proportional load sharing and hence is robust with respect to parameter drifts, component mismatches and disturbances.

In the steady state, the input to the integrator should be 0. Hence,

$$n_i P_i = K_e(E^* - V_o). \tag{19.16}$$

The right-hand side of the above equation is always the same for all inverters operated in parallel as long as K_e is chosen the same, which can easily be met. Hence,

$$n_i P_i = constant,$$

which guarantees accurate real power sharing without having the same E_i. This is more natural than the case with the conventional droop controller. The accuracy of real power sharing no longer depends on the inverter output impedances (including the feeder impedance) and is also immune to numerical computational errors and disturbances.

19.6.2 Error Due to Inaccurate Voltage Measurements

The only possible error in the real power sharing comes from the error in measuring the RMS value of the load voltage. From (19.16), the real power deviation ΔP_i due to the error ΔV_{oi} in the measurement of the RMS voltage is

$$\Delta P_i = -\frac{K_e}{n_i}\Delta V_{oi}.$$

For two inverters operated in parallel with $P_1 + P_2 = P_1^* + P_2^*$, the relative real power sharing error due to the error in the measurement of the RMS voltage $\Delta V_o = \Delta V_{o2} - \Delta V_{o1}$ is

$$e_P\% = \frac{P_1}{P_1^*} - \frac{P_2}{P_2^*} = \frac{\Delta P_1}{P_1^*} - \frac{\Delta P_2}{P_2^*} = \frac{K_e E^*}{n_i P_i^*}\frac{\Delta V_o}{E^*}.$$

This characterises the percentage error $e_P\%$ of the real power sharing with respect to the percentage error $\frac{\Delta V_o}{E^*}$ of the RMS voltage measurement. The term $\frac{K_e E^*}{n_i P_i^*}$ is the inverse of the voltage drop ratio with respect to the rated voltage at the rated power. If all inverters measure the voltage at the same point accurately, then the error ΔV_o can be made zero and exact proportional sharing can be achieved.

In order to achieve exact power sharing, all the inverters need to share and measure the same load voltage $v_o = V_o \angle 0°$ accurately. In order to measure the same load voltage, each inverter can provide an extra terminal, which is not internally connected to the output terminal, for voltage measurement. This voltage measurement terminal can be connected to the load terminal with a separate wire to measure the load voltage. In this way, the accuracy of the voltage measured is not affected by the current flowing through the feeder line and hence the effect of the feeder line on the power sharing accuracy and voltage regulation is eliminated. Moreover, the THD of the voltage at the load terminal instead of the inverter terminal is maintained, which reduces the THD of the load voltage. If needed, the voltage measurement terminal can be connected to the output terminal externally to measure the local voltage if the performance is acceptable. When the inverters and/or the loads are spatially distributed, it may be difficult to measure the voltage at the same point. This causes an error in the power sharing but in most cases this is acceptable because the resulting relative error of (real) power sharing is the relative error of the voltage measurement divided by the voltage drop ratio at the rated power, as characterised above. The voltage drop ratio at the rated power in these applications can be designed to be bigger than the ones in applications where the loads/inverters are close to each other so that the system is less sensitive to the error in voltage measurement.

19.6.3 Voltage Regulation

The strategy also reduces the load voltage drop. From (19.16), the load voltage is

$$V_o = E^* - \frac{n_i}{K_e}P_i = E^* - \frac{n_i P_i}{K_e E^*}E^*,$$

where $\frac{n_i P_i}{K_e E^*}$ is the voltage drop ratio. Note that this voltage drop ratio is the overall effective voltage drop ratio, which is much smaller than the drop ratio due to the droop effect and/or the load effect, but the voltage drop ratio in both the conventional droop controller and the controller in (Sao and Lehn 2005) is just the voltage drop ratio due to the droop effect and does not include the voltage drop ratio due to the load effect. Although the controller in (Sao and Lehn 2005) can compensate for the voltage drop due to the load effect, it cannot compensate for the voltage drop due to the droop effect. The robust droop control strategy can compensate for the voltage drop due to both effects and, hence, offers much better capability of voltage regulation. The voltage drop here is no longer determined by the output impedance originally designed as characterised in (19.1) but by the parameters n_i, K_e and the actual power P_i. It can be considerably reduced by using a large K_e. If there are errors in the RMS voltage measurement, then the trade-off between the voltage drop and the accuracy of power sharing has to be made because the voltage drop is proportional to $\frac{n_i}{K_e}$ but the sharing error is inverse proportional to $\frac{n_i}{K_e}$.

Here is a calculation example. Assume that the voltage drop ratio at the rated power is $\frac{n_i P_i^*}{K_e E^*} = 10\%$ and the error in the RMS voltage measurement is $\frac{\Delta V_o}{E^*} = 0.5\%$, whether because the local voltages of inverters are measured or because the sensors are not accurate. Then, the error in the real power sharing is $\frac{K_e E^*}{n_i S_i^*} \frac{\Delta V_o}{E^*} = \frac{0.5\%}{10\%} = 5\%$, which is reasonable.

It is worth noting that the robust droop control still contains the voltage droop function. What is different from the conventional droop control is that the voltage droop is applied to the output voltage V_o but not to the voltage set-point E, which is able to improve the performance significantly.

19.6.4 Error Due to the Global Settings for E^* and ω^*

This sub-section is devoted to the sensitivity analysis of the error in the global settings E^* and ω^* for the robust droop controller.

Any small error $\Delta \omega_i$ in ω_i^* would lead to the reactive power deviation (if still stable) of

$$\Delta Q_i = -\frac{1}{m_i} \Delta \omega_i,$$

according to (19.6). For two inverters operated in parallel with $Q_1 + Q_2 = Q_1^* + Q_2^*$, the relative reactive power sharing error due to the error $\Delta \omega = \omega_2^* - \omega_1^* = \Delta \omega_2 - \Delta \omega_1$ is

$$e_Q\% = \frac{Q_1}{Q_1^*} - \frac{Q_2}{Q_2^*} = \frac{\Delta Q_1}{Q_1^*} - \frac{\Delta Q_2}{Q_2^*} = \frac{\omega^*}{m_i Q_i^*} \frac{\Delta \omega}{\omega^*},$$

where $\frac{\omega^*}{m_i Q_i^*}$ is the inverse of the frequency boost ratio at the rated reactive power. The smaller the frequency boost ratio, the bigger the reactive power sharing error; the bigger the error $\Delta \omega$, the bigger the sharing error. For example, for a typical frequency boost ratio of $\frac{m_i Q_i^*}{\omega^*} = 1\%$, the error of $\frac{\Delta\omega}{\omega^*} = 1\%$ in the frequency setting would lead to a 100% of error in the reactive power sharing! Hence, the accuracy of reactive power sharing is very sensitive to the accuracy of the global setting for ω^*, which should be made very accurate.

Similarly, according to (19.16), the real power deviation ΔP_i due to the error ΔE_i^* in E_i^* is

$$\Delta P_i = \frac{K_e}{n_i} \Delta E_i^*.$$

For two inverters operated in parallel with $P_1 + P_2 = P_1^* + P_2^*$, the relative real power sharing error due to the error $\Delta E^* = E_2^* - E_1^* = \Delta E_2^* - \Delta E_1^*$ in the global settings of E^* is

$$e_P\% = \frac{P_1}{P_1^*} - \frac{P_2}{P_2^*} = \frac{\Delta P_1}{P_1^*} - \frac{\Delta P_2}{P_2^*} = -\frac{K_e E^*}{n_i P_i^*} \frac{\Delta E^*}{E^*}.$$

For a typical voltage drop ratio at the rated power of $\frac{n_i P_i^*}{K_e E^*} = 10\%$, a 10% error in $\frac{\Delta E^*}{E^*}$ would lead to a -100% error in the real power sharing. Although the error in E^* is less sensitive than the error in ω^*, it is still quite significant. Hence, in practice, it is very important to make sure that the global settings are accurate. Anyway, this is not a problem at all.

19.6.5 Experimental Results

The above strategy is verified in a laboratory set-up consisting of two single-phase inverters controlled by dSPACE kits and powered by separate 42 VDC voltage supplies. The inverters are connected to the AC bus via a circuit breaker CB and the load is assumed to be connected to the AC bus. The values of the inductors and capacitors are 2.35 mH and 22 μF, respectively. The switching frequency is 7.5 kHz and the frequency of the system is 50 Hz. The rated voltage is 12 VRMS and $K_e = 10$. The droop coefficients are: $n_1 = 0.4$ and $n_2 = 0.8$; $m_1 = 0.1$ and $m_2 = 0.2$. Hence, it is expected that $P_1 = 2P_2$ and $Q_1 = 2Q_2$. Due to the configuration of the hardware set-up, the voltage for Inverter 2 was measured by the controller of Inverter 1 and then sent out via a DAC channel, which was then sampled by the controller of Inverter 2. This brought some latency into the system but the effect was not noticeable.

19.6.5.1 Inverters Having Different per-unit Output Impedances with a Linear Load

Both K_i were chosen as 4 for the two inverters to intentionally make the per-unit output impedances of the two inverters significantly different. In reality, this could be due to different feeder impedances or component mismatches.

A linear load of about 9 Ω was connected to Inverter 2 initially. Inverter 1 was connected to the system at around $t = 3$ s and was then disconnected at around $t = 10.5$ s. The relevant curves from the experiment with the robust droop controller are shown in the left column of Figure 19.8 and the relevant curves from the experiment with the conventional droop controller are shown in the right column of Figure 19.8. In both cases, the reactive power was shared accurately (in the ratio of 2 : 1), although the actual values were different (because

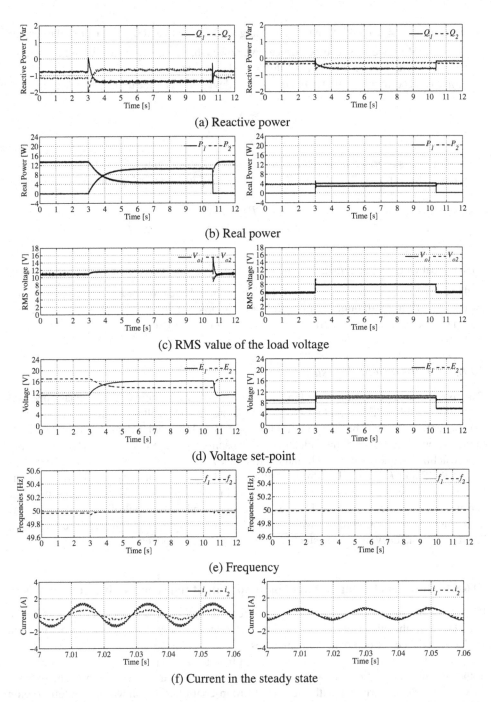

Figure 19.8 Experimental results for the case with a linear load when inverters have intentionally different per-unit output impedances: with the robust droop controller (left column) and with the conventional droop controller (right column)

the voltages were different). Inverter 1 was able to pick up the load, gradually in the case of the robust droop controller and very quickly in the case of the conventional droop controller. The robust droop controller could share the real power very accurately but the conventional one could not. The robust droop controller has considerably relaxed the trade-off between the sharing accuracy and the voltage drop. The voltage from the inverters equipped with the robust droop controller is very close to the rated voltage but the voltage from the inverters equipped with the conventional controller is only $\frac{2}{3}$ of the rated voltage. The voltage set-point of the conventional droop controller has to be lower than the rated voltage due to the droop effect. The bigger the voltage drop ratio, the lower the voltage set-point. It can also be clearly seen that $E_1 \neq E_2$ because the per-unit output impedance is different and also there are numerical errors and component mismatches etc. Because of the reduced deviation in the voltage, the reactive power becomes bigger. This leads to a slightly bigger deviation in the frequency but it is expected because of the $Q - \omega$ droop. The current sharing reflects the power sharing well. It is worth noting that there was no need to change the operation mode of Inverter 2 when connecting or disconnecting Inverter 1.

19.6.5.2 Inverters having the Same per-unit Output Impedance with a Linear Load

The current feedback gains were chosen as $K_{i1} = 2$ and $K_{i2} = 4$ so that the output impedance is consistent with the power sharing ratio 2 : 1. The results from the robust droop controller with $K_e = 10$ are shown in the left column of Figure 19.9 and the results from the conventional droop controller are shown in the right column of Figure 19.9. The robust droop controller was able to share the load according to the sharing ratio and considerably outperformed the conventional droop controller in terms of sharing accuracy and voltage drop. The difference between the voltage set-points can be clearly seen. This indicates the effect of numerical errors, parameter drifts and component mismatches etc. because the voltage set-points were supposed to be identical without these uncertain factors. Comparing the left columns of Figures 19.8 and 19.9, there were no noticeable changes in the performance for the robust droop controller but the difference in the voltage set-points was decreased. Comparing the right columns of Figures 19.8 and 19.9, the sharing accuracy and the voltage drop were improved slightly and the voltage set-points became closer to each other when the per-unit output impedances were the same.

Another experiment was carried out with $K_e = 1$ to demonstrate the role of K_e and the results are shown in the right column of Figure 19.10. The results with $K_e = 10$ shown in the left column of Figure 19.9 are shown in the left column of Figure 19.10 again for comparison. It can be seen that a large K_e helps speed up the response and reduce the voltage drop. However, a large K_e causes large ripples in the current.

19.6.5.3 With a Non-linear Load

A non-linear load, consisting of a rectifier loaded with an LC filter and the same rheostat used in the previous experiments, was connected to Inverter 2 initially. A similar procedure to connect/disconnect Inverter 1 was followed in the experiment. The relevant curves of the experiment are shown in the left column of Figure 19.11 for the case with the robust droop controller and in the right column of Figure 19.11 for the case with the conventional

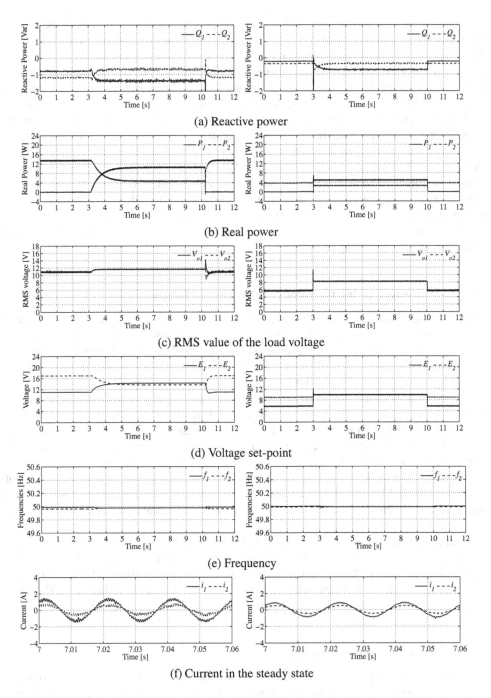

Figure 19.9 Experimental results for the case with a linear load when inverters have the same per-unit output impedance: with the robust droop controller (left column) and with the conventional droop controller (right column)

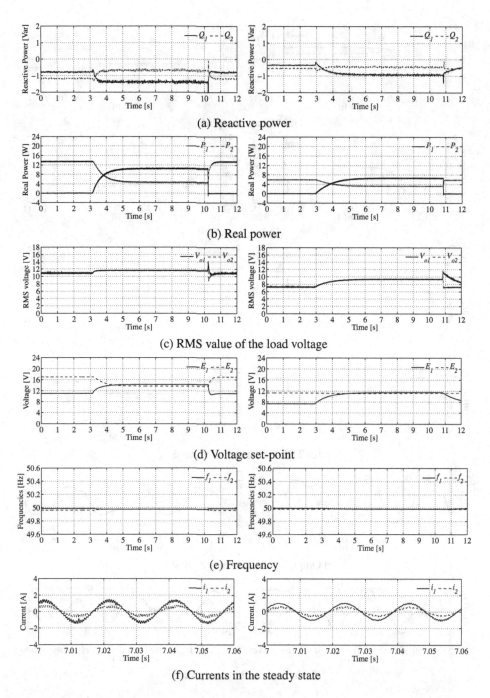

Figure 19.10 Experimental results for the case with the same per-unit output impedance using the robust droop controller: with $K_e = 10$ (left column) and with $K_e = 1$ (right column)

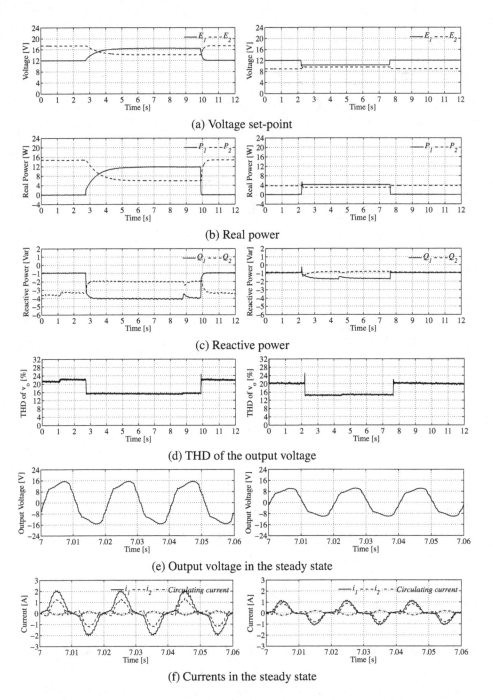

Figure 19.11 Experimental results with a non-linear load: with the robust droop controller (left column) and with the conventional droop controller (right column)

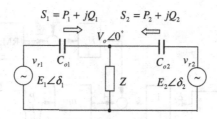

Figure 19.12 Two C-inverters operated in parallel

droop controller. It can be seen that the two inverters with the robust droop controllers were still able to share the load very accurately in the ratio of 2 : 1, although $E_1 \neq E_2$. The dynamic performance did not change much either. The circulating current is very small and does not contain noticeable fundamental component. It should be emphasised that the active power sharing is still very accurate although the output impedances of the inverters are not resistive over a wide enough frequency range and there is significant amount of harmonic current components.

The only drawback is that the THD of the output voltage v_o is not satisfactory (22% for one inverter and 16% for two inverters in parallel). However, this is expected because $R_{oi} = K_i = 4$ was used, which dominated the harmonic voltage drop on the output impedance and increased the THD. It can be improved by using smaller K_i while maintaining the output impedance resistive. A strategy to improve the output-voltage THD while maintaining accurate power sharing will be discussed in Chapter 20.

19.7 Robust Droop Control of C-inverters

19.7.1 Control Strategy

Figure 19.12 depicts the parallel operation of two C-inverters. As reported in (Zhong 2012c) and discussed above, the conventional droop control strategy is not able to accurately share both real power and reactive power at the same time because there is no mechanism to make sure that the voltage set-points are the same when numerical errors, noises and disturbances exist. Also it is impossible to make sure that the per-unit output impedance is the same because of component mismatches and parameter shifts. Hence, the voltage regulator bolted onto the conventional droop controllers for R-inverters should also be bolted onto the droop controller for C-inverters. This results in the robust droop controller shown in Figure 19.13. It is able to share both real power and reactive power accurately even if the per-unit output impedance is not the same and/or there are numerical errors, disturbances and noises because, at the steady state, there is

$$n_i Q_i + K_e(E^* - V_o) = 0. \tag{19.17}$$

This means

$$n_i Q_i = constant,$$

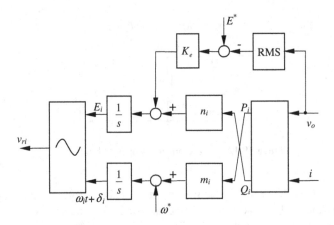

Figure 19.13 Robust droop controller for C-inverters

as long as K_e is the same for all inverters. This guarantees accurate sharing of reactive power in proportion to their ratings. As long as the system is stable, which leads to the same frequency, the real power can be guaranteed as well (Zhong 2012c).

According to (19.17), the output voltage is

$$V_o = E^* + \frac{n_i}{K_e} Q_i = E^* + \frac{n_i Q_i}{K_e E^*} E^*,$$

which can be maintained within the desired range via choosing a big K_e. Hence, the control strategy has very good capability of voltage regulation as well, in addition to accurate power sharing.

The droop coefficients n_i and m_i can be determined as usual by the desired voltage drop ratio $\frac{n_i Q_i^*}{K_e E^*}$ and the frequency boost ratio $\frac{m_i P_i^*}{\omega^*}$, respectively, at the rated reactive power Q^* and real power P^*.

19.7.2 Simulation Results

A system that consists of two single-phase inverters powered by two separate 42 VDC voltage supplies was used to carry out simulations to verify the design. The capacity of Inverter 1 is 25 VA and the capacity of Inverter 2 is 50 VA, with the rated power factor of 0.9. It is expected that $P_2 = 2P_1$ and $Q_2 = 2Q_1$. The switching frequency is 7.5 kHz and the frequency of the system is 50 Hz. The rated voltage is 12 V and $K_e = 20$. The filter inductor is $L = 2.35$ mH with a parasitic resistance of 0.1 Ω and the filter capacitance C is 22 μF.

Assume that the desired voltage drop ratio $\frac{n_i Q_i^*}{K_e E^*}$ is 10% and frequency boost ratio $\frac{m_i P_i^*}{\omega^*}$ is 1%, respectively, at the rated real power $P_i^* = 0.9 S_i^*$ and reactive power $Q_i^* = 0.436 S_i^*$. As a result, $n_1 = 2.2$ and $n_2 = 1.1$; $m_1 = 0.14$ and $m_2 = 0.07$.

The capacitor is chosen as $C_o = 479$ μF and the corresponding impedance at the fundamental frequency is $Z_o(j\omega^*) = -j6.65$ Ω, which is capacitive and is able to dominate the impedance between the voltage reference and the terminal. The results from the R-inverters with $K_i = 4$ are provided for comparison.

19.7.2.1 With a Linear Load

Simulations were carried out for a linear load with $R_L = 9\,\Omega$. The results for the C-inverters and R-inverters are shown in the left and right columns of Figure 19.14, respectively. The real power and reactive power were well shared in both cases and the output voltage was very close to the rated voltage. The THD of the output voltages in both cases was very low as well.

19.7.2.2 With a Non-linear Load

The same simulations were carried out for a full-bridge rectifier load with an LC filter $L = 150\,\mu H$, $C = 1000\,\mu F$ and $R_L = 9\,\Omega$. The results for C- and R-inverters are shown in the left and right columns of Figure 19.15, respectively, with the steady-state performance shown in Table 19.1. Again, both the C- and R-inverters performed very well: the real power and reactive power were accurately shared and the output voltage was close to the rated voltage. The voltage THD of the C-inverters was slightly better than that of the R-inverters, with considerably reduced low-order harmonics. It is worth noting that although the output impedances of the two inverters are the same, which means the per-unit output impedance is significantly different, the inverters shared both the real power and the reactive power accurately.

19.7.3 Experimental Results

Experiments were carried out on a test rig that consists of two single-phase inverters powered by two separate 42 VDC voltage supplies. The parameters of the system are the same as those given in the previous section. Experiments were carried out for C-inverters with $C_o = 479\,\mu F$ and R-inverters with $K_i = 4\,\Omega$.

19.7.3.1 With a Linear Load

Experiments were carried out for a linear load with $R_L = 9\,\Omega$. The results for C-inverters and R-inverters are shown in the left and right columns of Figure 19.16, respectively, for comparison. The system worked very well for both cases: with accurate sharing of real power and reactive power, good regulation of output voltage and low THD. With comparison to R-inverters, the voltage regulation performance is slightly better because this is related to the reactive power of the load, which is small, and the frequency variation is slightly higher because this is related to the real power.

19.7.3.2 With a Non-linear Load

The same experiments were carried out for a full-bridge rectifier load with an LC filter $L = 150\,\mu H$, $C = 1000\,\mu F$ and $R_L = 9\,\Omega$. The results for C-inverters and R-inverters are shown in the left and right columns of Figure 19.17, respectively. Again, the system worked very well for both cases with accurate sharing of real power and reactive power and good capability of voltage regulation. The C-inverters demonstrated much better THD than the

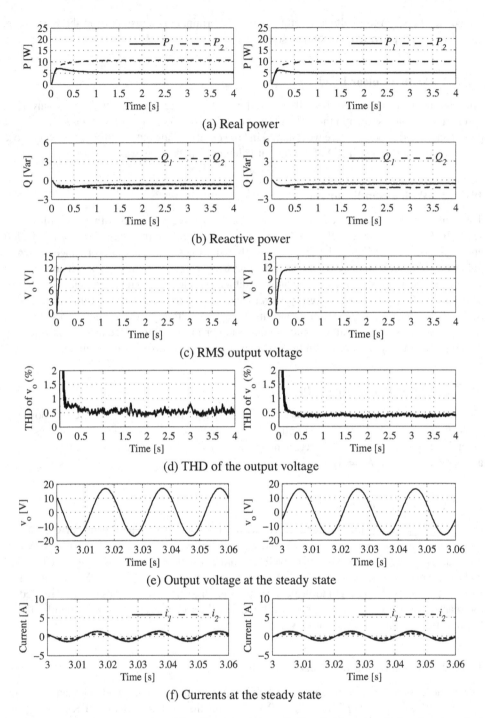

Figure 19.14 Simulation results with a linear load $R_L = 9\ \Omega$: of C-inverters (left column) and of R-inverters (right column)

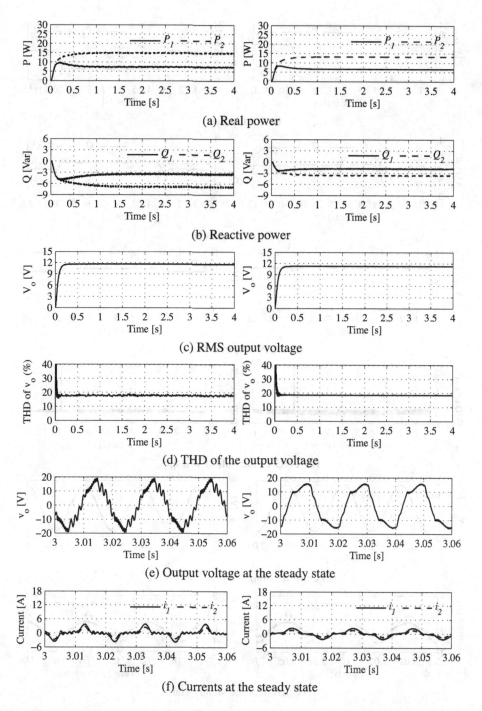

Figure 19.15 Simulation results with a non-linear load: of C-inverters (left column) and of R-inverters (right column)

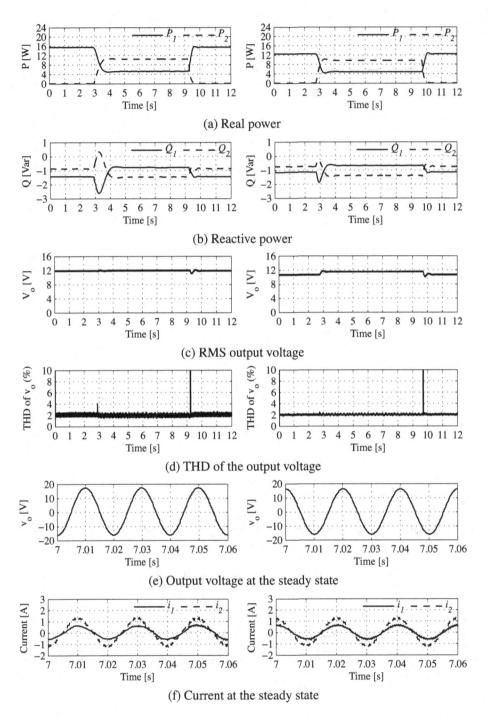

Figure 19.16 Experimental results with a linear load $R_L = 9\ \Omega$: of C-inverters (left column) and of R-inverters (right column)

Parallel Operation of Inverters

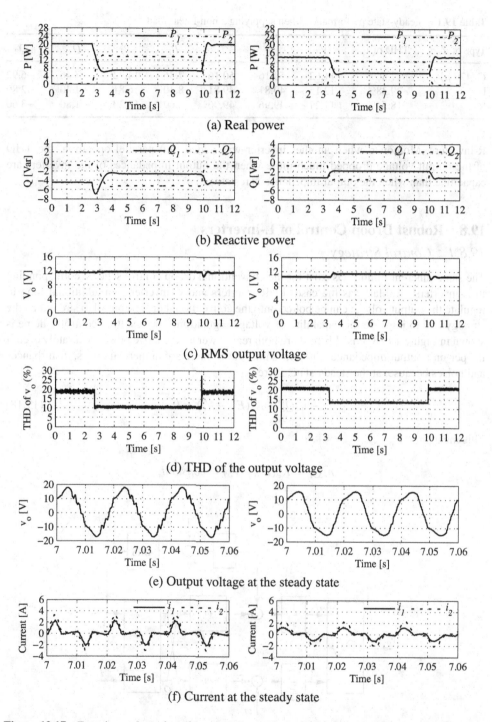

Figure 19.17 Experimental results with a non-linear load: of C-inverters (left column) and of R-inverters (right column)

Table 19.1 Steady-state performance when supplying a non-linear load

Type of Z_o	THD of v_o	V_o	f_1	f_2	P_1	P_2	Q_1	Q_2
C (479 μF)	17.86%	11.62	50.16	50.16	7.36	14.72	−3.46	−6.92
L	29.38%	12.15	49.81	49.81	8.40	16.81	−1.40	−2.80
R	18.54%	11.27	49.96	49.96	6.60	13.20	−1.80	−3.60

R-inverters. Moreover, when another inverter was added into parallel operation, the THD of the output voltage dropped much more when the output impedances of the inverters are capacitive than when the output impedance is resistive.

19.8 Robust Droop Control of L-inverters

19.8.1 Control Strategy

The same inherent limitations exist for the conventional droop control scheme corresponding to L-inverters. Similarly as the cases of R-inverters and C-inverters, an additional loop to regulate the output voltage can be bolted onto the conventional droop control to strengthen the strategy and to improve the capability of voltage regulation. The resulting control scheme is shown in Figure 19.18. It is able to share both real power and reactive power accurately even if the per-unit output impedance is not the same and/or there are numerical errors, disturbances and noises because, at the steady state, there is

$$-n_i Q_i + K_e(E^* - V_o) = 0. \tag{19.18}$$

This means

$$n_i Q_i = K_e(E^* - V_o) = \text{constant},$$

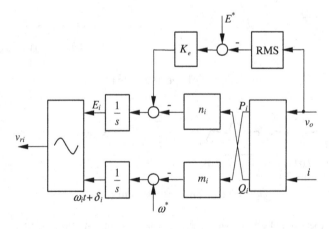

Figure 19.18 Robust droop controller for L-inverters

as long as K_e is the same for all inverters. This guarantees the accurate sharing of reactive power in proportion to their ratings. As long as the system is stable, which leads to the same frequency, the real power can be guaranteed as well (Zhong 2012c).

According to (19.18), the output voltage is

$$V_o = E^* - \frac{n_i}{K_e} Q_i = E^* - \frac{n_i Q_i}{K_e E^*} E^*,$$

which can be maintained within the desired range via choosing a big K_e. Hence, the control strategy has very good capability of voltage regulation as well, in addition to the accurate power sharing.

The droop coefficients n_i and m_i can then be determined as usual by the desired voltage drop ratio $\frac{n_i Q_i^*}{K_e E^*}$ and the frequency drop ratio $\frac{m_i P_i^*}{\omega^*}$, respectively, at the rated reactive power Q^* and real power P^*.

19.8.2 Simulation Results

A system that consists of two single-phase inverters powered by two separate 42 VDC voltage supplies was used to carry out simulations to verify the design. The capacity of Inverter 1 is 25 VA and the capacity of Inverter 2 is 50 VA, with the rated power factor of 0.9. It is expected that $P_2 = 2P_1$ and $Q_2 = 2Q_1$. The switching frequency is 7.5 kHz and the frequency of the system is 50 Hz. The rated voltage is 12 V and $K_e = 20$. The filter inductor is $L = 2.35$ mH with a parasitic resistance of 0.1 Ω and the filter capacitance C is 22 μF. Without any inner-loop controller, the output impedance of the inverter is about $Z_o(j\omega^*) = j0.74 \,\Omega$, which is inductive.

Assume that the desired voltage drop ratio $\frac{n_i Q_i^*}{K_e E^*}$ is 10% and the frequency drop ratio $\frac{m_i P_i^*}{\omega^*}$ is 1%, respectively, at the rated real power $P_i^* = 0.9 S_i^*$ and reactive power $Q_i^* = 0.436 S_i^*$. As a result, $n_1 = 2.2$ and $n_2 = 1.1$; $m_1 = 0.14$ and $m_2 = 0.07$.

19.8.2.1 With a Linear Load

Simulations were carried out for a linear load with $R_L = 9 \,\Omega$. The results for L-inverters and R-inverters are shown in the left and right columns of Figure 19.19, respectively. The system performed very well in both cases. It is worth noting that the output voltage of the L-inverters was slightly higher than the rated voltage because the reactive power is capacitive due to the filter capacitor.

19.8.2.2 With a Non-linear Load

The same simulations were carried out for a full-bridge rectifier load with an LC filter $L = 150 \,\mu\text{H}$, $C = 1000 \,\mu\text{F}$ and $R_L = 9 \,\Omega$. The results for L-inverters and R-inverters are shown in the left and right columns of Figure 19.20, respectively, with the steady-state performance shown in Table 19.1 with comparison to C-inverters. Again, the system performed well in both

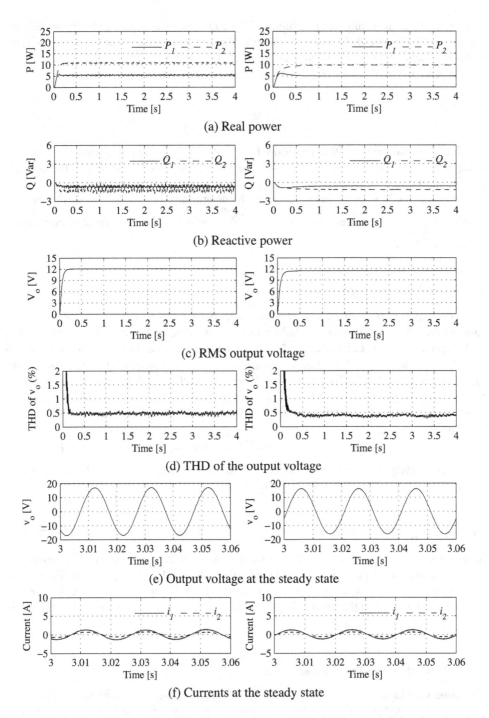

Figure 19.19 Simulation results with a linear load $R_L = 9\ \Omega$: of L-inverters (left column) and of R-inverters (right column)

Parallel Operation of Inverters

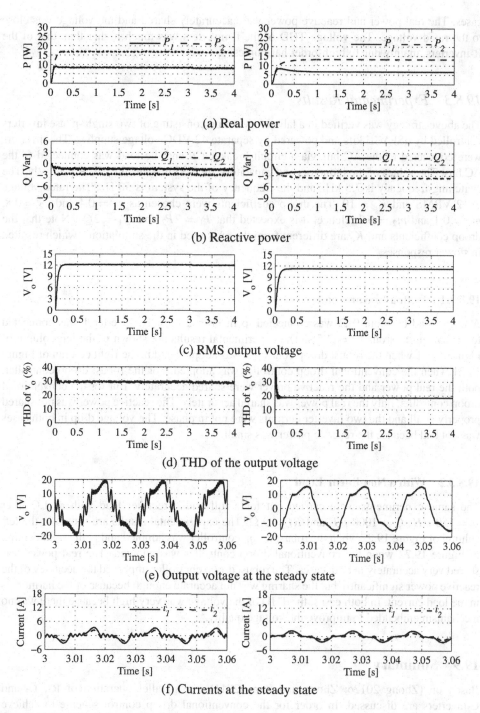

Figure 19.20 Simulation results with a non-linear load: of L-inverters (left column) and of R-inverters (right column)

cases. The real power and reactive power were accurately shared and the voltage was close to the rated voltage. The voltage THD of the L-inverters was much higher than that of the R-inverters, mainly due to high-order harmonics.

19.8.3 Experimental Results

The above strategy was verified in a laboratory set-up consisting of two single-phase inverters controlled by dSPACE kits and powered by separate 42 VDC voltage supplies. The inverters were connected to the AC bus via a circuit breaker CB and the load was connected to the AC bus. The values of the inductors and capacitors are 2.35 mH and 22 μF, respectively. The switching frequency is 7.5 kHz and the frequency of the system is 50 Hz. The rated voltage is 12 VRMS and $K_e = 10$. The droop coefficients were chosen as $n_1 = 0.4$ and $n_2 = 0.8$, $m_1 = 0.1$ and $m_2 = 0.2$. Hence, it is expected that $P_1 = 2P_2$ and $Q_1 = 2Q_2$. Note that the droop coefficients and K_e are different from the ones used in the simulations, which resulted in slower responses.

19.8.3.1 With a Linear Load

A linear load of about 9 Ω was connected to Inverter 2 initially. Inverter 1 was connected to the system at around $t = 2.5$ s. The experimental results are shown in the left column of Figure 19.21 when the robust droop controller was adopted and in the right column of Figure 19.21 when the conventional droop controller was adopted. For the robust droop controller, both the real power and the reactive power were shared accurately. But for the conventional droop controller, only the real power was shared accurately. The reactive power was not shared properly at all and the two inverter currents were not in phase. The voltage drop in both cases was not bad because the reactive power was small.

19.8.3.2 With a Non-linear Load

The same experiments were carried out for a full-bridge rectifier load with an LC filter $L = 150$ μH, $C = 1000$ μF and $R_L = 9$ Ω. The experimental results are shown in the left column of Figure 19.22 when the robust droop controller was adopted and in the right column of Figure 19.22 when the conventional droop controller was adopted. The real power was shared very accurately in both cases. The robust droop controller improved the accuracy of the reactive power significantly but the sharing was not accurate enough because of the harmonics in the load current. In both cases, the THD of the voltage was very high because there was no mechanism embedded to improve the voltage quality.

19.9 Summary

Based on (Zhong 2012c; Zhong and Zeng 2011), the parallel operation of R-, L- and C-inverters are discussed. In order for the conventional droop control scheme to achieve accurate proportional load sharing among parallel-operated inverters, the inverters should have the same per-unit output impedances and the voltage set-points (E_i) should be the same.

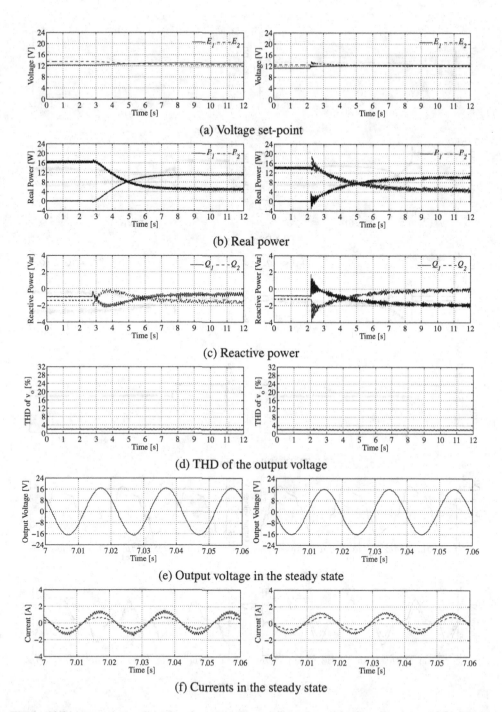

Figure 19.21 Experimental results of parallel-operated L-inverters with a linear load: with the robust droop controller (left column) and with the conventional droop controller (right column)

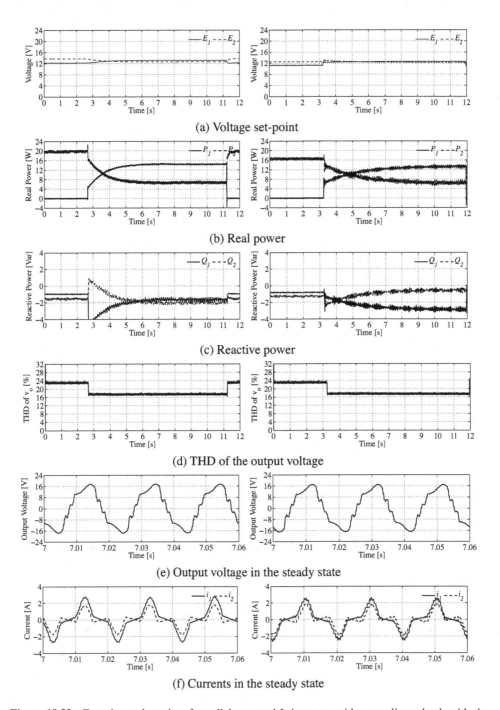

Figure 19.22 Experimental results of parallel-operated L-inverters with a non-linear load: with the robust droop controller (left column) and with the conventional droop controller (right column)

These conditions are almost impossible to meet in practical applications. A robust droop controller is then presented to obtain accurate proportional load sharing for inverters working in the stand-alone mode (and naturally also for inverters working in the grid-connected mode). This strategy does not require the above two conditions to be met in order to achieve accurate proportional sharing. The strategy is also able to compensate for the voltage drop due to the load effect and the droop effect and, hence, the load voltage can be maintained within the desired range around the rated value. Quantitative analysis of the error in power sharing is carried out thoroughly. Extensive simulation and experimental results are presented.

20

Robust Droop Control with Improved Voltage Quality

In Chapter 19, a robust droop controller is presented to achieve accurate sharing of both real power and reactive power for different types of inverters while maintaining excellent capability of voltage regulation. Apart from C-inverters, the THD of the output voltage is high when the load is non-linear. In this chapter, the strategy described in Chapter 8 is combined with the robust droop controller to improve the voltage quality. As a result, all the three major problems associated with the parallel operation of inverters, that is, sharing accuracy, voltage regulation and voltage THD, are solved with a compact controller. Experimental results are provided to verify the analysis and design. In this chapter, R-inverters are taken as an example but the concept can be extended to L- and C-inverters.

20.1 Control Strategy

For a single-phase inverter depicted in Figure 20.1, the controller presented in Chapter 8, as shown in Figure 20.2(a), is able to bypass the harmonic components in the load current and hence the output voltage THD is low even when the load is non-linear. It consists of a proportional feedback of the inductor current (Zhong 2012c), which is also widely referred to as a virtual resistor (Dahono 2003, 2004; Dahono et al. 2001; Li 2009) to dampen the resonance of the LC filter. The output impedance of the inverter is forced to be predominantly resistive with a large enough K_i. In addition to the current feedback through the virtual resistor K_i, the voltage error $v_r - v_o$ is added to v_r through a block $K_R(s)$, which is able to reduce the output impedance of the inverter at harmonic frequencies. As a result, the high voltage THD caused by a large K_i when nonlinear loads are present is reduced.

The block $K_R(s)$ in Figure 20.2(a) can be chosen as the resonant harmonic compensators (Castilla et al. 2009; Shen et al. 2010)

$$K_R(s) = \sum_{h=3,5,\ldots} \frac{2\xi h\omega s}{s^2 + 2\xi h\omega s + (h\omega)^2} \times K_h, \qquad (20.1)$$

Control of Power Inverters in Renewable Energy and Smart Grid Integration, First Edition.
Qing-Chang Zhong and Tomas Hornik.
© 2013 John Wiley & Sons, Ltd. Published 2013 by John Wiley & Sons, Ltd.

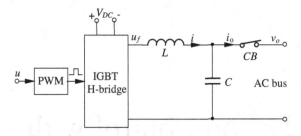

Figure 20.1 Single-phase inverter with a PWM block and an LC filter

of which the gain at frequency $h\omega$ is K_h with zero phase. It is almost 1 everywhere apart from at the frequencies around the harmonics. This is equivalent to reducing the output impedance of the inverter

$$Z_o(s) = \frac{sL + K_i}{1 + K_R(s)}$$

to $\frac{sL + K_i}{1 + K_h}$ at the frequency $h\omega$. This helps improve the THD of the output voltage v_o. The damping factor ξ can be chosen as $\xi = 0.01$ to accommodate frequency variations and h can

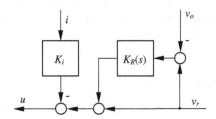

(a) The controller discussed in Chapter 8 to improve voltage THD

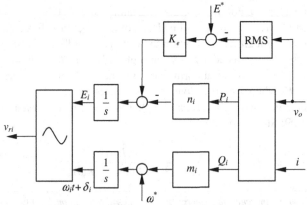

(b) Robust droop controller to generate the voltage reference

Figure 20.2 Robust droop control with improved voltage THD for R-inverters

be chosen to cover the major harmonic components in the current, e.g. the 3rd, 5th and 7th harmonics.

Since K_i has the capability of dampening oscillations, it can be chosen to be large so that the impedance at low frequencies is dominantly resistive. As a result, the robust droop controller for R-inverters shown in Figure 20.2(b), which is discussed in Chapter 19, can be adopted to generate the voltage reference v_r. The combination of both strategies is able to solve the three main problems associated with parallel-operated inverters, i.e. sharing accuracy, voltage regulation and voltage THD, with a compact controller.

20.2 Experimental Results

This strategy was verified with a laboratory set-up consisting of two single-phase inverters illustrated in Figure 20.1, which were controlled by dSPACE kits and powered by separate 42 VDC power supplies. The values of the inductors and capacitors were 2.35 mH and 22 μF, respectively. The switching frequency was 7.5 kHz and the fundamental frequency of the system was 50 Hz. The rated output voltage was 12 V RMS and $K_e = 10$. The inverters were designed to have resistive output impedance with the current feedback K_i. The coefficients for K_R were chosen as $K_3 = 14$, $K_5 = 10$ and $K_7 = 2.5$. Due to the configuration of the hardware set-up, the voltage for Inverter 2 was measured by the controller of Inverter 1 and then sent out via a DAC channel, which was then sampled by the controller of Inverter 2. This brought some latency but the effect was not noticeable. Inverter 1 was equipped with a synchronisation unit. It was synchronised with Inverter 2 when its output was not connected to that of Inverter 2 and was ready to be connected at any time.

20.2.1 1 : 1 Power Sharing

In this case, the droop coefficients were set as $n_1 = 0.8$ and $n_2 = 0.8$; $m_1 = 0.2$ and $m_2 = 0.2$. Hence, it was expected that $P_1 = P_2$. $K_1 = 4$ and $K_2 = 4$ were chosen and the output impedances of the inverters were resistive over a wide range of frequencies covering the fundamental frequency, as can be seen from the Bode plots of the output impedances of both inverters shown in Figure 20.3. Clearly, the impedances at the 3rd, 5th and 7th harmonic

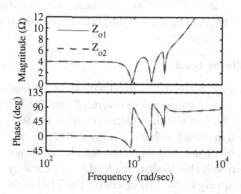

Figure 20.3 Bode plots of the output impedances of the inverters for 1 : 1 power sharing

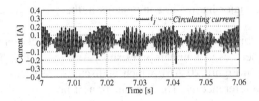

Figure 20.4 Circulating current at 1 : 1 power sharing

frequencies are much smaller than the impedance at the fundamental frequency. Both inverters had the same output impedance.

20.2.1.1 Without a Load

The current of each inverter contains two components: one is the load current component and the other is the circulating current component. Since the power sharing ratio is 1 : 1, the load current component is $\frac{i_1 + i_2}{2}$ and the circulating current component is $\pm\frac{i_1 - i_2}{2}$, which takes the positive sign for Inverter 1 and the negative sign for Inverter 2. When there was no load (apart from the capacitors in the filters) connected to the inverters, the circulating current is shown in Figure 20.4. Clearly, the circulating current is very small.

20.2.1.2 With a Linear Load

A linear load of 7.9 Ω was connected to Inverter 2 initially. Inverter 1 was connected to the system at around $t = 3$ s and was disconnected at around $t = 9.5$ s. The relevant curves are shown in the left column of Figure 20.5 when the K_R block was adopted and in the right column of Figure 20.5 when the K_R block was not adopted. There was no major change in the performance, apart from the reduced current ripples when the K_R block was used. The THD of the output voltage remained at about 2% before and after the connection.

Another experiment was done to see the effect of load change. Initially, two inverters were sharing a linear load of 5.7 Ω, which was then changed to 7.9 Ω at $t = 3.5$ s before being changed back to 5.7 Ω at $t = 9.2$ s. The two inverters shared the load accurately during the process, as shown in Figure 20.6. The THD of the output voltage remained almost unchanged before and after the load change. The voltage set-point decreased when the load was reduced.

20.2.1.3 With a Non-linear Load

A non-linear load, consisting of a rectifier loaded with an LC filter and the same rheostat used in the previous experiment, was connected to Inverter 2 initially. Inverter 1 was connected to the system at around $t = 3$ s and was then disconnected at $t = 9.5$ s. The relevant curves of the experiment are shown in the left column of Figure 20.7 when the K_R block was adopted and in the right column of Figure 20.7 when the K_R block was not adopted. It can be seen that the two inverters were still able to share the load very accurately in the ratio of 1 : 1. The load current was shared equally by the two inverters. The THD of the output voltage dropped

Robust Droop Control with Improved Voltage Quality 339

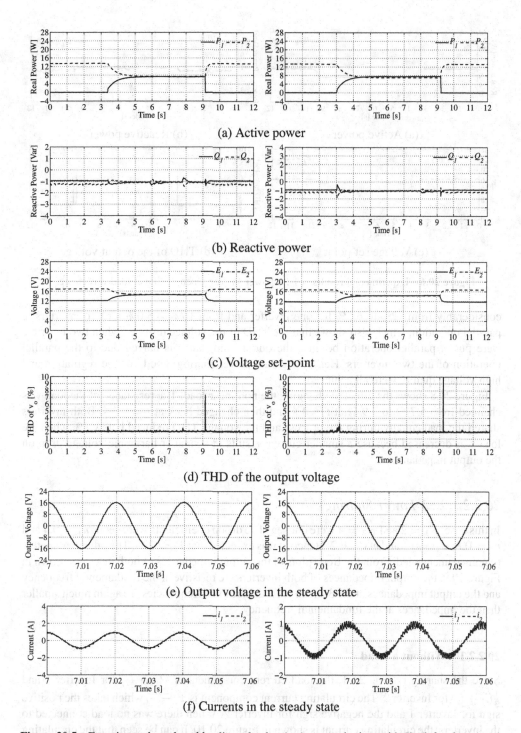

Figure 20.5 Experimental results with a linear load at 1 : 1 power sharing: with K_R (left column) and without K_R (right column)

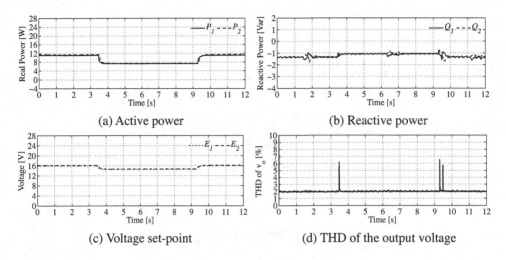

Figure 20.6 Experimental results at 1 : 1 power sharing when the linear load was changed

considerably from 22% to 6.5% for one inverter and from 16% to 5.0% for two inverters in parallel operation, which also shows that the THD of the voltage dropped when two inverters were put in parallel operation because the output impedance is halved due to the parallel operation of the two inverters. Hence, accurate power sharing, good voltage regulation and high voltage quality were all achieved.

Figure 20.8 shows the effect of changing the nonlinear load. The resistance of the load was changed from 5.7 Ω to 7.9 Ω at $t = 4$ s and then changed back at $t = 9.2$ s. Both inverters shared the load equally during the process and the voltage THD reduced below 5% when the load was decreased because a smaller current results in a smaller harmonic voltage drop on the output impedance.

20.2.2 2 : 1 Power Sharing

In this case, $K_1 = 2$ and $K_2 = 4$ were chosen. The droop coefficients were set as $n_1 = 0.4$ and $n_2 = 0.8$; $m_1 = 0.1$ and $m_2 = 0.2$. Hence, it was expected that $P_1 = 2P_2$. The other parameters remained unchanged. From the Bode plots of the output impedances of both inverters shown in Figure 20.9, the output impedances of both inverters are resistive at the fundamental frequency and the output impedances at the 3rd, 5th and 7th harmonic frequencies are again much smaller than the impedances at the fundamental frequency.

20.2.2.1 Without a Load

Since the sharing ratio is 2 : 1, the load current component is $\frac{2}{3}(i_1 + i_2)$ for Inverter 1 and $\frac{1}{3}(i_1 + i_2)$ for Inverter 2. The circulating current component is $\pm \frac{i_1 - 2i_2}{3}$, which takes the positive sign for Inverter 1 and the negative sign for Inverter 2. When there was no load connected to the inverters, the circulating current is shown in Figure 20.10. It can be seen that the circulating current is again very small.

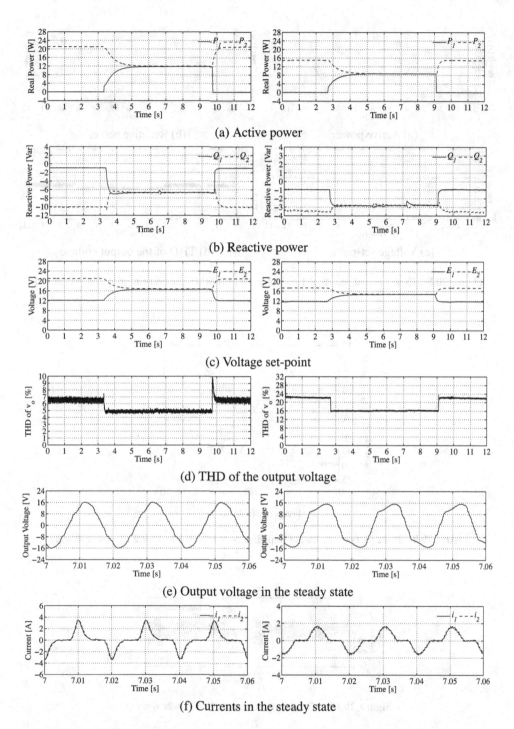

Figure 20.7 Experimental results with a non-linear load at a 1 : 1 power sharing: with K_R (left column) and without K_R (right column)

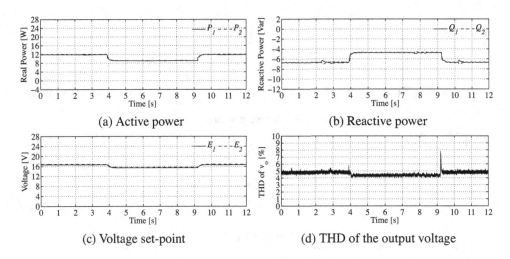

Figure 20.8 Experimental results at 1 : 1 power sharing when the non-linear load was changed

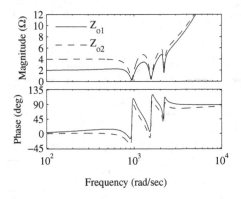

Figure 20.9 Bode plots of the output impedances of the inverters for 2 : 1 power sharing

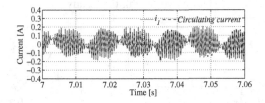

Figure 20.10 Circulating current at 2 : 1 power sharing

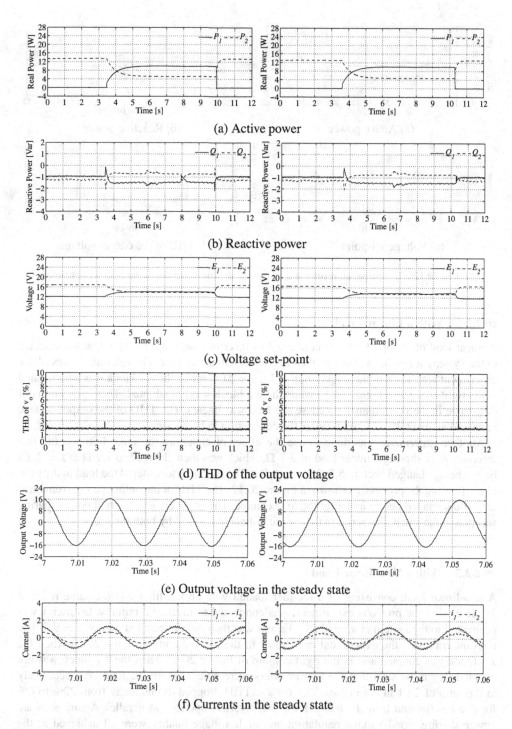

Figure 20.11 Experimental results with a linear load at 2 : 1 power sharing: with K_R (left column) and without K_R (right column)

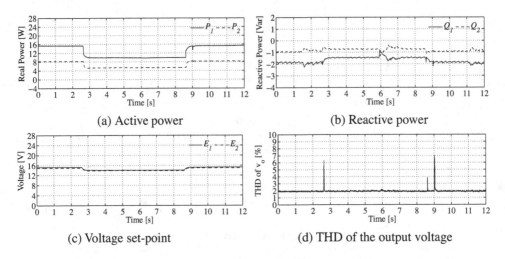

Figure 20.12 Experimental results at 2 : 1 power sharing when the linear load was changed

20.2.2.2 With a Linear Load

A linear load of about 7.9 Ω was connected to Inverter 2 initially. Inverter 1 was connected to the system at $t = 3.5$ s and was then disconnected at $t = 10$ s. The relevant curves of the experiment are shown in the left column of Figure 20.11 when the K_R block was adopted and in the right column of Figure 20.11 when the K_R block was not adopted. There was no major difference in the performance (because the load was linear). The THD of the output voltage remained at about 2% before and after the connection in both cases.

Another experiment was done to see the effect of changing the load. Initially, the two inverters were sharing a linear load of 5.7 Ω, which was then changed to 7.9 Ω at $t = 2.8$ s before being changed back to 5.7 Ω at $t = 8.8$ s. The two inverters shared the load in the ratio of 2 : 1 during the process, as shown in Figure 20.12. The THD of the output voltage remained almost unchanged before and after the load change. The small error in the real power sharing was due to the small difference in the output voltage measured by both inverters.

20.2.2.3 With a Non-linear Load

A non-linear load, consisting of a rectifier loaded with an LC filter and the same rheostat 7.9 Ω used in the previous experiment, was connected to Inverter 2 initially. Inverter 1 was connected to the system at around $t = 3$ s and was then disconnected at around $t = 9$ s. The relevant curves of the experiment are shown in the left column of Figure 20.13 when the K_R block was adopted and in the right column of Figure 20.13 when the K_R block was not adopted. It can be seen that the two inverters were able to share the load very accurately in the ratio of 2 : 1 in both cases. The voltage THD dropped dramatically from 22% to 6% for one inverter and from 16% to 5% for two inverters operated in parallel. Again, accurate power sharing, good voltage regulation and high voltage quality were all achieved at the same time.

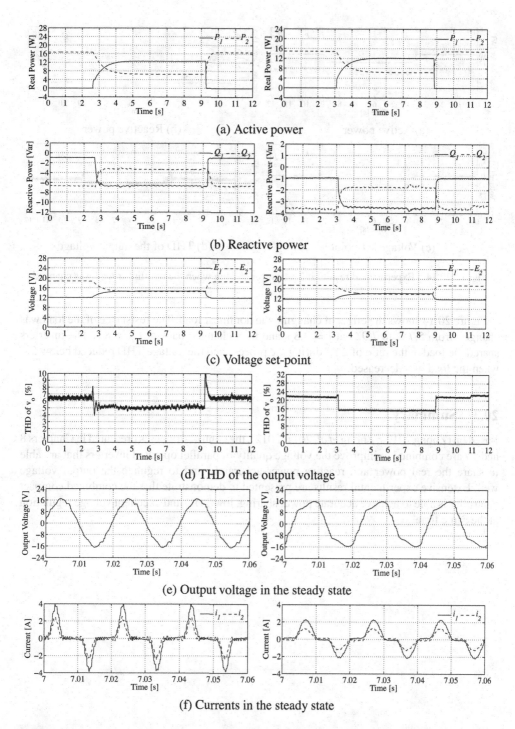

Figure 20.13 Experimental results with a non-linear load at 2 : 1 power sharing: with K_R (left column) and without K_R (right column)

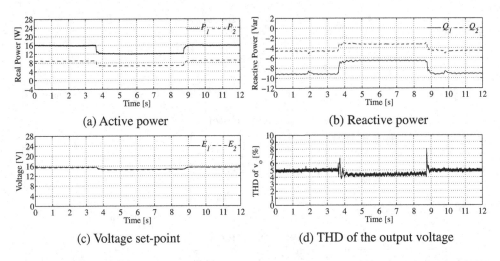

Figure 20.14 Experimental results at 2 : 1 power sharing when the non-linear load was changed

Figure 20.14 shows the effect of changing the nonlinear load. The resistance of the load was changed from 5.7 Ω to 7.9 Ω at $t = 3.8$ s and then changed back at $t = 8.8$ s. The inverters shared the load in the ratio of 2 : 1 during the process and the voltage THD reduced below 5% when the load was decreased.

20.3 Summary

Based on (Zhong 2012c; Zhong *et al.* 2011, 2012), the control strategies presented in Chapters 8 and 19 are combined to improve the voltage quality of parallel operated inverters that are able to share the real power and reactive power accurately and to regulate the output voltage well. Extensive experimental results are presented to demonstrate that the combined compact controller is able to achieve accurate power sharing, low THD and excellent voltage regulation at the same time.

21

Harmonic Droop Controller to Improve Voltage Quality

As discussed in Chapter 2, if the right amount of harmonic voltages is injected into the reference voltage of an inverter, then it is possible to cancel the harmonic voltages dropped on the output impedance due to the harmonic current components so that the THD of the output voltage is improved. In this chapter, this idea is developed further and a control strategy is presented. At first, the load and/or grid connected to an inverter is modelled as the combination of voltage sources and current sources at harmonic frequencies. As a result, the system can be analysed at each individual frequency, which avoids the difficulty in defining the reactive power for a system with different frequencies because it is now defined at each individual frequency. After developing a droop control strategy for systems delivering power to a constant current source, instead of a constant voltage source, a harmonic droop controller is developed so that the right amount of harmonic voltage is added to the inverter reference voltage. This forces the output voltage at the individual harmonic frequency close to zero and improves the THD of the output voltage considerably. Both simulation and experimental results are presented to demonstrate its capability of significantly improving the voltage THD.

21.1 Model of an Inverter System

It is widely known that a (linear) circuit having supplies/sinks with different frequencies can be analysed separately at each frequency according to the superposition theorem. Here, this is applied to inverter systems.

For an inverter, whether it is connected to a load, a grid or both, the mathematical model of the system can be illustrated as shown in Figure 21.1(a), where the inverter is modelled as a voltage reference v_r with an output impedance $Z_o(s)$ and the load is modelled as the combination of voltage sources and current sources. The terminal or output voltage is

$$v_o = v_{o1} + \Sigma_{h=2}^{\infty} v_{oh}$$

Control of Power Inverters in Renewable Energy and Smart Grid Integration, First Edition.
Qing-Chang Zhong and Tomas Hornik.
© 2013 John Wiley & Sons, Ltd. Published 2013 by John Wiley & Sons, Ltd.

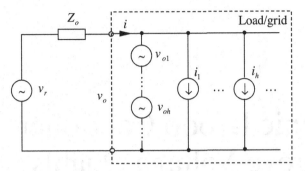

(a) One circuit including all harmonics

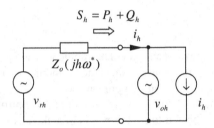

(b) The circuit at the h-th harmonic frequency

Figure 21.1 Model of an inverter connected to a load/grid in terms of harmonic voltage and current sources

with $v_{o1} = \sqrt{2}V_{o1}\sin(\omega^*t)$ and $v_{oh} = \sqrt{2}V_{oh}\sin(h\omega^*t + \psi_h)$, where ω^* is the rated fundamental angular frequency of the system, V_{o1} is the RMS value of the fundamental component and V_{oh} is the RMS value of the h-th harmonic component. The output or load current is described as

$$i = \Sigma_{h=1}^{\infty} i_h,$$

with $i_h = \sqrt{2}I_h \sin(h\omega^*t + \phi_h)$. This represents the effect of non-linear loads or harmonic currents and forces the current flowing through the series of voltage sources to be zero. The voltage reference v_r in the general case is described as

$$v_r = v_{r1} + \Sigma_{h=2}^{\infty} v_{rh}$$

with $v_{r1} = \sqrt{2}E \sin(\omega^*t + \delta)$ and $v_{rh} = \sqrt{2}E_h \sin(h\omega^*t + \delta_h)$. In many cases, in particular, when a droop controller is used in the inverter, E_h is often set to be zero. In this chapter, E_h is set to be non-zero to make v_{oh} close to zero.

This circuit can be analysed after decomposing it into multiple circuits at each harmonic frequency, according to the superposition theorem. The h-th harmonic circuit of the system is shown in Figure 21.1(b). The load of the circuit $\frac{V_{o1}}{I_1} \angle -\phi_1$ at the fundamental frequency[1] is expressed as the combination of the voltage source $V_{o1}\angle 0°$ and the current source $I_1\angle \phi_1$. If v_{oh} ($h \neq 1$) is close to zero, then what is left on the right-hand side of Figure 21.1(b) is a current source i_h.

The main function of an inverter (or a generator) is to supply the load with the real power and reactive power at the right voltage level and at the right frequency, which are regulated by industrial standards and/or law. To be more precise, this should be done *at the fundamental frequency*, without harmonics. When multiple inverters are connected in parallel, they should also share real power and reactive power in proportion to their capacity, again *at the fundamental frequency*. Then, what happens with harmonics? Ideally, the harmonics in the output voltage are expected to be 0, i.e., $v_{oh} = 0$ ($h = 2, 3, \cdots$), even when there are harmonics in the current i. This can be achieved when the voltage drop of the h-th harmonic current $\sqrt{2}I_h \sin(h\omega^* t + \phi_h)$ on the output impedance $Z_o(j\omega)$ is the same as the h-th harmonic component of the voltage reference $\sqrt{2}E_h \sin(h\omega^* t + \delta_h)$, i.e., when

$$E_h = I_h|Z_o(jh\omega^*)| \quad \text{and} \quad \delta_h = \phi_h + \angle Z_o(jh\omega^*). \tag{21.1}$$

In this chapter, this idea is exploited to design a controller for the inverter so that the harmonics in the output voltage are considerably reduced, after filling a gap in the theory of power delivery through an impedance.

21.2 Power Delivered to a Current Source

As discussed in Chapter 19, it is well understood how real power and reactive power are delivered through an impedance, whether it is inductive, resistive, capacitive or other types, when the terminal voltage is more or less maintained constant as a voltage source. Moreover, power sharing schemes, e.g., different droop control strategies (Guerrero et al. 2005, 2007, 2011; Mohamed and El-Saadany 2008a; Yao et al. 2011; Zhong 2012c; Zhong et al. 2011), have been developed for this case. Here, how real power and reactive power are delivered when the terminal is connected to a current source with a constant current is studied.

Figure 21.2 illustrates a voltage source v_r delivering power to a current source $I\angle 0°$ through an impedance $Z_o\angle \theta$. The terminal voltage is

$$\dot{V}_o = E\angle \delta - Z_o I\angle \theta$$
$$= E\cos\delta - Z_o I \cos\theta + j(E\sin\delta - Z_o I \sin\theta)$$

and the real power and reactive power delivered to the terminal are, respectively,

$$P = EI\cos\delta - Z_o I^2 \cos\theta, \tag{21.2}$$

$$Q = EI\sin\delta - Z_o I^2 \sin\theta. \tag{21.3}$$

---
[1] Note that the phasors in this chapter may be at different frequencies, which should be clear from the context.

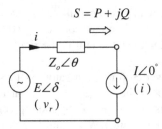

Figure 21.2 Power delivery to a current source through an impedance by a voltage source

Here, δ is the phase difference between the supply (voltage) and the terminal (current). When δ is small,

$$P \approx EI - Z_o I^2 \cos\theta,$$
$$Q \approx EI\delta - Z_o I^2 \sin\theta.$$

Hence, roughly,

$$P \sim E \quad \text{and} \quad Q \sim \delta$$

for any type of impedance $Z_o \angle \theta$. This is quite different from the case with a voltage source, where these relationships change with the type of the impedance. The conventional droop control strategy should then take the form

$$E_i = E^* - n_i P_i, \qquad (21.4)$$
$$\omega_i = \omega^* - m_i Q_i. \qquad (21.5)$$

This strategy is sketched in Figure 21.3. It is different from any of the droop control strategies when the power is delivered to a voltage source discussed in Chapter 19. Note that, in order to make sure that the $P - E$ loop and the $Q - \omega$ loop are of a negative feedback, respectively, so that the droop controller is able to regulate the frequency and the voltage, the signs before $n_i P_i$ and $m_i Q_i$ are all negative, which makes them droop terms. The main advantage of this

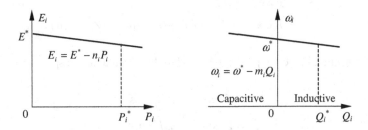

Figure 21.3 Droop control for inverters maintaining a constant output current (for any type of impedances)

scheme is that it does not depend on the type of the impedance and, hence, it can be used for any type of impedances. This facilitates the controller design, without the need for checking the impedance type at the corresponding harmonics, and will be applied to develop a strategy to improve the voltage THD in this chapter.

Note that $P = 0$ and $Q = 0$ when

$$E = Z_o I \quad \text{and} \quad \delta = \theta,$$

according to (21.2) and (21.3). This is another way to express (21.1) and can be used to reduce or even eliminate harmonics in the output voltage.

21.3 Reduction of Harmonics in the Output Voltage

As discussed above, in order to force the h-th harmonics in the output voltage of an inverter to be (nearly) zero, the voltage v_{oh} in Figure 21.1(b) needs to be zero. In other words, the real power and reactive power delivered to the current source i_h in Figure 21.1(b) should be 0. As a result, the voltage set-point E^* for the droop controller obtained in (21.4) should be 0 for the h-th harmonics ($h \neq 1$). The frequency set-point should simply be set as the h-th harmonic frequency. This leads to the following h-th harmonic droop controller:

$$E_h = -n_h P_h, \tag{21.6}$$

$$\omega_h = h\omega^* - m_h Q_h, \tag{21.7}$$

where P_h and Q_h are the real power and reactive power at the terminal for the h-th harmonic frequency, and n_h and m_h are the corresponding droop coefficients. Here, the subscripts of the relevant variables are changed to reflect the h-th harmonics. The reference voltage v_{rh} at the h-th harmonic frequency can then be formed with the RMS value E_h and the phase angle generated from the integration of ω_h. In practice, instead of generating a harmonic frequency ω_h from (21.7), it can be obtained from $h\omega t$ with the addition of δ_h, which is integrated from $-m_h Q_h$. Here, ωt is the phase of the voltage reference at the fundamental frequency. This leads to the h-th harmonic droop controller shown in Figure 21.4. As explained above, it does

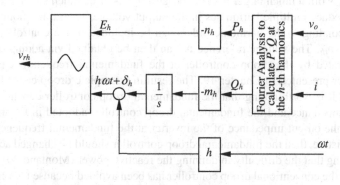

Figure 21.4 The h-th harmonic droop controller ($h \neq 1$)

not depend on the type of the output impedance at the h-th harmonic frequency, which could be resistive, inductive, capacitive or even complex.

Since the controller (21.6) in the voltage channel is a proportional controller, there is a static error and V_{oh} is not exactly zero (but close to zero). The V_{oh} can be calculated approximately via

$$V_{oh} \approx E_h - |Z_o(jh\omega^*)|I_h \approx -n_h V_{oh} I_h - |Z_o(jh\omega^*)|I_h.$$

That is,

$$V_{oh} \approx -\frac{|Z_o(jh\omega^*)| I_h}{n_h I_h + 1}.$$

Its contribution to the voltage THD is approximately $\frac{|Z_o(jh\omega^*)| I_h}{(n_h I_h + 1)E^*}$. The smaller the output impedance at the harmonic frequency $h\omega^*$, the smaller the THD. Hence, strategies like the one proposed in (Zhong et al. 2011) can be adopted to reduce $|Z_o(jh\omega^*)|$ and the voltage THD. The parameter n_h can be chosen to be large to make V_{oh} small as long as the system remains stable. As a rule of thumb, the tuning can be started with

$$n_h = \frac{|Z_o(jh\omega^*)|}{\gamma E^*},$$

with which the contribution of the h-harmonic component to the THD is about γ. If this causes instability, then it can be reduced. The parameter m_h can be determined in the same way as m_1 because $\frac{m_h Q_h^*}{h\omega^*}$ is the frequency drop ratio at the h-th harmonics, which should be the same as that at the fundamental frequency, i.e., $\frac{m_1 Q^*}{\omega^*}$. Hence,

$$m_h = m_1 \frac{h Q^*}{Q_h^*}.$$

As a result, m_h is often much larger than m_1 because Q_h^* is often much smaller than $h Q^*$.

In order to reduce multiple harmonics in the output voltage, several harmonic droop controllers corresponding to the harmonic orders can be included in the controller to generate the required $\Sigma_h v_{rh}$. The voltage reference v_r can then be obtained via adding $\Sigma_h v_{rh}$ to v_{r1}, which is generated by the droop controller at the fundamental frequency, e.g. the robust droop controller presented in Chapter 19. The resulting complete droop controller is shown in Figure 21.5. It is worth noting that the fundamental droop controller depends on the type of the output impedance and the fundamental droop controller adopted in Figure 21.5 is for R-inverters. If the output impedance of the inverter at the fundamental frequency is not predominantly resistive, then the fundamental droop controller should be changed accordingly. It is worth stressing that the difficulty in defining the reactive power (Montano 2011; Watanabe et al. 1993) for the conventional droop controller has been avoided because the reactive power in the this strategy is defined at the corresponding frequency.

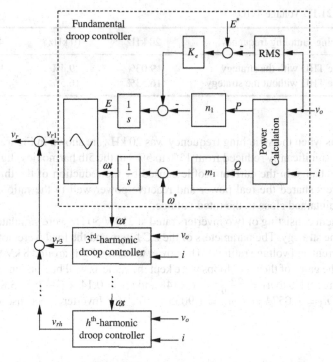

Figure 21.5 Droop controller consisting of a robust droop controller at the fundamental frequency for R-inverters and several harmonic droop controllers at individual harmonic frequencies

21.4 Simulation Results

Simulations were carried out with two inverters connected in parallel to verify the strategy. The values of the inductors and capacitors are 2.35 mH and 22 μF, respectively. The fundamental frequency of the system was 50 Hz and the rated output voltage was 12 VRMS. The inverter load was a rectifier bridge connected to a 9 Ω resistor after an LC filter with a 0.15 mH inductor and a 1000 μF capacitor. The inverters were designed, according to Chapter 7, to have resistive output impedances with a proportional current feedback of K_i. The harmonic droop controller shown in Figure 21.5 provides the voltage reference v_r to the inner-loop current controller. The droop controllers at the fundamental frequency were also designed according to (Zhong 2012c) with the droop coefficients $n_1 = 2.2$ and $m_1 = 0.14$ for Inverter 1 and $n_1 = 1.1$ and $m_1 = 0.07$ for Inverter 2. The current feedback gains were chosen as $K_{i1} = 4$ and $K_{i2} = 2$ and the parameter K_e was chosen as $K_e = 20$. The 3rd and 5th harmonic droop controllers were adopted with coefficients chosen as $n_h = 5$ and $m_h = 50$.

The switching/sampling frequency needs to be high enough in order to handle the 5th harmonics. Table 21.1 shows the voltage THD obtained with three switching frequencies. It shows that 4 kHz is not high enough, which resulted in a high THD. The results when the strategy was not used were also shown in Table 21.1 for comparison. The harmonic droop control strategy considerably improved the voltage THD, by 45%. The output voltage and the

Table 21.1 Voltage THD

Switching/sampling frequency	20 kHz	10 kHz	4 kHz
Voltage THD with the strategy	9.03%	9.4%	14.4%
Voltage THD without the strategy	16.53%	16.5%	20%

inverter currents when the switching frequency was 20 kHz are shown in Figure 21.6. The 3rd harmonics was significantly reduced from 15% to 5% but the 5th harmonics slightly increased because of the change in the current profile caused by the reduction of the third harmonics. The two inverters shared the real power and reactive power well in the ratio of 2:1 and the voltage was maintained close to the rated voltage.

Another system consisting of two inverters rated at 230 V 50 Hz were simulated to show the scalability of the strategy. The parameters of the LC filters and the load were not changed (but with higher current and voltage ratings). The power consumed was around 8 kVA with a power factor of 0.9. The gains of the inner loops were kept the same as well but the fundamental droop coefficients were scaled to $n_1 = \frac{2.2 \times 12}{230} = 0.1148$ and $m_1 = 0.14 \times (\frac{12}{230})^2 = 3.811 \times 10^{-4}$ for Inverter 1 and $n_1 = 0.0574$ and $m_1 = 1.9055 \times 10^{-4}$ for Inverter 2 because of the change

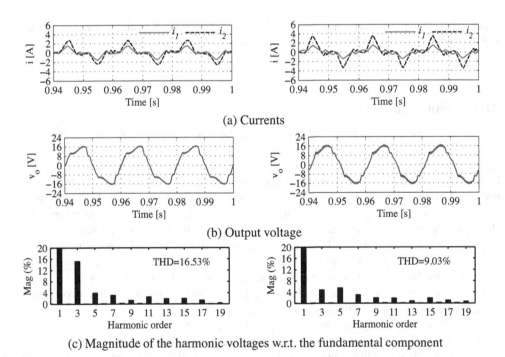

(a) Currents

(b) Output voltage

(c) Magnitude of the harmonic voltages w.r.t. the fundamental component

Figure 21.6 Simulation results for a 12 V system: without the harmonic droop controller (left column) and with a 3rd and 5th harmonics droop controller (right column)

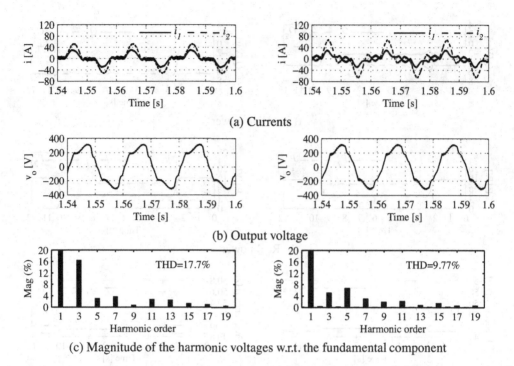

Figure 21.7 Simulation results for a 230 V system: without the harmonic droop controller (left column) and with a 3rd and 5th harmonics droop controller (right column)

of the rated voltage and power. The coefficients n_h were also scaled, correspondingly, to $n_h = \frac{5 \times 12}{230} = 0.2609$ and $m_h = 50 \times (\frac{12}{230})^2 = 0.1361$. The simulation results are shown in Figure 21.7. Again, the THD of the voltage was significantly improved by 45%, from 17.7% to 9.77%. The third harmonics was reduced from 16% to 5% but the 5th harmonics increased slightly.

21.5 Experimental Results

Experiments were carried out with a laboratory set-up, which consisted of two single-phase inverters controlled by dSPACE ACE1104 kits and was powered by separate 42 VDC power supplies. The parameters of the inverters are the same as those in the simulations for the 12 V system. The switching/sampling frequency was set at 4 kHz because of the hardware limitation, partially caused by the heavy computation involved in calculating the real power and reactive power at each frequency. Due to the configuration of the hardware set-up, the voltage of Inverter 2 was measured by the controller of Inverter 1 and then sent out via a DAC channel, which was then sampled by the controller of Inverter 2. This brought some latency into the system. The voltages were measured through a multiplexer, which resulted in a much lower sampling frequency for the voltage. This may have had an impact on the performance. Inverter 2 was equipped with a synchronisation unit. It was synchronised with Inverter 1 when its output was not connected to that of Inverter 1 so it was ready to be connected at any time.

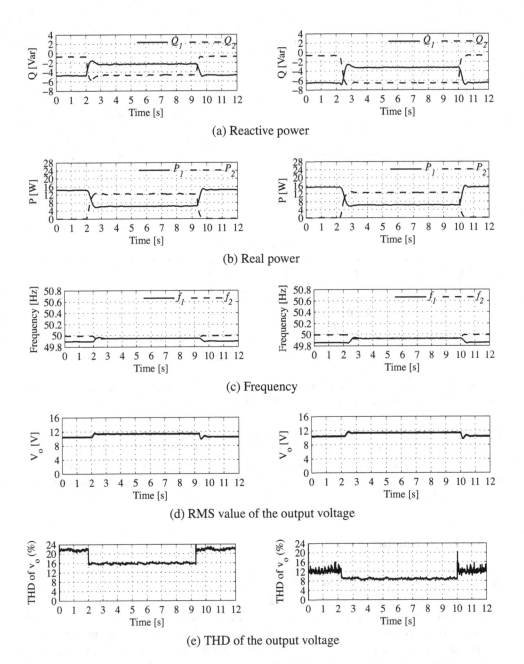

Figure 21.8 Experimental results (I): without the harmonic droop controller (left column) and with a 3rd and 5th harmonics droop controller (right column)

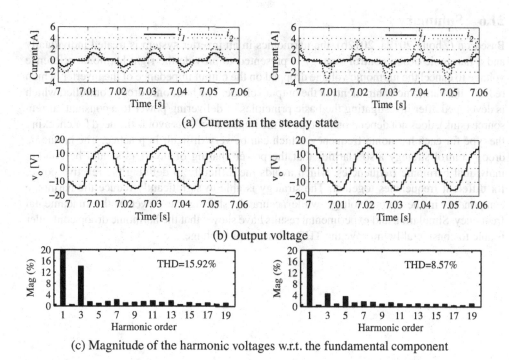

Figure 21.9 Experimental results (II): without the harmonic droop controller (left column) and with a 3rd and 5th harmonics droop controller (right column)

The experiments were carried out in three stages: (1) Inverter 1 was supplying the load; (2) Inverter 2 was put in parallel operation with Inverter 1; and (3) Inverter 2 was disconnected from the parallel operation. The experimental results are divided into two parts and shown in the left column of Figures 21.8 and 21.9 when the harmonic droop controller was not adopted and in the right column of Figures 21.8 and 21.9 when it was added for the 3rd and 5th harmonics. Since the dynamic process at the fundamental frequency is much lower than that at harmonic frequencies, the dynamics of the system is dominated by that of the fundamental droop controller and the harmonic droop controller does not bring noticeable change to the response speed for the real power, the reactive power, the frequency and the output voltage RMS.[2] There was no noticeable change in the steady-state performance of voltage regulation and the accuracy of power sharing, either. After the harmonic droop controller was introduced, there was significant improvement in the THD of the output voltage: from 22% to 12% for one inverter and from 15.92% to 8.57% for two inverters in parallel. This corresponds to the improvement of 46%. The 3rd harmonics was reduced from 14% to 4.5% although there was a slight increase in the 5th harmonics. The voltage THD is lower than that obtained from the simulation because of the filtering effect in the practical system.

---

[2] On the other hand, in order to handle high-order harmonics, the sampling speed of the controller needs to be increased.

21.6 Summary

Based on (Zhong 2012a, 2012b), the harmonics in an inverter system is treated individually and a harmonic droop control strategy is presented to provide the right harmonic reference voltage to cancel the harmonic voltage dropped on the output impedance of the inverter, which reduces the harmonic components in the output voltage. The harmonic droop controller, which is developed after investigating the basic principles of delivering power to a constant current source (sink), does not depend on the type of the impedance. This avoids the need for checking the type for each harmonic frequency, which can be very difficult in practice. The harmonic droop control strategy also clarifies that the power sharing of inverters should be done at individual harmonic frequencies, which avoids the difficulty in defining the reactive power for different frequencies together. The strategy is able to significantly reduce the harmonic components in the output voltage while accurately sharing the power at the fundamental frequency. Simulation and experimental results have shown that the harmonic droop controller is able to considerably improve the THD of the output voltage.

Part IV
Synchronisation

22

Conventional Synchronisation Techniques

One of the most important problems in renewable energy and smart grid integration is how to synchronise the inverters with the grid. There are two different scenarios: one is before connecting an inverter to the grid and the other is during the operation. If the inverter is not synchronised with the grid or another power source it is to be connected to, then large transient currents may appear at the time of connection, which may damage the equipment. During the normal operation, the inverter needs to be synchronised with the source it is connected to so that the system can work properly. In both cases, the grid information is needed accurately and in a timely manner. Depending on the control strategies adopted, the grid information needed can be any combination of the phase, the frequency and the voltage amplitude.

In this chapter, some synchronisation methods widely available in the literature are discussed. Simulation and experimental results are provided for two widely adopted methods, SOGI-PLL and STA.

22.1 Introduction

There are two categories of methods for the purpose of synchronisation: open-loop methods and close-loop methods. Typical open-loop methods include detecting the zero crossing of the grid voltage and directly filtering the grid voltage, etc. An example of the filtering method is the one used in Chapters 3, 15 and 16, where a phase-lead low-pass filter is adopted to provide the information of the grid voltage. Other filtering methods include the space-vector filter (SVF)-based method and the extended Kalman filter (EKF) method (Svensson 2001), etc. These methods are often found to be sluggish with high sensitivity to frequency deviations, voltage distortions and voltage imbalance (Timbus et al. 2005).

Closed-loop methods introduce a mechanism to make sure that the information obtained is accurate. Typical examples of closed-loop methods include the conventional PLL, which is widely used in single-phase applications, and the synchronously rotating reference frame PLL (SRF-PLL), which is widely used in three-phase applications. PLLs have been adopted as part

---

Control of Power Inverters in Renewable Energy and Smart Grid Integration, First Edition.
Qing-Chang Zhong and Tomas Hornik.
© 2013 John Wiley & Sons, Ltd. Published 2013 by John Wiley & Sons, Ltd.

of the controllers for most of the grid-connected applications nowadays (Barrena *et al.* 2008; Freijedo *et al.* 2009; Kesler and Ozdemir 2011; Santos Filho *et al.* 2008; Shen *et al.* 2009; Singh *et al.* 2009; Teodorescu and Blaabjerg 2004), e.g. in renewable energy applications (Shen *et al.* 2009; Teodorescu and Blaabjerg 2004), FACTS devices (Barrena *et al.* 2008; Singh *et al.* 2009), active power filters (Freijedo *et al.* 2009), UPS applications (Santos Filho *et al.* 2008) and power quality control (Kesler and Ozdemir 2011). The robustness and accuracy of the PLL are essential to the operation of these controllers (Barrena *et al.* 2008; Rolim *et al.* 2006). In recent years, the second-order generalised integrator (SOGI)-based PLL and the sinusoidal tracking algorithm (STA) (Karimi-Ghartemani and Ziarani 2003; Karimi-Ghartemani and Iravani 2001, 2002), also called the enhanced PLL (EPLL), have attracted a lot of attention.

One common issue with three-phase systems is the voltage imbalance. The SRF-PLL can achieve excellent performance for ideal balanced conditions but the performance degrades dramatically with unbalanced voltages due to the second-order harmonics appearing in the PLL output (Chung 2000; Rodriguez *et al.* 2007b). Reducing the PLL bandwidth could mitigate this problem but at the cost of lowering the dynamic performance (Rodriguez *et al.* 2007b). Another simple solution is to filter out the second-order harmonics by using a repetitive controller (Timbus *et al.* 2006b). There are also several more sophisticated techniques, e.g. the decoupled double synchronously rotating reference frame PLL (DDSRF-PLL) (Rodriguez *et al.* 2007a, 2007b, 2008), the delayed signal cancellation (Wang and Li 2011a,b), the fixed-reference-frame PLL (FRF-PLL) (Escobar *et al.* 2011), and the DSOGI-FLL (Rodriguez *et al.* 2006).

It is worth mentioning that there are also frequency-domain detection methods, e.g. Fourier transform methods (Lascu *et al.* 2009; McGrath *et al.* 2005) and space-vector discrete Fourier transform method (Neves *et al.* 2010, 2012). These methods require intensive data storage and computational resources (Wang and Li 2011b) and, thus, are not suitable for real-time control applications.

Some of the methods mentioned above are discussed in detail in this chapter. Simulation and experimental results are provided for SOGI-PLL and STA.

22.2 Zero-crossing Method

Zero-crossing is the simplest way to calculate the frequency and to provide the phase information of a sinusoidal signal (Timbus *et al.* 2005). For AC to DC converters with triacs or thyristors, zero-crossing is often used to calculate the firing angle for the distributed gating pulses (Valiviita 1999; Weidenbrug *et al.* 1993). For DC to AC converters, this can be used to detect the frequency and phase of the grid voltage so that the generated voltage can be synchronised with the grid voltage.

The method is depicted in Figure 22.1. A timer is restarted each time when the input signal crosses zero. The interval between the two crosses is multiplied by two (or added with the previous stored interval) to get the period T of the signal, from which the frequency of the signal can be easily derived as $f = \frac{1}{T}$. An integrator can also be reset when the signal crosses zero so that the phase of the signal can be obtained.

Although the method is very simple and therefore takes minimum computational resources, there are several well-known problems. One problem is the slow updating rate because the frequency information is only available every half cycle. Moreover, the frequency is assumed

Conventional Synchronisation Techniques

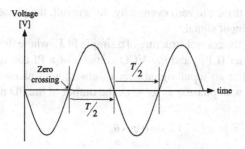

Figure 22.1 Zero-crossing method

to be constant in at least one half cycle, which is not always true. The method is thus very vulnerable to phase jumps, when large loads are switched on and off, and to variations in the grid frequency. Another problem is the multiple zero-crossing that could happen when the input signal is distorted by harmonic components (Wang and Li 2011a). The detection could also be affected by the bias in the analog to digital converter (ADC) or the comparator circuit and hence the system should be well calibrated.

22.3 Basic Phase-locked Loops (PLL)

The block diagram to demonstrate the operating concept of a PLL is shown in Figure 22.2(a). It consists of a phase (error) detection (PD) unit, a loop filter (LF) and a voltage controlled oscillator (VCO). The PD unit measures the phase difference between the input signal and the reproduced output signal and then passes it through the loop filter to extract the DC component obtained from the phase error. The DC component is amplified and then passed to the VCO, which could be a PI controller to generate the frequency of the output signal. The frequency is integrated to form the phase of the output signal. If the frequency of the output signal is locked with the input frequency, then the phase difference between the input and output signals, i.e.

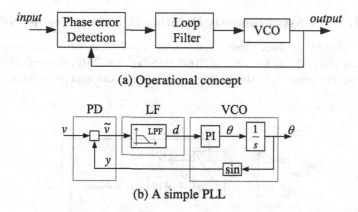

Figure 22.2 Block diagrams of a conventional PLL

the output of the PD, is driven to zero eventually. As a result, the phase of the output signal is locked with that of the input signal.

Figure 22.2(b) shows the control structure of a simple PLL, where the PD unit is a multiplier, the LF is a low-pass filter (LPF) and the VCO consists of a PI controller, an integrator and a sinusoidal function. For an input signal $v = V_m \cos\theta_g$ with phase $\theta_g = \omega_g t + \phi_g$ and an output signal $y = \sin\theta$ with phase $\theta = \omega t + \phi$, the output of the PD unit is

$$\tilde{v} = vy = V_m \sin\theta \cos\theta_g$$
$$= \frac{V_m}{2}\sin(\theta - \theta_g) + \frac{V_m}{2}\sin(\theta + \theta_g). \qquad (22.1)$$

The two components in (22.1) can be rewritten as $\frac{V_m}{2}\sin[(\omega - \omega_g)t + (\phi - \phi_g)]$ and $\frac{V_m}{2}\sin[(\omega + \omega_g)t + (\phi + \phi_g)]$. It is obvious that the first term is a low frequency component that contains the phase difference between v and y and the second term is a high frequency component, which is out of interest and can be filtered out with the loop filter. The output d of the LF is

$$d = \frac{V_m}{2}\sin[(\omega - \omega_g)t + (\phi - \phi_g)],$$

which is then fed into a PI controller to generate the estimated frequency $\omega = \dot\theta$ until $d = 0$. The estimated frequency is integrated to form the phase of the output signal $y = \sin\theta$, which is sent back to the PD unit to complete the loop. In the steady state, d is driven to zero and $\theta = \theta_g$, i.e. $\omega = \omega_g$ and $\phi = \phi_g$. The phase of the output signal y is said to be locked with that of the input signal v.

It is worth noting that the input signal and the output signal are actually 90° shifted in order for the DC component d to be zero when $\theta = \theta_g$. Indeed, in the case shown here, v is a cosine curve and y is a sine curve. This is not a problem because a constant can be added to θ to obtain any phase angle needed.

22.4 PLL in the Synchronously Rotating Reference Frame (SRF-PLL)

A common technique in three-phase applications is a PLL in the synchronously rotating reference frame (SRF-PLL) (Amuda et al. 2000; da Silva et al. 2010; Kaura and Blasko 1997), which is shown in Figure 22.3. Similar operating concepts can also be found, e.g. in (Chung 2000).

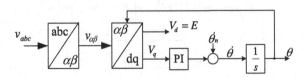

Figure 22.3 Three-phase PLL in the synchronously rotating reference frame (SRF-PLL)

Conventional Synchronisation Techniques

As discussed in Chapter 2, a three-phase voltage vector $v_{abc} = [\,v_a\ v_b\ v_c\,]^T$ in the natural frame can be transformed into a vector $[\,V_d\ V_q\,]^T$ in the synchronously rotating reference frame by first using the Clarke transformation

$$v_{\alpha\beta} = \begin{bmatrix} v_\alpha \\ v_\beta \end{bmatrix} = \frac{2}{3}\begin{bmatrix} 1 & -\dfrac{1}{2} & -\dfrac{1}{2} \\ 0 & -\dfrac{\sqrt{3}}{2} & \dfrac{\sqrt{3}}{2} \end{bmatrix} v_{abc} = T_{\alpha\beta} \times v_{abc}, \tag{22.2}$$

and then the Park transformation

$$\begin{bmatrix} V_d \\ V_q \end{bmatrix} = \begin{bmatrix} \cos\theta & -\sin\theta \\ \sin\theta & \cos\theta \end{bmatrix} v_{\alpha\beta} = T_{dq}\, v_{\alpha\beta}. \tag{22.3}$$

For a voltage vector

$$\begin{bmatrix} v_a \\ v_b \\ v_c \end{bmatrix} = \begin{bmatrix} E\cos(\theta_g) \\ E\cos\left(\theta_g - \dfrac{2\pi}{3}\right) \\ E\cos\left(\theta_g + \dfrac{2\pi}{3}\right) \end{bmatrix},$$

there is

$$\begin{bmatrix} V_d \\ V_q \end{bmatrix} = \begin{bmatrix} E\cos(\theta - \theta_g) \\ E\sin(\theta - \theta_g) \end{bmatrix}.$$

Hence, $[\,V_d\ V_q\,]^T$ is a vector with two DC components V_d and V_q in the SRF. In order to lock into the phase of the input signal, i.e. to achieve $\theta = \theta_g$, the V_q component can be fed into a PI controller to achieve $V_q = 0$ in the steady state. The output of the PI controller is actually the estimated frequency, which is integrated to obtain the estimated phase angle θ, as shown in Figure 22.3. The estimated amplitude E of the voltage vector can also be obtained, via

$$E = \sqrt{V_d^2 + V_q^2}. \tag{22.4}$$

When the phase is locked, $E = V_d$. Hence, the frequency, the amplitude and the phase are all available from the SRF-PLL.

Comparing Figure 22.3 to the basic PLL shown in Figure 22.2(a), the transformation from abc to dq with the estimated θ plays the role of the phase error detection. Because V_q is already a DC component, the loop filter is simply a unity gain. The PI controller and the integrator actually play the role of the VCO to generate the frequency and the phase.

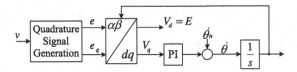

Figure 22.4 Typical structure of a single-phase PLL

The SRF-PLL is popular and widely applied in power control applications. It is also available in MATLAB®/Simulink® SimPowerSystems™ blocksets. In the ideal balanced condition, the SRF-PLL is able to eliminate the steady state error in tracking the phase and frequency and to achieve high bandwidth, which delivers fast and accurate tracking performance. However, it is very sensitive to harmonics or imbalance in the voltage and improved control schemes are necessary in these applications (Escobar *et al.* 2011; Rodriguez *et al.* 2007a, 2007b).

The SRF-PLL concept can also be applied to single-phase applications although there are no three balanced voltages available and an instantaneous single-phase voltage v cannot be transformed into a space vector as is done in three-phase systems (Shinnaka 2008). The idea (Ciobotaru *et al.* 2006; Shinnaka 2008; Yuan *et al.* 2002) is to generate two perpendicular components e and e_q and then treat these two components as v_α and v_β to form $v_{\alpha\beta}$ to be used in (22.3). The component e is the estimated version of v and the component e_q is a quadrature component, that is 90° shifted from e. The resulting single-phase PLL is sketched in Figure 22.4. The second-order generalised integrator (SOGI) to be described in detail in the next section is one way to generate the two perpendicular components e and e_q. It is worth noting that there are other single-phase synchronisation methods that are not based on the generation of a quadrature signal too; see e.g. (da Silva *et al.* 2010; Freijedo *et al.* 2011; Mojiri *et al.* 2007).

22.5 Second-order Generalised Integrator-based PLL (SOGI-PLL)

As mentioned above, if two perpendicular components e and e_q are generated from a single-phase voltage v, then these two components can be treated as v_α and v_β to further obtain two DC components V_d and V_q. The simplest way is to use a transport delay block or a lag compensator to shift the input signal v by $-90°$. However, this method suffers from high sensitivity to frequency variations and harmonics, which makes the DC components polluted (Ciobotaru *et al.* 2006).

The SOGI is able to generate both in-phase e and quadrature e_q components that are filtered and contain only the fundamental component. This is referred to as the SOGI-based quadrature-signal generator (SOGI-QSG) (Yuan *et al.* 2002). It consists of feedback loops involving two integrators, as shown in Figure 22.5(a). The transfer function $G_d(s)$ from v to e is

$$G_d(s) = \frac{k\omega s}{s^2 + k\omega s + \omega^2}, \tag{22.5}$$

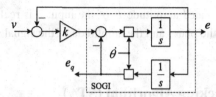

(a) SOGI-based quadrature-signal generator (SOGI-QSG)

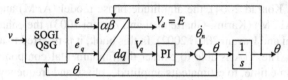

(b) Single-phase PLL equipped with the SOGI-QSG

Figure 22.5 SOGI-based single-phase PLL

where $\omega = \dot{\theta}$ is the resonant frequency of the SOGI-QSG, and the transfer function $G_q(s)$ from v to e_q is

$$G_q(s) = \frac{k\omega^2}{s^2 + k\omega s + \omega^2}. \tag{22.6}$$

Both $G_d(s)$ and $G_q(s)$ are resonant filters for $0 \leq k < 2$ and are able to select the component of v at the resonant frequency ω. For $s = j\omega$, $G_d = 1$ and $G_q = -j$, which means $e = v$ and e_q has the same amplitude as v but the phase is 90° delayed. When the frequency moves away from ω, $|G_d|$ and $|G_q|$ decrease, depending on the gain k. As a result, only the fundamental frequency component can pass the SOGI-QSG. A smaller gain k leads to better selectivity and offers better attenuation to other frequency components but it takes longer to settle down.

Another important characteristics of SOGI-QSG is that e_q is always 90° delayed, at any frequency, from e because

$$G_d(s) = \frac{s}{\omega} G_q(s),$$

which means the SOGI-QSG always generates two perpendicular components e and e_q, at the frequency of ω.

The SOGI-QSG can be plugged into the typical single-phase PLL shown in Figure 22.4 to form a SOGI-PLL as shown in Figure 22.5(b). The components e and e_q are treated as v_α and v_β, which are further transformed into DC components V_d and V_q using the Park transformation (22.3). A PI controller is used to drive V_q to zero. The output of the PI controller is added with the nominal frequency $\dot{\theta}_n$ to form the estimated frequency $\dot{\theta}$, which is then integrated to obtain the estimated phase angle θ. Note that the estimated frequency $\dot{\theta}$ is fed back to the SOGI-QSG so that it is able to select the component at the right frequency. When the phase is locked,

$V_q = 0$. Moreover, the estimated amplitude E of the input v is realised as $E = \sqrt{V_d^2 + V_q^2}$. As a result, the frequency, the phase and the amplitude of the signal v are all available.

22.6 Sinusoidal Tracking Algorithm (STA)

This method was introduced with several different names, e.g. the sinusoidal tracking algorithm (STA) (Ziarani and Konrad 2004), the amplitude phase model (APM) and amplitude phase frequency model (APFM) (Karimi-Ghartemani and Ziarani 2003), the enhanced PLL (EPLL) (Karimi-Ghartemani and Iravani 2001, 2002). In this book, it is referred to as the STA in order to avoid confusion. The STA is able to extract the fundamental component of a sinusoidal signal and, at the same time, to estimate its amplitude, phase and frequency.

A typical voltage $v(t)$ has the general form of

$$v(t) = \sum_{i=0}^{\infty} V_{mi} \sin \theta_{gi} + n(t)$$

where V_{mi} and $\theta_{gi} = \omega_{gi} t + \delta$ are the amplitude and phase of the i-th harmonic component of the voltage, and $n(t)$ represents the noise on the signal. The objective of a PLL can be regarded as extracting the component $e(t)$ of interest, which is usually the fundamental component, from the input signal $v(t)$. Assume that the estimated signal is

$$e(t) = E(t) \sin(\int_0^t \omega(\tau) d\tau + \delta(t)),$$

where $E(t)$ is the estimated amplitude, $\omega(t) = \dot{\theta}(t)$ is the estimated frequency and $\theta(t) = \int_0^t \omega(\tau) d\tau + \delta(t)$ is the estimated phase of $e(t)$. Define the state vector of the PLL to be $\psi(t) = [E(t) \; \omega(t) \; \delta(t)]^T$. Then the problem of designing a PLL can be formulated as finding the optimal vector $\psi(t)$ that minimises the cost function

$$J(\psi(t), t) = d^2(t) = [v(t) - e(t)]^2,$$

where $d(t) = v(t) - e(t)$ is the tracking error. There are several methods to solve this problem. The method applied in (Karimi-Ghartemani and Ziarani 2003; Ziarani and Konrad 2004) is the gradient descent method (Giordano and Hsu 1985), which is outlined below. In order to solve the optimisation problem, formulate

$$\frac{d\psi(t)}{dt} = -\mu \frac{\partial [J(\psi(t), t)]}{\partial \psi(t)}$$

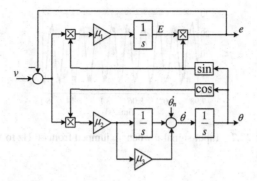

Figure 22.6 Sinusoidal tracking algorithm (STA) or the enhanced PLL (EPLL)

where μ is a diagonal matrix chosen to minimise J along the direction of $-\frac{\partial [J(\psi(t),t)]}{\partial \psi(t)}$. The resulting set of differential equations are then (Ziarani and Konrad 2004; Karimi-Ghartemani and Ziarani 2003)

$$\begin{cases} \dfrac{dE(t)}{dt} = 2\mu_1 d \sin\theta, \\ \dfrac{d\omega(t)}{dt} = 2\mu_2 E d \cos\theta, \\ \dfrac{d\theta(t)}{dt} = \omega + \mu_3 \dfrac{d\omega}{dt}. \end{cases} \quad (22.7)$$

This is the basis for implementing the STA as shown in Figure 22.6. Since the variation of E is relatively small compared to the variation of d, the major dynamics of ω is from d and the effect of E can then be combined with the proper selection of μ_2. In order to speed up the response, the nominal frequency $\dot{\theta}_n$ is added to the estimated frequency $\dot{\theta}$.

The behaviour of the frequency loop is determined by the two gains μ_2 and μ_3, which forms a PI controller $\mu_2(\mu_3 + \frac{1}{s})$, and μ_2 directly affects the bandwidth of the loop. The larger the μ_2, the higher the bandwidth and therefore the faster the response. However, if a large phase jump occurs, a large gain μ_2 can cause oscillations and it takes long time for the loop to reach the steady state. Too large a μ_2 could even cause instability if the phase jump is too large.

22.7 Simulation Results with SOGI-PLL and STA

22.7.1 With a Noisy Distorted Signal having a Variable Frequency

The input signal v has the amplitude $V_m = 20\sqrt{2}$ and contains a noise-added harmonic component

$$v_h(t) = 2\sqrt{2}\sin(3\omega_g t + 1.5) + 2\sqrt{2}\sin(5\omega_g t + 2.5) + n(t), \quad (22.8)$$

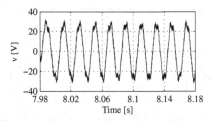

Figure 22.7 Input signal v when f_g jumped from 40 Hz to 50 Hz

where $n(t)$ is a uniform random noise with the amplitude of $2\sqrt{2}$ V. The frequency $f_g = \frac{\omega_g}{2\pi}$ of the signal, which varied from 40 Hz to 60 Hz periodically with the cycle of 8 s, is expressed for the first period as

$$f_g(t) = 50 + \sin(2\pi t) + \begin{cases} 0, & (0 < t \le 2) \\ 10(t-2), & (2 < t \le 3) \\ 10, & (3 < t \le 3.5) \\ 10 - 18(t - 3.5), & (3.5 < t \le 4) \\ 1 - (t - 4), & (4 < t \le 5) \\ -5(t - 5), & (5 < t \le 7) \\ -10, & (7 < t \le 8). \end{cases} \quad (22.9)$$

The signal when the frequency f_g jumped from 40 Hz to 50 Hz is shown in Figure 22.7.

The SOGI-PLL shown in Figure 22.5(b) and the STA shown in Figure 22.6 were simulated with the parameters given in Table 22.1, which were optimised to compromise the dynamic performance and stability. The relevant signals from the simulations are shown in the left column of Figure 22.8 for the SOGI-PLL and in the right column of Figure 22.8 for the STA. It is worth mentioning that, although the amplitude of the fundamental component does not change, the amplitude of v does change because of $v_h(t)$. However, the estimated amplitude E should track that of the fundamental component, which is V_m. Although the phase of the reference signal was tracked well with both methods, the frequency variations were noticeable. The result is compatible with the simulation results in (Ciobotaru *et al.* 2006; Ziarani and Konrad 2004). The SOGI-PLL offered better performance than the STA in tracking the frequency.

Table 22.1 Parameters of SOGI-PLL and STA for simulations and experiments

For the SOGI-PLL	Values	For the STA	Values
k	1	μ_1	200
K_p	2.5	μ_2	500
K_i	50	μ_3	0.01

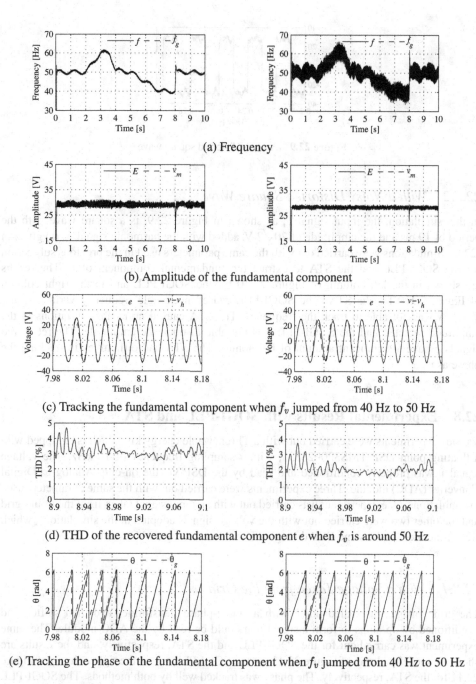

Figure 22.8 Tracking a distorted noisy signal with a variable frequency: Simulation results with the SOGI-PLL (left column) and with the STA (right column)

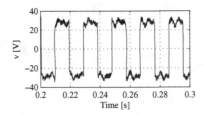

Figure 22.9 A noisy distorted square wave

22.7.2 With a Noisy Distorted Square Wave

In this simulation, the input signal v, as shown in Figure 22.9, is a square wave with the period of 19 ms and the amplitude of $20\sqrt{2}$ V, added with the harmonic signal $v_h(t)$ given in (22.8). Simulations were carried out with the same parameters used in the previous subsection for the SOGI-PLL and the STA to extract the fundamental component of v. The results are shown in the left column of Figure 22.10 for the SOGI-PLL and in the right column of Figure 22.10 for the STA. The SOGI-PLL caused noticeable frequency variations and the STA was not able to track the frequency. The STA caused noticeable variations in the amplitude detected and the SOGI-PLL was not able to track the amplitude. The recovered signals e have a significant amount of harmonics. The STA had difficulties in tracking the phase as well.

22.8 Experimental Results with SOGI-PLL and STA

Various experiments were carried out with a TI renewable energy kit, which is equipped with a floating point DSP TMS320F28335, with a sampling frequency of 10 kHz. The voltage signal v was properly conditioned and read by the DSP via the on-chip Analog to Digital Converter (ADC) module. Three experiments were carried out with the same parameters used in simulations. One experiment was carried out with an input signal taken from the utility grid and the other two were carried out with the voltage signals adopted in the simulations, which were generated by the same DSP through a Digital to Analog (DAC) module.

22.8.1 With a Voltage Taken from the Grid

The grid voltage was scaled down with a single-phase transformer and then shifted and conditioned by op-amps to form a signal that could be read by the ADC module. The same experiment was carried out for the SOGI-PLL and the STA, respectively, and the results are shown in the left column of Figure 22.11 for the SOGI-PLL and in the right column of Figure 22.11 for the STA, respectively. The phase was tracked well by both methods. The SOGI-PLL took about one cycle to produce the correct amplitude and a half cycle to produce the correct frequency while the STA took about two cycles to produce the correct amplitude and one cycle to produce the correct frequency, all with noticeable overshoots.

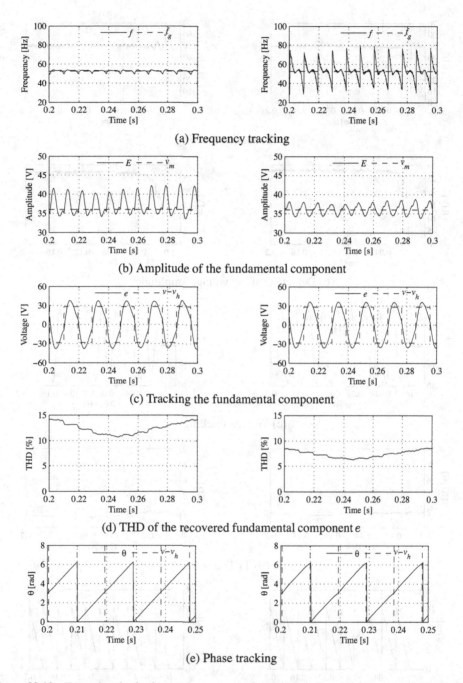

Figure 22.10 Extracting the fundamental component from a noisy distorted square wave: Simulation results with the SOGI-PLL (left column) and with the STA (right column)

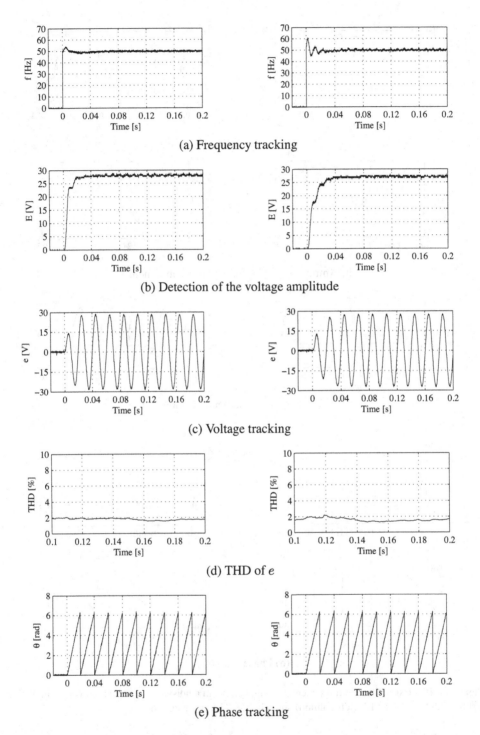

Figure 22.11 Tracking the grid voltage: Experimental results with the SOGI-PLL (left column) and with the STA (right column)

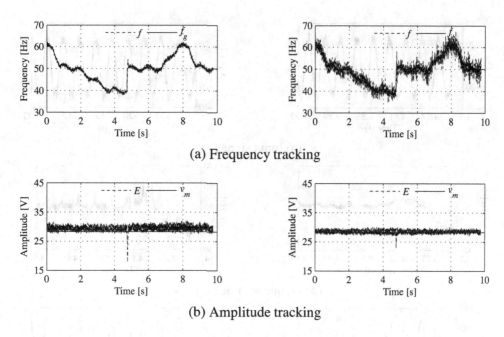

Figure 22.12 Frequency and amplitude tracking for a distorted noisy signal with a variable frequency: Experimental results with the SOGI-PLL (left column) and with the STA (right column)

22.8.2 With a Noisy Distorted Signal having a Variable Frequency

The voltage signal used in Section 22.7.1 was generated and sent out via a DAC channel as the voltage signal v. The experiment was carried out for the SOGI-PLL and the STA and the results are shown in Figure 22.12. The experimental results matched very well with the simulation results shown in Figure 22.8. The SOGI-PLL was able to track the frequency and the amplitude but with much bigger ripples in the amplitude. The frequency produced by the STA varied in a wide range. The experimental results when the frequency changed from 40 to 50 Hz are shown in Figure 22.13. The SOGI-PLL tracked the frequency well but there were noticeable variations in the frequency after the change. The STA was not able to deal with the change in the frequency.

22.8.3 With a Noisy Distorted Square Wave

The voltage signal used in Section 22.7.2 was generated and sent out via a DAC channel as the voltage signal v. The experiment was carried out for the SOGI-PLL and the STA and the results are shown in Figure 22.14. The experimental results matched the simulation results. The STA was not able to track the amplitude and the frequency although the phase was tracked well. The voltages recovered contain significant harmonics.

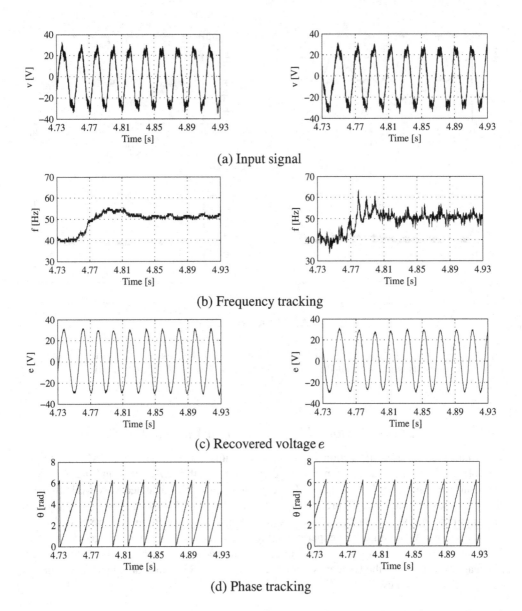

Figure 22.13 Tracking a distorted noisy signal with a variable frequency when the frequency jumped from 40 Hz to 50 Hz: Experimental results with the SOGI-PLL (left column) and with the STA (right column)

Conventional Synchronisation Techniques

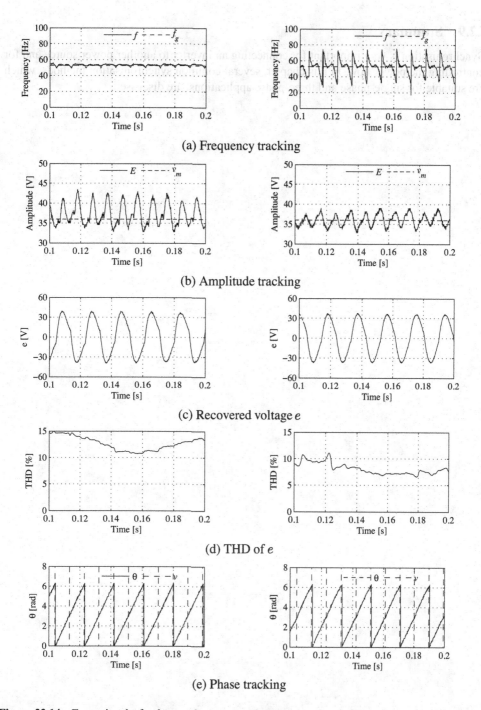

Figure 22.14 Extracting the fundamental component from a noisy distorted square wave: Experimental results with the SOGI-PLL (left column) and with the STA (right column)

22.9 Summary

Synchronisation is very important for connecting an inverter to another power source and for controlling the power flow. In this chapter, several common synchronisation methods, which are suitable for single-phase and three-phase applications, are discussed.

23

Sinusoid-locked Loops

Several conventional synchronisation methods are discussed in Chapter 22. Simulation and experimental results have shown that these methods have difficulties in providing the information on the voltage when there are frequency variations and/or harmonics, etc. Moreover, the response is not fast enough. In this chapter, a sinusoid-locked loop (SLL) that is able to quickly track the amplitude, frequency and phase of the fundamental component of a signal is presented. This is based on the idea of mimicking a grid-connected synchronous machine that does not exchange power with the grid because such a machine generates the same instantaneous voltage as the grid voltage. That is, the voltage amplitude, the phase and the frequency of the generated voltage are the same as those of the grid voltage. Following the idea of synchronverters (Zhong and Weiss 2011) discussed in Chapter 18, the mathematical model of a synchronous machine can be adopted as the core of an SLL to lock with the fundamental component of the voltage. The control objective of the SLL is then to zero the real power and reactive power exchanged with the input voltage. Because the dynamics of a tiny machine can be very fast, the SLL is able to provide very fast response. Both simulation and experimental results are presented.

23.1 Single-phase Synchronous Machine (SSM) Connected to the Grid

The simplified model of an SSM connected to the grid is depicted in Figure 23.1, where the grid voltage is $v = v_m \sin\theta_g$ and the SSM is modelled as a voltage source $e = E\sin\theta$, which represents the generated voltage, in series with the synchronous reactance X_s.

The real power P and reactive power Q flowing out of the SSM are (Singh et al. 2009; Wildi 2005)

$$P = \frac{v_m E}{2X_s} \sin(\theta - \theta_g), \qquad (23.1)$$

and

$$Q = \frac{v_m}{2X_s}\left[E\cos(\theta - \theta_g) - v_m\right]. \qquad (23.2)$$

Control of Power Inverters in Renewable Energy and Smart Grid Integration, First Edition.
Qing-Chang Zhong and Tomas Hornik.
© 2013 John Wiley & Sons, Ltd. Published 2013 by John Wiley & Sons, Ltd.

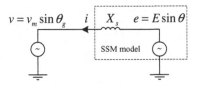

Figure 23.1 Model of an SSM connected to the grid

The factor 2 in the denominator is because E and v_m are peak amplitude values instead of RMS values. The SSM is considered to be synchronised and floating on the grid (Wildi 2005) if and only if

$$\begin{cases} E = v_m, \\ \theta = \theta_g. \end{cases} \tag{23.3}$$

In this case, $P = 0$ and $Q = 0$. In other words, if P and Q are driven to zero, then the condition (23.3) is satisfied and the generated voltage e is the same as the input (terminal) voltage v.

23.2 Structure of a Sinusoid-locked Loop (SLL)

As mentioned above, the idea of the SLL is to operate a virtual (single-phase) synchronous generator with $P = 0$ and $Q = 0$ so that the generated voltage e is the same as the fundamental component of the terminal (or input) voltage v. Hence, the controller of the synchronverter shown in Figure 18.4 can be adopted to implement it after making some necessary changes. The resulting SLL is shown in Figure 23.2(a), which is able to provide the frequency $\dot{\theta}$, phase θ, voltage amplitude E and a recovered voltage e for the voltage v.

As discussed above, the desired real power and reactive power should all be set as 0. For single-phase applications, the instantaneous values of T_e and Q given in (18.7) and (18.9) are pulsating. Hence, their average values

$$T_e = \frac{1}{T} \int_{t-T}^{t} M_f i_f i \sin\theta \, dt, \tag{23.4}$$

$$Q = -\frac{1}{T} \int_{t-T}^{t} \dot{\theta} M_f i_f i \cos\theta \, dt, \tag{23.5}$$

where $T = \frac{2\pi}{\dot{\theta}}$ is the period of voltage v, should be used. The amplitude of the generated voltage e is

$$E = \dot{\theta} M_f i_f$$

and the instantaneous value of the generated voltage e is

$$e = \dot{\theta} M_f i_f \sin\theta = E \sin\theta, \tag{23.6}$$

which should track that of the input voltage v.

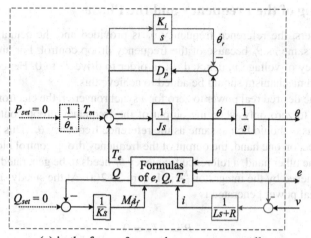

(a) in the form of a synchronverter controller

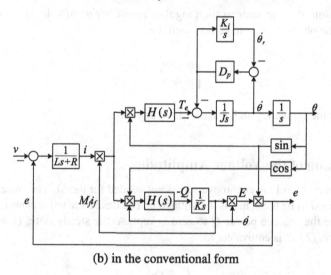

(b) in the conventional form

Figure 23.2 Structure of the sinusoid-locked loop (SLL)

The stator current i, which is the inductor current in the case of a synchronverter, should be generated internally as the voltage difference between e and v divided by the virtual synchronous reactance $X_s(s) = sL + R$, i.e.,

$$i = \frac{e - v}{sL + R}.$$

This forms the (virtual) current feedback loop, which is crucial for the synchronisation process.

23.3 Tracking of the Frequency and the Phase

For synchronverters, the reference frequency $\dot{\theta}_r$ is provided and the actual frequency $\dot{\theta}$ is normally not the same as $\dot{\theta}_r$ because of the frequency droop control. For an SLL, $\dot{\theta}$ should track the frequency of voltage v, denoted $\dot{\theta}_g$, in order to drive T_e to 0. Hence, it is expected that $\dot{\theta}_r = \dot{\theta}_g$ and a mechanism should be added to achieve this.

After setting the desired real power to zero for a synchronverter, the electromagnetic torque T_e can be driven to zero only when the output of the frequency droop control block D_p is zero, which means $\dot{\theta}$ should be the same as the reference frequency $\dot{\theta}_r$. This is actually what is expected. Hence, on one hand, the output of the frequency droop control block D_p needs to be zero and, on the other hand, a reference frequency $\dot{\theta}_r$ needs to be generated so that $\dot{\theta}_r = \dot{\theta}_g$. These can be achieved by the integrator $\frac{K_i}{s}$ in Figure 23.2(a). At the steady state, T_e is driven to zero and the real power generated is

$$P = \dot{\theta} T_e = 0 \tag{23.7}$$

as well. As a result, $\dot{\theta}$ is the same as the (angular) frequency $\dot{\theta}_g$ of voltage v and the phase θ is the same as the phase θ_g of voltage v as well, i.e.,

$$\begin{cases} \dot{\theta} = \dot{\theta}_g, \\ \theta = \theta_g. \end{cases} \tag{23.8}$$

The SLL tracks the frequency and phase without any error.

23.4 Tracking of the Voltage Amplitude

The voltage droop control in synchronverters is not needed for the SLL because the generated voltage is expected to be the same as the voltage v. The desired reactive power is set to 0 and the loop to drive the reactive power Q to zero is kept. At the steady state, $Q = 0$ in addition to $P = 0$. Hence, (23.2) is equivalent to

$$E = v_m. \tag{23.9}$$

Together with (23.8), the condition (23.3) is satisfied and the SLL is synchronised with the voltage v.

It is worth noting that one advantage of the SLL is that the frequency, phase, voltage amplitude and the recovered signal are all directly available internally without any extra calculation.

23.5 Tuning of the Parameters

The SLL mainly contains an amplitude loop to regulate the reactive power (and the voltage), a frequency loop to regulate the real power (and the frequency $\dot{\theta}$) and a loop to generate the reference frequency $\dot{\theta}_r$ for the frequency loop.

Sinusoid-locked Loops

The time constant of the frequency loop is

$$\tau_f = \frac{J}{D_p}. \tag{23.10}$$

The choice of τ_f determines the dynamic response of the loop. It is proportional to the moment of inertia J. A large τ_f is equivalent to having a large J, which makes the SLL less sensitive to variations in the grid frequency and also makes the system more stable. However, the response is slow. A small τ_f is equivalent to having a small J, which leads to fast frequency tracking. As a general rule of thumb, τ_f can be chosen to be much smaller than the period of the voltage v so that the frequency can be tracked very quickly.

The time constant of the amplitude loop is proportional to

$$\tau_q = \frac{K}{\dot{\theta}_n} \tag{23.11}$$

and the amplitude loop generates $M_f i_f$, which directly affects the amplitude of e in (23.6). Hence the choice of τ_q affects the dynamic response of the amplitude tracking. Generally, the frequency loop should be tuned much faster than the amplitude loop, which is normally the case because τ_f is often chosen to be much smaller than the period of the voltage v. This allows the voltage to be established. Otherwise $M_f i_f$ and eventually the voltage amplitude E would be driven to zero, which is also an equilibrium point of the system, before the frequency and phase could be synchronised. However, if a very large τ_q is chosen, it would take a long time for the voltage amplitude E to track v_m.

The inductance L and resistance R of the virtual synchronous reactance X_s can be chosen to be small to enable a large transient current i, which helps speed up the tracking process. However, too small L and R may cause oscillations in the frequency estimated. Moreover, the ratio $\frac{R}{L}$ is the cut-off frequency of the filter $\frac{1}{sL+R}$, which determines the capability of filtering out the harmonics from the voltage v.

The loop to generate the reference frequency $\dot{\theta}_r$ is an outer loop for the frequency loop so it should be tuned much slower than the frequency loop. Its time constant is

$$\tau_{fn} = \frac{1}{D_p K_i}$$

and can be tuned as $\tau_{fn} = (10 \sim 100)\tau_f$.

23.6 Equivalent Structure

The SLL shown in Figure 23.2(a) can be redrawn as shown in Figure 23.2(b) to demonstrate the differences from conventional PLLs. The hold filter

$$H(s) = \frac{1 - e^{-Ts}}{Ts}$$

with $T = \frac{2\pi}{\dot{\theta}}$ is adopted to take the average value of an incoming signal.

With comparison to the STA or EPLL shown in Figure 22.6, the SLL has the following features:

1. The error signal $v - e$ is passed through a low-pass filter $\frac{1}{Ls + R}$, which reduces the impact of the harmonics in v and also amplifies the difference to speed up the tracking process because R and L are often chosen to be small.
2. The hold filter $H(s)$ is applied to both channels, which removes the ripples in the signal entering the integrators and reduces the variations in frequency $\dot{\theta}$ and amplitude E.
3. The frequency $\dot{\theta}$ is forwarded into the voltage channel, which speeds up the tracking process when the frequency changes.
4. $M_f i_f$ is forwarded into the frequency channel, which speeds up the tracking process when the voltage changes.
5. A local feedback loop consisting of D_p and $\frac{K_i}{s}$ is added to the frequency loop, which is equivalent to cascading $\frac{s + D_p K_i}{s + D_p K_i + D_p/J} = \frac{s + 1/\tau_{fn}}{s + 1/\tau_{fn} + 1/\tau_f}$ to $\frac{1}{Js}$. Since $\tau_{fn} \gg \tau_f$, $\frac{s + D_p K_i}{s + D_p K_i + D_p/J}$ is a lead compensator. It provides phase lead at all frequencies and enhances the responsiveness and stability of the system.

23.7 Simulation Results

The same simulations carried out in MATLAB®/Simulink® for the SOGI-PLL and the STA presented in Chapter 22 were repeated to verify the SLL with the parameters given in Table 23.1. The results of the SOGI-PLL are reproduced here for comparison.

23.7.1 With a Noisy Distorted Signal having a Variable Frequency

The distorted noisy voltage v used for simulations in Section 22.7.1, of which the frequency varies in a wide range, was adopted to test the SLL. The results are shown in the left column of Figure 23.3. It is worth mentioning that, although the amplitude of the fundamental component does not change, the amplitude of v does change because of $v_h(t)$. As a result, the estimated amplitude E is not a straight line. Although there was significant amount of

Table 23.1 Parameters of the SLL for simulations and experiments

Parameters	Values	Parameters	Values
f_n	50 Hz	J	2.0264×10^{-5}
τ_f	0.0005 s	K	4809.6
τ_{fn}	0.049 s	K_i	100
τ_q	18.37 s	L	0.3 mH
D_p	0.2026	R	0.01 Ω

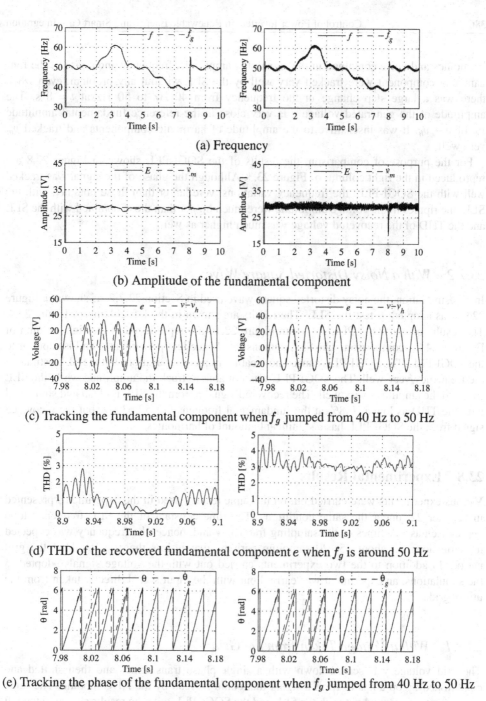

Figure 23.3 Tracking a distorted noisy signal with a variable frequency: Simulation results with the SLL (left column) and with the SOGI-PLL (right column)

harmonics and noise contained in the signal, the amplitude, frequency and phase of the fundamental component were tracked very well by the SLL without any problem, even when there was a large step change in the frequency from 40 Hz to 50 Hz at $t = 8$ s. The amplitude of the generated voltage e is very close to the reference fundamental amplitude v_m of $v - v_h$. It was insensitive to the amplitude of harmonic components and tracked v_m very well.

For the purpose of comparison, the results of the SOGI-PLL shown in Figure 22.8 are reproduced in the right column of Figure 23.3. Although the phase of the signal was tracked well with the SOGI-PLL, the frequency variations were larger than those obtained with the SLL; the ripples in the voltage amplitude were much bigger than those obtained with the SLL and the THD of the recovered voltage was much higher as well.

23.7.2 With a Noisy Distorted Square Wave

In this simulation, the noisy distorted square wave used in Section 22.7.2, as shown in Figure 22.9, was adopted to test the SLL. The results are shown in the left column of Figure 23.4. The results of the SOGI-PLL shown in Figure 22.10 are reproduced in the right column of Figure 23.4 for comparison. The SLL demonstrated excellent performance and is superior to the SOGI-PLL. The SOGI-PLL caused noticeable frequency variations but the SLL estimated the frequency very well. The SOGI-PLL was not able to track the amplitude while the SLL detected the amplitude very well. The recovered signal e from the SLL is clean and sinusoidal with the THD as low as 0.8% at the fundamental frequency of 52.63 Hz, but the recovered signal from the SOGI-PLL has a significant amount of harmonics.

23.8 Experimental Results

Various experiments were carried out on the same test rig used in the experiments presented in Chapter 22, under the same conditions. The time constant of the SLL frequency loop was chosen as five times of the sampling frequency and, hence, the frequency was expected to settle down in about 15~20 sampling periods, that is less than one-tenth of the grid period. In addition to the two experiments carried out with the voltage signals adopted in the simulations, an experiment was carried out with the input signal directly taken from the utility grid.

23.8.1 With a Voltage Taken from the Grid

The grid voltage was scaled down with a single-phase transformer and then shifted and conditioned by op-amps to form a signal that could be read by the ADC module. The same experiment was carried out with the SLL and the SOGI-PLL with the results shown in the left column and the right column of Figure 23.5, respectively. The phase was tracked well by both methods. The SLL tracked the voltage almost immediately, producing accurate frequency, amplitude and phase. Although the grid signal was not clean, the SLL did not have any difficulty in tracking it. Because there is no phase delay, the SLL can also be used as a filter. The SOGI-PLL took about one cycle to produce the correct amplitude and a half cycle to

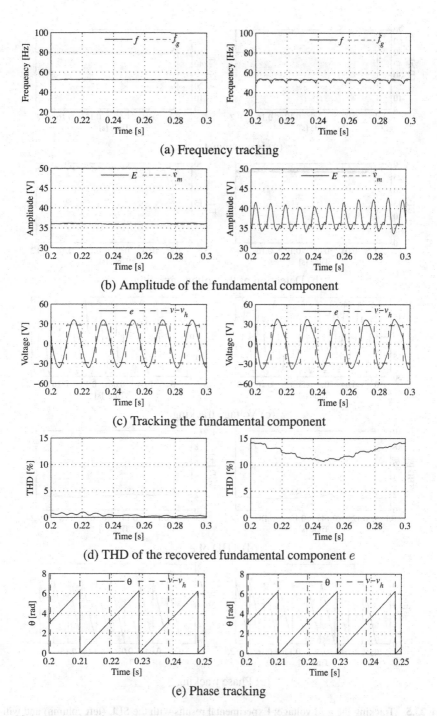

Figure 23.4 Extracting the fundamental component from a noisy distorted square wave: Simulation results with the SLL (left column) and with the SOGI-PLL (right column)

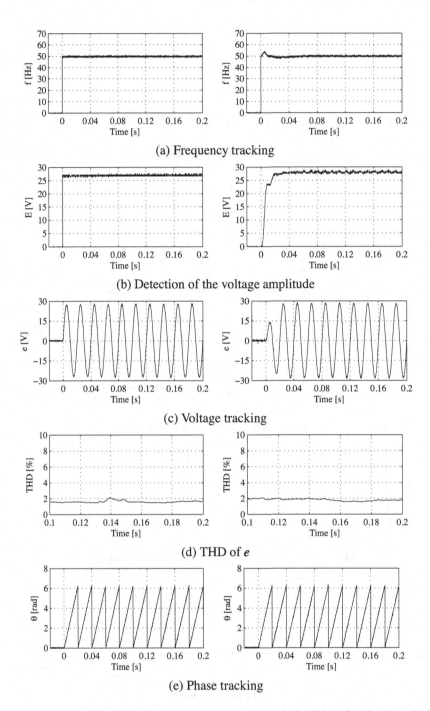

Figure 23.5 Tracking the grid voltage: Experimental results with the SLL (left column) and with the SOGI-PLL (right column)

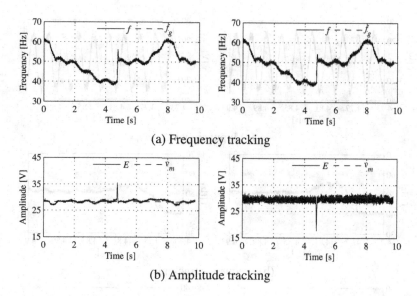

Figure 23.6 Frequency and amplitude tracking for a distorted noisy signal with a variable frequency: Experimental results with the SLL (left column) and with the SOGI-PLL (right column)

produce the correct frequency, all with noticeable overshoot. Apparently, the SLL outperforms the SOGI-PLL significantly.

23.8.2 With a Noisy Distorted Signal having a Variable Frequency

The voltage signal used in Section 23.7.1 was generated and sent out via a DAC channel as the voltage signal v. Again, the experiment was carried out for the SLL and the SOGI-PLL with the results shown in Figure 23.6. The experimental results matched the simulation results shown in Figure 23.3. The SLL tracked the frequency and the amplitude very well; the SOGI-PLL was able to track the frequency and the amplitude but with much bigger ripples. The experimental results when the frequency changed from 40 to 50 Hz are shown in Figure 23.7. The SLL tracked the frequency quickly. The SOGI-PLL tracked the frequency well but there were noticeable variations in the frequency after the jump.

23.8.3 With a Noisy Distorted Square Wave

The voltage signal used in Section 23.7.2 was generated and sent out via a DAC channel as the voltage signal. Again, the experiment was carried out for the SLL and the SOGI-PLL. The results are shown in Figure 23.8. The experimental results matched with the simulation results. The SLL was able to track the phase, frequency and amplitude but the SOGI-PLL cannot track the amplitude. The voltage recovered by the SLL is very clean but the voltage recovered by the SOGI-PLL contains significant harmonics.

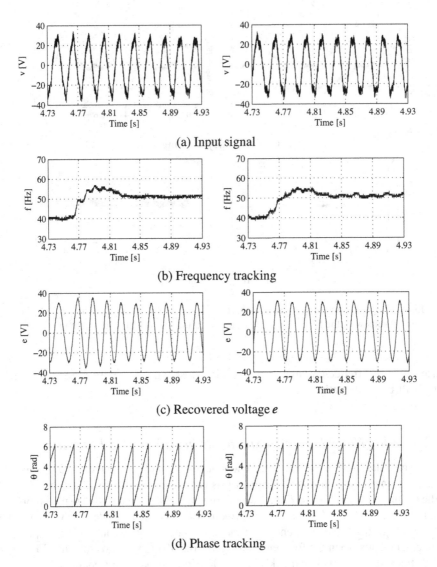

Figure 23.7 Tracking a distorted noisy signal with a variable frequency when the frequency jumped from 40 Hz to 50 Hz: Experimental results with the SLL (left column) and with the SOGI-PLL (right column)

23.9 Summary

Based on (Zhong and Nguyen 2012), a sinusoid-locked loop is discussed in this chapter to track the fundamental component of a periodic signal following the idea of synchronverters (Zhong and Weiss 2011). The underlying principle is that a grid-connected synchronous machine that does not exchange any power with the grid generates the same voltage as the grid voltage. Both

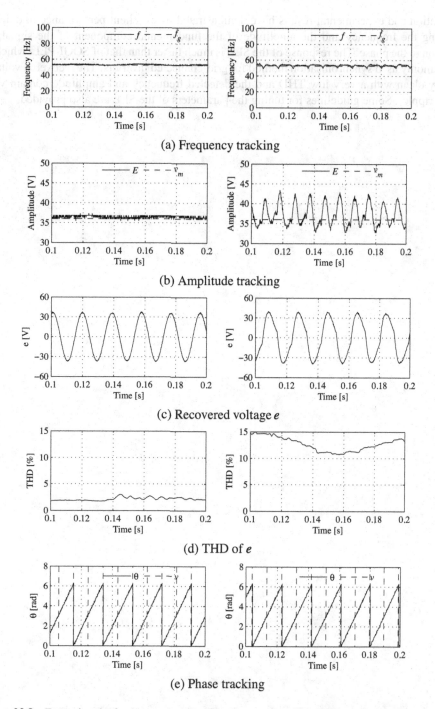

Figure 23.8 Extracting the fundamental component from a noisy distorted square wave: Experimental results with the SLL (left column) and with the SOGI-PLL (right column)

simulation and experimental results have demonstrated its excellent performance in quickly tracking the frequency and the amplitude of the fundamental component of the signal, in addition to the phase. The response of the SLL is much faster than that of SOGI-PLL, which is a commonly used synchronisation method for grid-connected inverters. The recovered voltage is very clean with a very low THD and the detected frequency and amplitude contain very small ripples. Some guidelines for tuning the parameters of the SLL are also provided.

References

ABB 2010 SVCs for load balancing and trackside voltage control. Technical report. http://www05.abb.com/global/scot/scot221.nsf/veritydisplay/c9f996d4f6c5c601c12570c9004b1754/$file/A02–0196%20E%20LR.pdf.
Ackerman T 2005 *Wind Power in Power Systems*. John Wiley & Sons, Ltd., Chichester.
Amuda L, Cardoso Filho B, Silva S, Silva S and Diniz A 2000 Wide bandwidth single and three-phase PLL structures for grid-tied PV systems, in *Proceedings of the 28th IEEE Photovoltaic Specialists Conference (PVSC)*, pp. 1660–1663.
Araujo S, Engler A, Sahan B and Antunes F 2007 LCL filter design for grid-connected NPC inverters in offshore wind turbines, in *Proceedings of the 7th International Conference on Power Electronics (ICPE)*, pp. 1133–1138.
Asiminoaei L, Rodriguez P and Blaabjerg F 2008 Application of discontinuous PWM modulation in active power filters. *IEEE Transactions on Power Electronics* 23(4), 1692–1706.
Aström K and Hägglund T 1988 *Automatic Tuning of PID Controllers*. Instrument Society of America, Research Triangle Park, NC.
Aström K and Hägglund T 1995 *PID Controllers: Theory, Design, and Tuning*, 2nd edn. Instrument Society of America, Research Triangle Park, NC.
Aström K and Hägglund T 2006 *Advanced PID Control*. Instrument Society of America, Research Triangle Park, NC.
Bansal RC 2005 Three-phase self-excited induction generators: An overview. *IEEE Transactions of Energy Conversion* 20(2), 292–299.
Barklund E, Pogaku N, Prodanovic M, Hernandez-Aramburo C and Green T 2008 Energy management in autonomous microgrid using stability-constrained droop control of inverters. *IEEE Transactions on Power Electronics* 23(5), 2346–2352.
Baroudi J, Dinavahi V and Knight A 2005 A review of power converter topologies for wind generators, in *Proceedings of IEEE International Conference on Electric Machines and Drives*, pp. 458–465.
Barrena J, Marroyo L, Vidal M and Apraiz J 2008 Individual voltage balancing strategy for PWM cascaded H-Bridge converter-based STATCOM. *IEEE Transactions on Industrial Electronics* 55(1), 21–29.
Bendre A, Venkataramanan G, Rosene D and Srinivasan V 2006 Modeling and design of a neutral-point voltage regulator for a three-level diode-clamped inverter using multiple-carrier modulation. *IEEE Transactions on Industrial Electronics* 53(3), 718–726.
Bianchi F, Battista H and Mantz R 2007 *Wind Turbine Control Systems: Principles, Modelling and Gain Scheduling Design*. Springer-Verlag.
Blaabjerg F and Chen Z 2006 *Power Electronics for Modern Wind Turbines*. Morgan and Claypool Publishers, San Rafael, CA.
Blaabjerg F, Freysson S, Hansen HH and Hansen S 1997 A new optimized space-vector modulation strategy for a component-minimized voltage source inverter. *IEEE Transactions on Power Electronics* 12(4), 704–714.
Blaabjerg F, Teodorescu R, Liserre M and Timbus A 2006 Overview of control and grid synchronization for distributed power generation systems. *IEEE Transactions on Industrial Electronics* 53(5), 1398–1409.
Blooming T and Carnovale D 2007 Harmonic convergence. *IEEE Industry Applications Magazine* 13(1), 21–27.

Control of Power Inverters in Renewable Energy and Smart Grid Integration, First Edition.
Qing-Chang Zhong and Tomas Hornik.
© 2013 John Wiley & Sons, Ltd. Published 2013 by John Wiley & Sons, Ltd.

Bollen M and Hassan F 2011 *Integration of Distributed Generation in the Power System*. Wiley-IEEE Press.
Bollen MH 2000 *Understanding Power Quality Problems: Voltage Sags and Interruptions*. Wiley-IEEE Press.
Bolsens B, De Brabandere K, Van den Keybus J, Driesen J and Belmans R 2006 Model-based generation of low distortion currents in grid-coupled PWM-inverters using an LCL output filter. *IEEE Transactions on Power Electronics* 21(4), 1032–1040.
Borup U, Blaabjerg F and Enjeti P 2001 Sharing of nonlinear load in parallel-connected three-phase converters. *IEEE Transactions on Industry Applications* 37(6), 1817–1823.
Bose B 2001 *Modern Power Electronics and AC Drives*. Prentice-Hall, Englewood Cliffs, NJ.
Bose B 2009 Power electronics and motor drives: recent progress and perspective. *IEEE Transactions on Industrial Electronics* 56(2), 581–588.
Brabandere KD, Bolsens B, den Keybus JV, Woyte A, Driesen J and Belmans R 2007 A voltage and frequency droop control method for parallel inverters. *IEEE Transactions on Power Electronics* 22(4), 1107–1115.
Brenna M, Foiadelli F and Zaninelli D 2011 New stability analysis for tuning PI controller of power converters in railway application. *IEEE Transactions on Industrial Electronics* 58(2), 533–543.
Burton T 2001 *Wind energy: Handbook*. John Wiley & Sons.
Busco B, Marino P, Porzio M, Schiavo R and Vasca F 2003 Digital control and simulation for power electronic apparatus in dual voltage railway locomotive. *IEEE Transactions on Power Electronics* 18(5), 1146–1157.
Buso S, Fasolo S and Mattavelli P 2001 Uninterruptible power supply multiloop control employing digital predictive voltage and current regulators. *IEEE Transactions on Industry Applications* 37(6), 1846–1854.
Busquets-Monge S, Ortega J, Bordonau J, Beristain J and Rocabert J 2008a Closed-loop control of a three-phase neutral-point-clamped inverter using an optimized virtual-vector-based pulse width modulation. *IEEE Transactions on Industrial Electronics* 55(5), 2061–2071.
Busquets-Monge S, Rocabert J, Rodriguez P, Alepuz S and Bordonau J 2008b Multilevel diode-clamped converter for photovoltaic generators with independent voltage control of each solar array. *IEEE Transactions on Industrial Electronics* 55(7), 2713–2723.
Carrasco J, Franquelo L, Bialasiewicz J, Galvan E, Guisado R, Prats M, Leon J and Moreno-Alfonso N 2006a Power-electronic systems for the grid integration of renewable energy sources: A survey. *IEEE Transactions on Industry Applications* 53(4), 1002–1016.
Carrasco J, Franquelo L, Bialasiewicz J, Galvan E, Portillo-Guisado R, Prats M, Leon J and Moreno-Alfonso N 2006b Power-electronic systems for the grid integration of renewable energy sources: A survey. *IEEE Transactions on Industrial Electronics* 53(4), 1002–1016.
Castilla M, Miret J, Matas J, Garcia de Vicuna L and Guerrero J 2009 Control design guidelines for single-phase grid-connected photovoltaic inverters with damped resonant harmonic compensators. *IEEE Transactions on Industrial Electronics* 56(11), 4492–4501.
Celanovic N and Boroyevich D 2000 A comprehensive study of neutral-point voltage balancing problem in three-level neutral-point-clamped voltage source PWM inverters. *IEEE Transactions on Power Electronics* 15(2), 242–249.
Cetin A and Ermis M 2009 VSC-Based D-STATCOM with selective harmonic elimination. *IEEE Transactions on Industry Applications* 45(3), 1000–1015.
Chandorkar M, Divan D and Adapa R 1993 Control of parallel connected inverters in standalone AC supply systems. *IEEE Transactions on Industry Applications* 29(1), 136–143.
Chang G, Chu SY and Wang HL 2006 A new method of passive harmonic filter planning for controlling voltage distortion in a power system. *IEEE Transactions on Power Delivery* 21(1), 305–312.
Chang G, Lin HW and Chen SK 2004 Modeling characteristics of harmonic currents generated by high-speed railway traction drive converters. *IEEE Transactions on Power Delivery* 19(2), 766–773.
Chen BK and Guo BS 1996 Three phase models of specially connected transformers. *IEEE Transactions on Power Delivery* 11(1), 323–330.
Chen CL, Wang Y, Lai JS, Lee YS and Martin D 2010 Design of parallel inverters for smooth mode transfer microgrid applications. *IEEE Transactions on Power Electronics* 25(1), 6–15.
Chen MW, Li QZ and Wei G 2009 Optimized design and performance evaluation of new cophase traction power supply system, in *Proceedings of Asia-Pacific Power and Energy Engineering Conference (APPEEC)*, pp. 1–6.
Chen S, Lai Y, Tan SC and Tse C 2008 Analysis and design of repetitive controller for harmonic elimination in PWM voltage source inverter systems. *IET Proceedings on Power Electronics* 1(4), 497–506.
Chen SL, Li RJ and Hsi PH 2004 Traction system unbalance problem-analysis methodologies. *IEEE Transactions on Power Delivery* 19(4), 1877–1883.

Chen TH, Yang WC and Hsu YF 1998 A systematic approach to evaluate the overall impact of the electric traction demands of a high-speed railroad on a power system. *IEEE Transactions of Vehicular Technology* **47**(4), 1378–1384.

Chen Y, Mwinyiwiwa B, Wolanski Z and Ooi BT 2000 Unified power flow controller (UPFC) based on chopper stabilized diode-clamped multilevel converters. *IEEE Transactions on Power Electronics* **15**(2), 258–267.

Chen Z and Spooner E 2001 Grid power quality with variable speed wind turbines. *IEEE Transactions of Energy Conversion* **16**(2), 148–154.

Chung SK 2000 A phase tracking system for three phase utility interface inverters. *IEEE Transactions on Power Electronics* **15**(3), 431–438.

Ciobotaru M, Teodorescu R and Blaabjerg F 2006 A new single-phase PLL structure based on second order generalized integrator, in *Proceedings of the 37th IEEE Power Electronics Specialists Conference (PESC)*, pp. 1–6.

Cipriano dos Santos E, Jacobina C, Dias J and Rocha N 2011 Single-phase to three-phase universal active power filter. *IEEE Transactions on Power Delivery* **26**(3), 1361–1371.

Coelho E, Cortizo P and Garcia P 2002 Small-signal stability for parallel-connected inverters in stand-alone AC supply systems. *IEEE Transactions on Industry Applications* **38**(2), 533–542.

Corless R, Gonnet G, Hare D, Jeffrey D and Knuth D 1996 On the Lambert W function. *Advances in Computational Mathematics* **5**, 329–359. Available at http://www.apmaths.uwo.ca/~rcorless/frames/papers.htm, accessed on 3/9/2005.

Costa-Castello R, Grino R and Fossas E 2004 Odd-harmonic digital repetitive control of a single-phase current active filter. *IEEE Transactions on Power Electronics* **19**(4), 1060–1068.

Costa-Castello R, Grino R, Cardoner R and Fossas E 2007 High performance control of a single-phase shunt active filter, in *Proceedings of IEEE International Symposium on Industrial Electronics (ISIE)*, pp. 3350–3355.

da Silva C, Pereira R, da Silva L, Lambert-Torres G, Bose B and Ahn S 2010 A digital PLL scheme for three-phase system using modified synchronous reference frame. *IEEE Transactions on Industrial Electronics* **57**(11), 3814–3821.

Dahono P 2003 A method to damp oscillations on the input LC filter of current-type AC-DC PWM converters by using a virtual resistor, in *Proceedings of the 25th International Telecommunications Energy Conference (INTELEC'03)*, pp. 757–761.

Dahono P 2004 A control method for DC-DC converter that has an LCL output filter based on new virtual capacitor and resistor concepts, in *Proceedings of the 35th Annual Power Electronics Specialists Conference*, pp. 36–42.

Dahono P, Bahar Y, Sato Y and Kataoka T 2001 Damping of transient oscillations on the output LC filter of PWM inverters by using a virtual resistor, in *Proceedings of the 4th IEEE International Conference on Power Electronics and Drive Systems*, pp. 403–407.

Dai M, Marwali M, Jung JW and Keyhani A 2008a Power flow control of a single distributed generation unit. *IEEE Transactions on Power Electronics* **23**(1), 343–352.

Dai NY, Wong MC, Ng F and Han YD 2008b A FPGA-based generalized pulse width modulator for three-leg center-split and four-leg voltage source inverters. *IEEE Transactions on Power Electronics* **23**(3), 1472–1484.

Das J 2004 Passive filters-potentialities and limitations. *IEEE Transactions on Industry Applications* **40**(1), 232–241.

De D and Ramanarayanan V 2010 Decentralized parallel operation of inverters sharing unbalanced and non-linear loads. *IEEE Transactions on Power Electronics* **25**(12), 3015–3025.

Dewan S and Ziogas P 1979 Optimum filter design for a single-phase solid-state UPS system. *IEEE Transactions on Industry Applications* **IA-15**(6), 664–669.

Dewan SB 1981 Optimum input and output filters for a single-phase rectifier power supply. *IEEE Transactions on Industry Applications* **IA-17**(3), 282–288.

Diaz G, Gonzalez-Moran C, Gomez-Aleixandre J and Diez A 2010 Scheduling of droop coefficients for frequency and voltage regulation in isolated microgrids. *IEEE Transactions on Power Systems* **25**(1), 489–496.

DOE 2009a The smart grid: An introduction. Technical report, The U.S. Department of Energy. http://energy.gov/sites/prod/files/oeprod/DocumentsandMedia/DOE_SG_Book_Single_Pages.pdf.

DOE 2009b Smart grid system report. Technical report, The U.S. Department of Energy. http://energy.gov/sites/prod/files/oeprod/DocumentsandMedia/SGSR_Annex_A-B_090707_lowres.pdf.

Driesen J and Craenenbroeck TV 2002 Voltage disturbances: Introduction to unbalance. Technical report. http://www.apqi.org/file/attachment/2008716/95537.pdf.

Driesen J and Visscher K 2008 Virtual synchronous generators, in *Proceedings of IEEE Power and Energy Society General Meeting*, pp. 1–3.

Ekanayake J, Holdsworth L and Jenkins N 2003 Control of DFIG wind turbines. *Power Engineer* **17**(1), 28–32.

Ekanayake J, Jenkins N, Liyanage K, Wu J and Yokoyama A 2012 *Smart Grid: Technology and Applications*. John Wiley & Sons.

Ellison A 1965 *Electromechanical Energy Conversion*. George G. Harrap Co. Ltd, London.

Enslin J, Wolf M, Snyman D and Swiegers W 1997 Integrated photovoltaic maximum power point tracking converter. *IEEE Transactions on Industrial Electronics* **44**(6), 769–773.

EPIA 2010 Photovoltaic energy, electricity from the sun. Technical report, European Photovoltaic Industry Association. http://www.epia.org.

EPIA 2011 Global market outlook for photovoltaics until 2015. Technical report, European Photovoltaic Industry Association. http://www.epia.org.

Erickson R and Maksimović D 2001 *Fundamentals of Power Electronics*. Kluwer Academic.

Escobar G, Hernandez-Briones P, Martinez P, Hernandez-Gomez M and Torres-Olguin R 2008 A repetitive-based controller for the compensation of harmonic components. *IEEE Transactions on Industrial Electronics* **55**(8), 3150–3158.

Escobar G, Martinez-Montejano M, Valdez A, Martinez P and Hernandez-Gomez M 2011 Fixed-reference-frame phase-locked loop for grid synchronization under unbalanced operation. *IEEE Transactions on Industrial Electronics* **58**(5), 1943–1951.

Etezadi-Amoli M and Choma K 2001 Electrical performance characteristics of a new micro-turbine generator, in *Proceedings of IEEE Power Engineering Society Winter Meeting*, pp. 736–740.

Farhangi H 2010 The path of the smart grid. *IEEE Power Energy Magazine* **8**(1), 18–28.

Fisher M 1991 *Power Electronics*. PWS-KENT publishing company.

Fitzgerald A, Kingsley C and Umans S 2003 *Electric Machinery*. McGraw-Hill, New York, NY.

Francis B and Wonham W 1975 The internal model principle for linear multivariable regulators. *Applied Mathematics and Optimization* **2**(2), 170–194.

Franklin G, Powell J and Workman M 1990 *Digital Control of Dynamic Systems* 2nd edn. Addison-Wesley, Reading, MA.

Freijedo F, Doval-Gandoy J, Lopez O, Fernandez-Comesana P and Martinez-Penalver C 2009 A signal-processing adaptive algorithm for selective current harmonic cancellation in active power filters. *IEEE Transactions on Industrial Electronics* **56**(8), 2829–2840.

Freijedo F, Yepes A, Malvar J, Lóandpez O, Fernandez-Comesañ A P, Vidal A and Doval-Gandoy J 2011 Frequency tracking of digital resonant filters for control of power converters connected to public distribution systems. *IET Proceedings on Power Electronics* **4**(4), 454–462.

FUJI 2004 *FUJI IGBT-IPM Application Manual* Fuji Electric Device Technology Co. Ltd. http://www.fujielectric.com/products/semiconductor/technical/application/pdf/REH983a/REH983a.pdf.

Garcia-Cerrada A, Pinzon-Ardila O, Feliu-Batlle V, Roncero-Sanchez P and Garcia-Gonzalez P 2007 Application of a repetitive controller for a three-phase active power filter. *IEEE Transactions on Power Electronics* **22**(1), 237–246.

Ghennam T, Berkouk E and Francois B 2010 A novel space-vector current control based on circular hysteresis areas of a three-phase neutral-point-clamped inverter. *IEEE Transactions on Industrial Electronics* **57**(8), 2669–2678.

Giordano AA and Hsu FM 1985 *Least Square Estimation with Applications to Digital Signal Processing*. John Wiley & Sons, New York.

Grainger J and Stevenson W 1994 *Power System Analysis*. McGraw-Hill, New York, NY.

Green M and Limebeer D 1995 *Linear Robust Control*. Prentice-Hall, Englewood Cliffs, NJ.

Green T and Prodanović M 2003 Control of inverter-based micro-grids, submitted to Electric Power Systems Research, special issue on Distributed Generation.

Grino R, Cardoner R, Costa-Castello R and Fossas E 2007 Digital repetitive control of a three-phase four-wire shunt active filter. *IEEE Transactions on Industrial Electronics* **54**(3), 1495–1503.

Guerrero J, Berbel N, de Vicuna L, Matas J, Miret J and Castilla M 2006a Droop control method for the parallel operation of online uninterruptible power systems using resistive output impedance, in *Proceedings of the 21st IEEE Applied Power Electronics Conference and Exposition*, pp. 1716–1722.

Guerrero J, de Vicuna L, Miret J, Matas J and Cruz J 2004 Output impedance performance for parallel operation of UPS inverters using wireless and average current-sharing controllers, in *Proceedings of the 35th IEEE Power Electronics Specialists Conference*, pp. 2482–2488.

Guerrero J, Garcia de Vicuna L, Matas J, Castilla, M. and Miret J 2005 Output impedance design of parallel-connected UPS inverters with wireless load-sharing control. *IEEE Transactions on Industrial Electronics* **52**(4), 1126–1135.

Guerrero J, Hang L and Uceda J 2008 Control of distributed uninterruptible power supply systems. *IEEE Transactions on Industrial Electronics* **55**(8), 2845–2859.

Guerrero J, Matas J, de Vicuna L, Castilla M and Miret J 2006b Wireless-control strategy for parallel operation of distributed-generation inverters. *IEEE Transactions on Industrial Electronics* **53**(5), 1461–1470.

Guerrero J, Matas J, de Vicuna LG, Castilla M and Miret J 2007 Decentralized control for parallel operation of distributed generation inverters using resistive output impedance. *IEEE Transactions on Industrial Electronics* **54**(2), 994–1004.

Guerrero J, Vasquez J, Matas J, Castilla M and de Vicuna L 2009 Control strategy for flexible microgrid based on parallel line-interactive UPS systems. *IEEE Transactions on Industrial Electronics* **56**(3), 726–736.

Guerrero JM, Vasquez JC, Matas J, Garcia de Vicuna L and Castilla M 2011 Hierarchical control of droop-controlled AC and DC microgrids: a general approach towards standardization. *IEEE Transactions on Industrial Electronics* **58**(1), 158–172.

Guo S and Liu D 2011 Analysis and design of output LC filter system for dynamic voltage restorer, in *Proceedings of the 26th IEEE Applied Power Electronics Conference and Exposition*, pp. 1599–1605.

Guo W, Xiao L and Dai S 2012 Enhancing low-voltage ride-through capability and smoothing output power of DFIG with a superconducting fault-current limiter–Magnetic energy storage system. *IEEE Transactions of Energy Conversion* **27**(2), 277–295.

Habetler T 1993 A space vector-based rectifier regulator for AC/DC/AC converters. *IEEE Transactions on Power Electronics* **8**(1), 30–36.

Hamadi A, Rahmani S and Al-Haddad K 2010 A hybrid passive filter configuration for VAR control and harmonic compensation. *IEEE Transactions of Energy Conversion* **57**(7), 2419–2434.

Hara S, Yamamoto Y, Omata T and Nakano M 1988 Repetitive control system: A new type servo system for periodic exogenous signals. *IEEE Transactions on Automatic Control* **33**, 659–668.

Hatua K, Jain A, Banerjee D and Ranganathan V 2012 Active damping of output LC filter resonance for vector-controlled VSI-Fed AC motor drives. *IEEE Transactions on Industrial Electronics* **59**(1), 334–342.

Hatziargyriou N, Asano H, Iravani R and Marnay C 2007 Microgrids. *IEEE Power Energy Magazine* **5**(4), 78–94.

Heier S 2006 *Grid integration of wind energy conversion systems*. John Wiley & Sons.

Holmes D and Martin D 1996 Implementation of a direct digital predictive current controller for single and three phase voltage source inverters, in *Proceedings of IEEE Industry Applications Conference (IAS)*, pp. 906–913.

Holmes D, Lipo T and Lipo T 2003 *Pulse Width Modulation for Power Converters: Principles and Practice* IEEE Press series on power engineering. IEEE Press.

Holtz J 1992 Pulse width modulation: a survey. *IEEE Transactions on Industrial Electronics* **39**(5), 410–420.

Holtz J 1994 Pulsewidth modulation for electronic power conversion. *Proceedings of the IEEE* **82**(8), 1194–1214.

Hopkins DC and Safiuddin M 2010 Power electronics in a smart-grid distribution system. Technical report, State University of New York. http://www.dchopkins.com/professional/open_seminars/PowerElectronics_SmartGrid.pdf.

Horita Y, Morishima N, Kai M, Onishi M, Masui T and Noguchi M 2010 Single-phase STATCOM for feeding system of Tokaido Shinkansen, in *Proceedings of International Power Electronics Conference (IPEC)*, pp. 2165–2170.

Hornik T and Zhong QC 2009 H^∞ repetitive current controller for grid-connected inverters, in *Proceedings of the 35th IEEE Annual Conference of Industrial Electronics (IECON)*, pp. 554–559.

Hornik T and Zhong QC 2010a H^∞ repetitive current-voltage control of inverters in microgrids, in *Proceedings of the 36th Annual IEEE Conference of Industrial Electronics (IECON)*, Phoenix, USA.

Hornik T and Zhong QC 2010b H^∞ repetitive voltage control of grid-connected inverters with frequency adaptive mechanism. *IET Proceedings on Power Electronics* **3**(6), 925–935.

Hornik T and Zhong QC 2010c Voltage control of grid-connected inverters based on h^∞ and repetitive control, in *Proceedings of the 8th World Congress on Intelligent Control and Automation (WCICA)*, pp. 270–275.

Hornik T and Zhong QC 2011a A current control strategy for voltage-source inverters in microgrids based on H^∞ and repetitive control. *IEEE Transactions on Power Electronics* **26**(3), 943–952.

Hornik T and Zhong QC 2011b H^∞ current control strategy for the neutral point of a three-phase inverter, in *Proceedings of the 50th IEEE Conference on Decision and Control and European Control Conference*, pp. 520–525.

Hossain Z, Olejniczak KJ, Burgers KC and Balda JC 1997a Design of RCD snubbers based upon approximations to the switching characteristics. I. theoretical development, in *Proceedings IEEE International Electric Machines and Drives Conf. Record*, pp. TC2–6.1–TC2–6.3.

Hossain Z, Olejniczak KJ, Burgers KC and Balda JC 1997b Design of RCD snubbers based upon approximations to the switching characteristics. II. simulation and experimental results, in *Proceedings IEEE International Electric Machines and Drives Conf. Record*, pp. TC2–5.1–TC2–5.3.

Huang SR and Chen BN 2002 Harmonic study of the Le Blanc transformer for Taiwan railway's electrification system. *IEEE Transactions on Power Delivery* **17**(2), 495–499.

IR 2004 *IR 2130 Datasheet* International Rectifier. http://www.irf.com/product-info/datasheets/data/ir2130.pdf.

Irwin JD 1996 *The Industrial Electronics Handbook* 2 edn. CRC Press.

Iwase H, Kawasumi K, Tachibana K and Shioda T 2003 WT1600 digital power meter. Technical Report 35, Yokogawa Ltd. http://www.yokogawa.com/rd/pdf/TR/rd-tr-r00035-006.pdf.

IXYS 1998 *Inverter Interface and Digital Deadtime Generator for 3 Phase PWM Controls* IXYS Corporation. http://ixapps.ixys.com/DataSheet/98568.pdf.

Iyer SV, Belur MN and Chandorkar MC 2010 A generalized computational method to determine stability of a multi-inverter microgrid. *IEEE Transactions on Power Electronics* **25**(9), 2420–2432.

Jacobson MZ 2009 Review of solutions to global warming, air pollution, and energy security. *Energy Environment, Science* **2**, 148–173.

Jahns T, Doncker RD, Radun A, Szczesny P and Turnbull F 1993 System design considerations for a high-power aerospace resonant link converter. *IEEE Transactions on Power Electronics* **8**(4), 663–672.

Jenkins N, Allan R, Crossley P, Kirshen D and Strbac G 2000 *Embedded Generation*. IEE Books.

Jouanne AV, Dai S and Zhang H 2002 A multilevel inverter approach providing DC-link balancing, ride-through enhancement, and common-mode voltage elimination. *IEEE Transactions on Industrial Electronics* **49**(4), 739–745.

Karady G and Holbert K 2004 *Electrical Energy Conversion and Transport: An Interactive Computer-Based Approach IEEE Press Series on Power Engineering*. Wiley–IEEE Press.

Karimi-Ghartemani M and Iravani M 2001 A new phase-locked loop (PLL) system, in *Proceedings of the 44th IEEE 2001 Midwest Symposium on Circuits and Systems (MWSCAS)*, pp. 421–424.

Karimi-Ghartemani M and Iravani M 2002 A nonlinear adaptive filter for online signal analysis in power systems: Applications. *IEEE Transactions on Power Delivery* **17**(2), 617–622.

Karimi-Ghartemani M and Ziarani A 2003 Periodic orbit analysis of two dynamical systems for electrical engineering applications. *Journal of Engineering Mathematics* **45**, 135–154.

Katiraei F, Iravani R, Hatziargyriou N and Dimeas A 2008 Microgrids management. *IEEE Power Energy Magazine* **6**(3), 54–65.

Kaura V and Blasko V 1997 Operation of a phase locked loop system under distorted utility conditions. *IEEE Transactions on Industry Applications* **33**(1), 58–63.

Kawabata T, Miyashita T and Yamamoto Y 1990 Dead beat control of three phase PWM inverter. *IEEE Transactions on Power Electronics* **5**(1), 21–28.

Kawabata T, Miyashita T and Yamamoto Y 1991 Digital control of three-phase PWM inverter with LC filter. *IEEE Transactions on Power Electronics* **6**(1), 62–72.

Kesler M and Ozdemir E 2011 Synchronous-reference-frame-based control method for UPQC under unbalanced and distorted load conditions. *IEEE Transactions on Industrial Electronics* **58**(9), 3967–3975.

Khalil HK 2001 *Nonlinear Systems*. Prentice Hall.

Kim J, Choi J and Hong H 2000 Output LC filter design of voltage source inverter considering the performance of controller, in *Proceedings on Power System Technology Conference*, pp. 1659–1664.

Kimmel WD and Gerke DD 1995 *Electromagnetic Compatibility in Medical Equipment: A Guide for Designers and Installers*. Taylor & Francis US.

Kjaer S, Pedersen J and Blaabjerg F 2005 A review of single-phase grid-connected inverters for photovoltaic modules. *IEEE Transactions on Industry Applications* **41**(5), 1292–1306.

Kneschke TA 1985 Control of utility system unbalance caused by single-phase electric traction. *IEEE Transactions on Industry Applications* **IA-21**(6), 1559–1570.

Kojabadi H, Yu B, Gadoura I, Chang L and Ghribi M 2006 A novel DSP-based current-controlled PWM strategy for single phase grid-connected inverters. *IEEE Transactions on Power Electronics* **21**(4), 985–993.

Kundur P 1994 *Power System Stability and Control*. McGraw-Hill, New York, NY.

Kuo B 1992 *Digital Control Systems* 2nd edn. Saunders College Publishing, Ft. Worth, TX.

Kwak S and Toliyat H 2005 A hybrid solution for load-commutated-inverter-fed induction motor drives. *IEEE Transactions on Industry Applications* **41**(1), 83–90.

Lascu C, Asiminoaei L, Boldea I and Blaabjerg F 2007 High performance current controller for selective harmonic compensation in active power filters. *IEEE Transactions on Power Electronics* **22**(5), 1826–1835.

Lascu C, Asiminoaei L, Boldea I and Blaabjerg F 2009 Frequency response analysis of current controllers for selective harmonic compensation in active power filters. *IEEE Transactions on Industrial Electronics* **56**(2), 337–347.

Lasseter R 2002 Microgrids, in *Proceedings of IEEE Power Engineering Society Winter Meeting*, pp. 305–308.

Ledwich G and George T 1994 Using phasors to analyze power system negative phase sequence voltages caused by unbalanced loads. *IEEE Transactions on Power Systems* **9**(3), 1226–1232.

Lee CT, Chu CC and Cheng PT 2010 A new droop control method for the autonomous operation of distributed energy resource interface converters, in *Proceedings of IEEE Energy Conversion Congress and Exposition (ECCE)*, pp. 702–709.

Lee D, Lee S and Lee F 1998 An analysis of midpoint balance for the neutral-point-clamped three-level VSI, in *Proceedings on Power Electronics Specialists Conference (PESC)*, pp. 193–199.

Lee DC and Kim YS 2007 Control of single-phase-to-three-phase AC/DC/AC PWM converters for induction motor drives. *IEEE Transactions on Industrial Electronics* **54**(2), 797–804.

Lee H, Lee C, Jang G and Kwon SH 2006 Harmonic analysis of the Korean high-speed railway using the eight-port representation model. *IEEE Transactions on Power Delivery* **21**(2), 979–986.

Lee TL and Cheng PT 2007 Design of a new cooperative harmonic filtering strategy for distributed generation interface converters in an islanding network. *IEEE Transactions on Power Electronics* **22**(5), 1919–1927.

LEM n.d.a *LAH 25-NP Datasheet* LEM. http://www.lem.com/docs/products/lah 25-np e.pdf.

LEM n.d.b *LV 25-P Datasheet* LEM. http://www.lem.com/docs/products/lv 25-p.pdf.

Lewicki A, Krzeminski Z and Abu-Rub H 2011 Space-vector pulse width modulation for three-level npc converter with the neutral point voltage control. *IEEE Transactions on Industrial Electronics* **58**(11), 5076–5086.

Li Q and Wolfs P 2008 A review of the single phase photovoltaic module integrated converter topologies with three different DC link configurations. *IEEE Transactions on Power Electronics* **23**(3), 1320–1333.

Li Y 2009 Control and resonance damping of voltage-source and current-source converters with LC filters. *IEEE Transactions on Industrial Electronics* **56**(5), 1511–1521.

Li Y and Kao CN 2009 An accurate power control strategy for power-electronics-interfaced distributed generation units operating in a low-voltage multibus microgrid. *IEEE Transactions on Power Electronics* **24**(12), 2977–2988.

Li Y, Vilathgamuwa D and Loh PC 2004 Design, analysis, and real-time testing of a controller for multibus microgrid system. *IEEE Transactions on Power Electronics* **19**(5), 1195–1204.

Li YW, Vilathgamuwa D and Loh PC 2006 A grid-interfacing power quality compensator for three-phase three-wire microgrid applications. *IEEE Transactions on Power Electronics* **21**(4), 1021–1031.

Li YW, Vilathgamuwa D and Loh PC 2007 Robust control scheme for a microgrid with PFC capacitor connected. *IEEE Transactions on Industry Applications* **43**(5), 1172–1182.

Liang J, Green T, Feng C and Weiss G 2009 Increasing voltage utilization in split-link, four-wire inverters. *IEEE Transactions on Power Electronics* **24**(6), 1562–1569.

Lindgren M 1996 Analysis and simulation of digitally-controlled grid-connected PWM-converters using the space-vector average approximation, in *Proceedings of IEEE Workshop on Computers in Power Electronics*, pp. 85–89.

Lindgren M and Svensson J 1998 Control of a voltage-source converter connected to the grid through an LCL-filter-application to active filtering, in *Proceedings of the 29th Annual IEEE Power Electronics Specialists Conference (PESC)*, pp. 229–235.

Liserre M, Blaabjerg F and Hansen S 2005 Design and control of an LCL-filter-based three-phase active rectifier. *IEEE Transactions on Industry Applications* **41**(5), 1281–1291.

Logemann H and Pandolfi L 1994 A note on stability and stabilizability of neutral systems. *IEEE Transactions on Automatic Control* **39**, 138–143.

Logemann H and Townley S 1996 The effect of small delays in the feedback loop on the stability of neutral systems. *Systems & Control Letters* **27**(5), 267–274.

Loh PC and Holmes D 2005 Analysis of multiloop control strategies for LC/CL/LCL-filtered voltage-source and current-source inverters. *IEEE Transactions on Industry Applications* **41**(2), 644–654.

Loix T, Brabandere KD, Driesen J and Belmans R 2007 A three-phase voltage and frequency droop control scheme for parallel inverters, in *Proceedings of the 33rd Annual Conference of the IEEE Industrial Electronics Society (IECON)*, pp. 1662–1667.

Luo A, Wu CP, Shen J, Shuai ZK and Ma FJ 2011 Railway static power conditioners for high-speed train traction power supply systems using three-phase V/V transformers. *IEEE Transactions on Power Electronics* **26**(10), 2844–2856.

Machado R, Buso S and Pomilio J 2006 A line-interactive single-phase to three-phase converter system. *IEEE Transactions on Power Electronics* **21**(6), 1628–1636.

Majumder R, Chaudhuri B, Ghosh A, Ledwich G and Zare F 2010 Improvement of stability and load sharing in an autonomous microgrid using supplementary droop control loop. *IEEE Transactions on Power Systems* **25**(2), 796–808.

Manwell J, McGowan J and Rogers A 2009 *Wind Energy Explained: Theory, Design and Application*. John Wiley & Sons.

Marwali M, Jung JW and Keyhani A 2004 Control of distributed generation systems-part II: Load sharing control. *IEEE Transactions on Power Electronics* **19**(6), 1551–1561.

Mathew S 2006 *Wind Energy: Fundamentals, Resource Analysis and Economics*. Springer-Verlag.

Mathew S and Philip G 2011 *Advances in Wind Energy and Conversion Technology*. Springer-Verlag.

McGrath B, Holmes D and Galloway J 2005 Power converter line synchronization using a discrete Fourier transform (DFT) based on a variable sample rate. *IEEE Transactions on Power Electronics* **20**(4), 877–884.

Michels L, de Camargo R, Botteron F, Grudling H and Pinheiro H 2006 Generalised design methodology of second-order filters for voltage-source inverters with space-vector modulation. *IEE Electronic Letters* **153**(2), 219–226.

Middleton R and Goodwin G 1990 *Digital Control and Estimation: A Unified Approach*. Prentice-Hall, Inc.

Ming-Li D, Guang-Ning W, Xue-Yan Z, Chun-Lei F, Chang-Hong H and Qiang Y 2008 The simulation analysis of harmonics and negative sequence with Scott wiring transformer, in *Proceedings of International Conference on Condition Monitoring and Diagnosis*, pp. 513–516.

Mohamed YR and El-Saadany E 2007 An improved deadbeat current control scheme with a novel adaptive self-tuning load model for a three-phase PWM voltage-source inverter. *IEEE Transactions on Industrial Electronics* **54**(2), 747–759.

Mohamed Y and El-Saadany E 2008a Adaptive decentralized droop controller to preserve power sharing stability of paralleled inverters in distributed generation microgrids. *IEEE Transactions on Power Electronics* **23**(6), 2806–2816.

Mohamed YR and El-Saadany E 2008b A control scheme for PWM voltage-source distributed-generation inverters for fast load-voltage regulation and effective mitigation of unbalanced voltage disturbances. *IEEE Transactions on Industrial Electronics* **55**(5), 2072–2084.

Mohan N 2003 *Power Electronics: Converters, Applications, and Design*. Hoboken, NJ : John Wiley and Sons.

Mojiri M, Karimi-Ghartemani M and Bakhshai A 2007 Estimation of power system frequency using an adaptive notch filter. *IEEE Transactions on Instrumentation and Measurement* **56**(6), 2470–2477.

Momoh J 2012 *Smart Grid: Fundamentals of Design and Analysis*. John Wiley & Sons.

Montano J 2011 Reviewing concepts of instantaneous and average compensations in polyphase systems. *IEEE Transactions on Industrial Electronics* **58**(1), 213–220.

Moreno L 2011 Concentrated solar power (CSP) in DESERTEC– Analysis of technologies to secure and affordable energy supply, in *Proceedings of the 6th IEEE International Conference on Intelligent Data Acquisition and Advanced Computing Systems (IDAACS)*, **vol. 2**, pp. 923–926.

Morimoto H, Uzuka T, Horiguchi A and Akita T 2009 New type of feeding transformer for AC railway traction system, in *Proceedings of International Conference on Power Electronics and Drive Systems*, pp. 800–805.

Nabae A, Takahashi I and Akagi H 1981 A new neutral-point clamped PWM inverter. *IEEE Transactions on Industry Applications* **17**(5), 518–523.

Naim R, Weiss G and Ben-Yaakov S 1997 H^∞ control applied to boost power converters. *IEEE Transactions on Power Electronics* **12**(4), 677–683.

Nakano M, Inoue T, Yamamoto Y and Hara S 1989 *Repetitive Control*. SICE Publications.

Neves F, de Souza H, Bradaschia F, Cavalcanti M, Rizo M and Rodriguez F 2010 A space-vector discrete fourier transform for unbalanced and distorted three-phase signals. *IEEE Transactions on Industrial Electronics* **57**(8), 2858–2867.

Neves F, Souza H, Cavalcanti M, Bradaschia F and Bueno E 2012 Digital filters for fast harmonic sequence components separation of unbalanced and distorted three-phase signals. *IEEE Transactions on Industrial Electronics* **59**(10), 3847–3859.

Nikkhajoei H and Lasseter R 2009 Distributed generation interface to the CERTS microgrid. *IEEE Transactions on Power Delivery* **24**(3), 1598–1608.

Partington J and Bonnet C 2004 H^∞ and BIBO stabilization of delay systems of neutral type. *Systems & Control Letters* **52**, 283–288.

Pasterczyk R, Guichon JM, Schanen JL and Atienza E 2009 PWM inverter output filter cost-to-losses tradeoff and optimal design. *IEEE Transactions on Industry Applications* **45**(2), 887–897.

Piagi P and Lasseter R 2006 Autonomous control of microgrids, in *Proceedings of IEEE Power Engineering Society General Meeting*, p. 8.

POWEREX 2000 *General Considerations for IGBT and Intelligent Power Modules* POWEREX Inc. http://www.pwrx.com/pwrx/app/IGBT-Intelligent-PwrMods.pdf.

Prodanovic M and Green T 2006 High-quality power generation through distributed control of a power park microgrid. *IEEE Transactions on Industrial Electronics* **53**(5), 1471–1482.

Quinn C and Mohan N 1992 Active filtering of harmonic currents in three-phase, four-wire systems with three-phase and single-phase nonlinear loads, in *Proceedings of IEEE Applied Power Electronics Conference and Exposition*, pp. 829–836.

Ragheb M 2009 Wind power systems. Technical report, University of Illinois at Urbana-Champaign. Available at https://netfiles.uiuc.edu/mragheb/www/NPRE

Rashid M 1993 *Power Electronics: Circuits, Devices and Applications* 2nd edn. Prentice-Hall International, Inc.

Rashid M 2010 *Power Electronics Handbook: Devices, Circuits, and Applications* Academic Press. Elsevier.

Ratnayake K, Murai Y and Watanabe T 1999 Novel PWM scheme to control neutral point voltage variation in three-level voltage source inverter, in *Proceedings of Conference Record of the IEEE Industry Applications Conference*, **vol. 3**, pp. 1950–1955.

Rockhill AA, Liserre M, Teodorescu R and Rodriguez P 2011 Grid-filter design for a multimegawatt medium-voltage voltage-source inverter. *IEEE Transactions on Industrial Electronics* **58**(4), 1205–1217.

Rodriguez J, Lai JS and Peng F 2002 Multilevel inverters: A survey of topologies, controls, and applications. *IEEE Transactions on Industrial Electronics* **49**(4), 724–738.

Rodriguez P, Luna A, Ciobotaru M, Teodorescu R and Blaabjerg F 2006a Advanced grid synchronization system for power converters under unbalanced and distorted operating conditions, in *Proceedings of the 32nd IEEE Annual Conference on Industrial Electronics (IECON)*, pp. 5173–5178.

Rodriguez P, Luna A, Teodorescu R and Blaabjerg F 2008 Grid synchronization of wind turbine converters under transient grid faults using a double synchronous reference frame PLL *IEEE Energy 2030 Conference*, pp. 1–8.

Rodriguez P, Luna A, Teodorescu R, Iov F and Blaabjerg F 2007a Fault ride-through capability implementation in wind turbine converters using a decoupled double synchronous reference frame PLL, *European Conference on Power Electronics and Applications*, pp. 1–10.

Rodriguez P, Pou J, Bergas J, Candela J, Burgos R and Boroyevich D 2007b Decoupled double synchronous reference frame PLL for power converters control. *IEEE Transactions on Power Electronics* **22**(2), 584–592.

Rodriguez J, Rivera M, Kolar J and Wheeler P 2012 A review of control and modulation methods for matrix converters. *IEEE Transactions on Industrial Electronics* **59**(1), 58–70.

Rodriguez P, Teodorescu R, Candela I, Timbus A, Liserre M and Blaabjerg F 2006b New positive-sequence voltage detector for grid synchronization of power converters under faulty grid conditions, in *Proceedings of the 37th IEEE Power Electronics Specialists Conference (PESC)*, pp. 1–7.

Rolim L, da Costa D and Aredes M 2006 Analysis and software implementation of a robust synchronizing PLL circuit based on the pq theory. *IEEE Transactions on Industrial Electronics* **53**(6), 1919–1926.

Routimo M, Salo M and Tuusa H 2007 Comparison of voltage-source and current-source shunt active power filters. *IEEE Transactions on Power Electronics* **22**(2), 636–643.

Salaet J, Gilabert A, Bordonau J, Alepuz S, Cano A and Gimeno L 2006 Nonlinear control of neutral point in three-level single-phase converter by means of switching redundant states. *Electronics Letters* **42**(5), 304–306.

Sankaran C 2002 *Power Quality The Electric Power Engineering Series*. CRC Press.

Santos Filho R, Seixas P, Cortizo P, Torres L and Souza A 2008 Comparison of three single-phase PLL algorithms for UPS applications. *IEEE Transactions on Industrial Electronics* **55**(8), 2923–2932.

Sao C and Lehn P 2005 Autonomous load sharing of voltage source converters. *IEEE Transactions on Power Delivery* **20**(2), 1009–1016.

Sera D, Kerekes T, Lungeanu M, Nakhost P, Teodorescu R, Andersen G and Liserre M 2005 Low-cost digital implementation of proportional-resonant current controllers for PV inverter applications using delta operator, in *Proceedings of the 31st Annual IEEE Conference of Industrial Electronics (IECON)*.

Severns R n.d. Design of snubbers for power circuits. Technical report, Cornell Dubilier Electronics, Inc. http://www.cde.com/tech/design.pdf.

Shen B, Mwinyiwiwa B, Zhang Y and Ooi BT 2009 Sensorless maximum power point tracking of wind by DFIG using rotor position phase lock loop (PLL). *IEEE Transactions on Power Electronics* **24**(4), 942–951.

Shen G, Zhu X, Zhang J and Xu D 2010 A new feedback method for PR current control of LCL-filter-based grid-connected inverter. *IEEE Transactions on Industrial Electronics* **57**(6), 2033–2041.

Shinnaka S 2008 A robust single-phase PLL system with stable and fast tracking. *IEEE Transactions on Industry Applications* **44**(2), 624–633.

Shu ZL, Xie SF and Li QZ 2011 Single-phase back-to-back converter for active power balancing, reactive power compensation, and harmonic filtering in traction power system. *IEEE Transactions on Power Electronics* **26**(2), 334–343.

Singh B, Saha R, Chandra A and Al-Haddad K 2009 Static synchronous compensators (STATCOM): A review. *IET Proceedings on Power Electronics* **2**(4), 297–324.

Singh B and Singh S 2010 Single-phase power factor controller topologies for permanent magnet brushless DC motor drives. *IET Proceedings on Power Electronics* **3**(2), 147–175.

Skvarenina T 2002 *The Power Electronics Handbook* Industrial Electronics Series. CRC Press.

Smith M 2002 Synthesis of mechanical networks: The inerter. *IEEE Transactions on Automatic Control* **47**(10), 1648–1662.

Song W and Huang A 2010 Fault-tolerant design and control strategy for cascaded h-bridge multilevel converter-based STATCOM. *IEEE Transactions on Industrial Electronics* **57**(8), 2700–2708.

Soto-Sanchez D and Green T 2001 Voltage balance and control in a multi-level unified power flow controller. *IEEE Transactions on Power Delivery* **16**(4), 732–738.

Spera D 2009 *Wind Turbine Technology: Fundamental Concepts of Wind Turbine Engineering*. ASME Press.

Srikanthan S and Mishra M 2010 DC capacitor voltage equalization in neutral clamped inverters for DSTATCOM application. *IEEE Transactions on Industrial Electronics* **57**(8), 2768–2775.

Strom JP, Korhonen J, Tyster J and Silventoinen P 2011 Active du/dt-new output-filtering approach for inverter-fed electric drives. *IEEE Transactions on Industrial Electronics* **58**(9), 3840–3847.

Sun Z, Jiang XJ, Zhu DQ and Zhang GX 2004 A novel active power quality compensator topology for electrified railway. *IEEE Transactions on Power Electronics* **19**(4), 1036–1042.

Svensson J 2001 Synchronisation methods for grid-connected voltage source converters. *IEE Proceedings Generation, Transmission and Distribution* **148**(3), 229–235.

Tan K, Wang Q and Hang C 1999 *Advances in PID Control*. Springer, Berlin.

Tan PC, Morrison RE and Holmes DG 2003 Voltage form factor control and reactive power compensation in a 25–kV electrified railway system using a shunt active filter based on voltage detection. *IEEE Transactions on Industry Applications* **39**(2), 575–581.

Tazil M, Kumar V, Bansal RC, Kong S, Dong ZY, Freitas W and Mathur HD 2010 Three-phase doubly fed induction generators: An overview. *IET Proceedings of Electric Power Application* **4**(2), 75–89.

Teodorescu R and Blaabjerg F 2004 Flexible control of small wind turbines with grid failure detection operating in stand-alone and grid-connected mode. *IEEE Transactions on Power Electronics* **19**(5), 1323–1332.

Teodorescu R, Blaabjerg F, Liserre M and Loh P 2006 Proportional-resonant controllers and filters for grid-connected voltage-source converters. *IEE Proceedings of Electric Power Application* **153**(5), 750–762.

Teodorescu R, Liserre M and Rodríguez P 2011 *Grid Converters for Photovoltaic and Wind Power Systems*. Wiley-IEEE Press.

Thongam JS and Ouhrouche M 2011 MPPT control methods in wind energy conversion systems *Fundamental and Advanced Topics in Wind Power* **number** 341–360 InTech.

Thorborg K 1988 *Power Electronics*. Prentice Hall.

TI 2002 *TMS320F28335 controlCARD* Texas Instruments. http://www.ti.com/tool/tmdscncd28335.

TI 2007 *TMS320F28335, TMS320F28334, TMS320F28332, TMS320F28235, TMS320F28234, TMS320F28232 Digital Signal Controllers (DSCs) Data Manual*, revised apr. 2012 edn Texas Instruments. http://www.ti.com/lit/ds/symlink/tms320f28335.pdf.

TI 2011 *Hardware Design Guidelines for TMS320F28xx and TMS320F28xxx DSCs*. http://www.ti.com/lit/an/spraas1b/spraas1b.pdf.

Tilli A and Tonielli A 1998 Sequential design of hysteresis current controller for three-phase inverter. *IEEE Transactions on Industrial Electronics* **45**(5), 771–781.

Timbus A, Ciobotaru M, Teodorescu R and Blaabjerg F 2006a Adaptive resonant controller for grid-connected converters in distributed power generation systems, in *Proceedings of the 21st Annual IEEE Applied Power Electronics Conference and Exposition (APEC)*, pp. 1601–1606.

Timbus A, Liserre M, Teodorescu R, Rodriguez P and Blaabjerg F 2009 Evaluation of current controllers for distributed power generation systems. *IEEE Transactions on Power Electronics* **24**(3), 654–664.

Timbus A, Teodorescu R, Blaabjerg F and Liserre M 2005 Synchronization methods for three phase distributed power generation systems: An overview and evaluation, in *Proceedings of IEEE Annual Power Electronics Specialists Conference (PESC)*, pp. 2474–2481.

Timbus A, Teodorescu T, Blaabjerg F, Liserre M and Rodriguez P 2006b PLL algorithm for power generation systems robust to grid voltage faults, in *Proceedings of the 37th IEEE Power Electronics Specialists Conference (PESC)*, pp. 1–7.

Timbus AV, Teodorescu R, Blaabjerg F, Liserre M and Rodriguez P 2006c Linear and nonlinear control of distributed power generation systems, in *Proceedings of IEEE Industry Applications Conference*, pp. 1015–1023.

Todd PC 2001 Snubber circuits: theory, design and application. Technical report, Texas Instruments. http://www.ti.com/lit/an/slup100/slup100.pdf.

Toshiba 2002 *TLP550 Datasheet* Toshiba Corporation. http://www.datasheetcatalog.org/datasheet/toshiba/2120.pdf.

Tuladhar A, Jin H, Unger T and Mauch K 1997 Parallel operation of single phase inverter modules with no control interconnections, in *Proceedings of the 12th IEEE Applied Power Electronics Conference and Exposition*, pp. 94–100.

Tuladhar A, Jin H, Unger T and Mauch K 2000 Control of parallel inverters in distributed AC power systems with consideration of line impedance effect. *IEEE Transactions on Industry Applications* **36**(1), 131–138.

Twining E and Holmes D 2003 Grid current regulation of a three-phase voltage source inverter with an LCL input filter. *IEEE Transactions on Power Electronics* **18**(3), 888–895.

Tzou YY, Jung SL and Yeh HC 1999 Adaptive repetitive control of PWM inverters for very low THD AC-voltage regulation with unknown loads. *IEEE Transactions on Power Electronics* **14**(5), 973–981.

Valiviita S 1999 Zero-crossing detection of distorted line voltages using 1-b measurements. *IEEE Transactions on Industrial Electronics* **46**(5), 917–922.

van der Schaft A 1996 L_2-*Gain and Passivity Techniques in Nonlinear Control*, vol. 218 of *Lecture Notes in Control and Information Sciences*. Springer-Verlag, London.

Venkataramanan G and Illindala M 2002 Microgrids and sensitive loads, in *Proceedings of IEEE Power Engineering Society Winter Meeting*, pp. 315–322.

Verdelho P and Marques G 1998 Four-wire current-regulated PWM voltage converter. *IEEE Transactions on Industrial Electronics* **45**(5), 761–770.

Vilathgamuwa D, Loh PC and Li Y 2006 Protection of microgrids during utility voltage sags. *IEEE Transactions on Industrial Electronics* **53**(5), 1427–1436.

Visioli A 2006 *Practical PID Control*. Springer.

Vithayathil J 1995 *Power Electronics: Principles and Applications*. McGraw-Hill.

Wagner HJ and Mathur J 2009 *Introduction to Wind Energy Systems: Basics, Technology and Operation*. Springer-Verlag.

Walker J 1981 *Large Synchronous Machines: Design, Manufacture and Operation*. Oxford University Press.

Wang C and Li Y 2010 Analysis and calculation of zero-sequence voltage considering neutral-point potential balancing in three-level NPC converters. *IEEE Transactions on Industrial Electronics* **57**(7), 2262–2271.

Wang HQ, Tian YJ and Gui QC 2009 Evaluation of negative sequence current injecting into the public grid from different traction substation in electrical railways, in *Proceedings of the 20th International Conference and Exhibition on Electricity Distribution-Part 1*, pp. 1–4.

Wang X 2008 *Design and Implementation of Internal Model Based Controllers for DC-AC Power Converters* PhD thesis. Imperial College London, UK.

Wang Y and Li Y 2011a Grid synchronization PLL based on cascaded delayed signal cancellation. *IEEE Transactions on Power Electronics* **26**(7), 1987–1997.

Wang Y and Li Y 2011b Three-phase cascaded delayed signal cancellation PLL for fast selective harmonic detection. *IEEE Transactions on Industrial Electronics* **PP**(99), 1.

Wang Y, Wang D, Zhang B and Zhou K 2007 Fractional delay based repetitive control with application to PWM DC/AC converters, in *Proceedings of IEEE International Conference on Control Applications*.

Watanabe E, Stephan R and Aredes M 1993 New concepts of instantaneous active and reactive powers in electrical systems with generic loads. *IEEE Transactions on Power Delivery* **8**(2), 697–703.

Weidenbrug R, Dawson F and Bonert R 1993 New synchronization method for thyristor power converters to weak ac-systems. *IEEE Transactions on Industrial Electronics* **40**(5), 505–511.

Weiss G 1997 Repetitive control systems: Old and new ideas In *Systems and Control in the Twenty-first Century* (ed. Byrnes C, Datta B, Gilliam D and Martin C) vol 22, Birkhäuser, Boston pp. 389–404.

Weiss G and Hafele M 1999 Repetitive control of MIMO systems using H^∞ design. *Automatica* **35**(7), 1185–1199.

Weiss G, Neuffer D and Owens DH 1998 A simple scheme for internal model based control, in *Proceedings of UKACC International Conference on Control*, pp. 630–634.

Weiss G, Zhong QC, Green T and Liang J 2004 H^∞ repetitive control of DC-AC converters in micro-grids. *IEEE Transactions on Power Electronics* **19**(1), 219–230.

Wheeler P, Rodriguez J, Clare J, Empringham L and Weinstein A 2002 Matrix converters: a technology review. *IEEE Transactions on Industrial Electronics* **49**(2), 276–288.

Wildi T 2005 *Electrical Machines, Drives and Power Systems* 6 edn. Prentice-Hall.

Wong MC, Zhao ZY, Han YD and Zhao LB 2001 Three-dimensional pulse-width modulation technique in three-level power inverters for three-phase four-wired system. *IEEE Transactions on Power Electronics* **16**(3), 418–427.

Wu CP, Luo A, Shen J, Ma FJ and Peng SJ 2012 A negative sequence compensation method based on two-phase three-wire converter for high-speed railway traction power supply system. *IEEE Transactions on Power Electronics* **27**(2), 706–717.

Wu E and Lehn P 2006 Digital current control of a voltage source converter with active damping of LCL resonance. *IEEE Transactions on Power Electronics* **21**(5), 1364–1373.

Xiarnay C, Asano H, Papathanassiou S and Strbac G 2008 Policymaking for microgrids. *IEEE Power Energy Magazine* **6**(3), 66–77.

Yao W, Chen M, Matas J, Guerrero J and Qian ZM 2011 Design and analysis of the droop control method for parallel inverters considering the impact of the complex impedance on the power sharing. *IEEE Transactions on Industrial Electronics* **58**(2), 576–588.

Yao Z, Xiao L and Yan Y 2010 Seamless transfer of single-phase grid-interactive inverters between grid-connected and stand-alone modes. *IEEE Transactions on Power Electronics* **25**(6), 1597–1603.

Ye Y, Zhang B, Zhou K, Wang D and Wang Y 2006 High-performance repetitive control of PWM DC-AC converters with real-time phase lead FIR filter. *IEEE Transactions of Circuits Syst* **53**(8), 768–772.

Ye Y, Zhang B, Zhou K, Wang D and Wang Y 2007 High-performance cascade-type repetitive controller for CVCF PWM inverter: Analysis and design. *IET Proceedings of Electric Power Application* **1**(1), 112–118.

Yousefpoor N, Fathi S, Farokhnia N and Abyaneh HA 2012 THD minimization applied directly on the line to line voltage of multi-level inverters. *IEEE Transactions on Industrial Electronics* **59**(1), 373–380.

Yu CC 1999 *Autotuning of PID Controllers: Relay Feedback Approach*. Springer Verlag, London UK.

Yuan X, Merk W, Stemmler H and Allmeling J 2002 Stationary-frame generalized integrators for current control of active power filters with zero steady-state error for current harmonics of concern under unbalanced and distorted operating conditions. *IEEE Transactions on Industry Applications* **38**(2), 523–532.

Zaragoza J, Pou J, Ceballos S, Robles E, Jaen C and Corbalan M 2009 Voltage-balance compensator for a carrier-based modulation in the neutral-point-clamped converter. *IEEE Transactions on Industrial Electronics* **56**(2), 305–314.

Zeng Q and Chang L 2005 Study of advanced current control strategies for three-phase grid-connected PWM inverters for distributed generation, in *Proceedings of IEEE Conference on Control Applications*, pp. 1311–1316.

Zeng Q and Chang L 2008 An advanced SVPWM-based predictive current controller for three-phase inverters in distributed generation systems. *IEEE Transactions on Industrial Electronics* **55**(3), 1235–1246.

Zhang B, Wang D, Zhou K and Wang Y 2008 Linear phase lead compensation repetitive control of a CVCF PWM inverter. *IEEE Transactions on Industrial Electronics* **55**(4), 1595–1602.

Zhang R, Boroyevich D, Prasad V, Mao HC, Leeand F and Dubovsky S 1997 A three-phase inverter with a neutral leg with space vector modulation, in *Proceedings of IEEE Applied Power Electronics Conference and Exposition*, vol 2, pp. 857–863.

Zhao YL, Zhao LP and Li QZ 2010 Some key problems on cophase traction power supply device, in *Proceedings of International Forum on Information Technology and Applications(IFITA)*, pp. 444–449.

Zhong QC 2006 *Robust Control of Time-delay Systems*. Springer.

Zhong QC 2012a Harmonic droop controller to reduce the voltage harmonics of inverters. *IEEE Transactions on Industrial Electronics*. Early Access.

Zhong QC 2012b Power delivery to a current source and reduction of voltage harmonics for inverters, in *Proceedings of the American Control Conference (ACC)*, Montreal, Canada.

Zhong QC 2012c Robust droop controller for accurate proportional load sharing among inverters operated in parallel. *IEEE Transactions on Industrial Electronics*.

Zhong QC, Blaabjerg F, Guerrero J and Hornik T 2011 Reduction of voltage harmonics for parallel-operated inverters, in *Proceedings of IEEE Energy Conversion Congress and Exposition (ECCE)*, pp. 473–478.

Zhong QC, Blaabjerg F, Guerrero J and Hornik T 2012 Improving the voltage quality of an inverter via bypassing the harmonic current components, in *Proceedings of IEEE Energy Conversion Congress and Exposition (ECCE)*, Raleigh, North Carolina.

Zhong QC, Green T, Liang J and Weiss G 2002a H^{∞} control of the neutral leg for 3–phase 4–wire DC-AC converters, in *Proceedings of the 28th Annual IEEE Conference of Industrial Electronics (IECON)*.

Zhong QC, Green T, Liang J and Weiss G 2002b Robust repetitive control of grid-connected DC-AC converters, in *Proceedings of the 41th IEEE Conf. on Decision & Control*, vol 3, pp. 2468–2473.

Zhong QC, Hobson L and Jayne M 2005a Classical control of the neutral point in 4–wire 3–phase DC-AC converters. *Journal of Electrical Power Quality and Utilisation* **11**(2), 111–119.

Zhong QC, Hobson L and Jayne M 2005b Generating a neutral point for 3-phase 4-wire DC-AC converters, in *Proceedings of IEEE Compatibility in Power Electronics (CPE)*, pp. 126–133.

Zhong QC and Hornik T 2012 Cascaded current-voltage control to improve the power quality for a grid-connected inverter with a local load. *IEEE Transactions on Industrial Electronics*. Early Access.

Zhong QC, Liang J, Weiss G, Feng C and Green T 2006 H^{∞} control of the neutral point in 4–wire 3–phase DC-AC converters. *IEEE Transactions on Industrial Electronics* **53**(5), 1594–1602.

Zhong QC and Nguyen PL 2012 Sinusoid-locked loops to detect the frequency, the ampitude and the phase of the fundamental component of a periodic signal, in *Proceedings of the 24th Chinese Control and Decision Conference (CCDC)*, Taiyuan, China.

Zhong QC and Weiss G 2009 Static synchronous generators for distributed generation and renewable energy, in *Proceedings of IEEE PES Power Systems Conference & Exhibition (PSCE)*.

Zhong QC and Weiss G 2011 Synchronverters: Inverters that mimic synchronous generators. *IEEE Transactions on Industrial Electronics* **58**(4), 1259–1267.

Zhong QC and Zeng Y 2011 Can the output impedance of an inverter be designed capacitive? *Proceedings of the 37th Annual IEEE Conference of Industrial Electronics (IECON)*.

Zhou D and Rouaud D 2001 Experimental comparisons of space vector neutral point balancing strategies for three-level topology. *IEEE Transactions on Power Electronics* **16**(6), 872–879.

Zhou K and Doyle J 1998 *Essentials of Robust Control*. Prentice-Hall.

Zhou K, Doyle J and Glover K 1996 *Robust and Optimal Control*. Prentice-Hall, Englewood Cliffs, NJ.

Zhou K and Wang D 2003 Digital repetitive controlled three-phase PWM rectifier. *IEEE Transactions on Power Electronics* **18**(1), 309–316.

Zhou K, Wang D, Zhang B and Wang Y 2009 Plug-in dual-mode-structure repetitive controller for CVCF PWM inverters. *IEEE Transactions on Industrial Electronics* **56**(3), 784–791.

Ziarani AK and Konrad A 2004 A method of extraction of nonstationary sinusoids. *Signal Processing* **84**(8), 1323–1346.

Zmood D, Holmes D and Bode G 1999 Frequency domain analysis of three phase linear current regulators, in *Proceedings of IEEE Industry Applications Conference*, pp. 818–825.

Zmood D, Holmes D and Bode G 2001 Frequency-domain analysis of three-phase linear current regulators. *IEEE Transactions on Industry Applications* **37**(2), 601–610.

Index

H^∞ repetitive control
    cascaded current-voltage, 127
    current, 81
    voltage, 109
    voltage and current, 93
H^∞ norm, 222
$\alpha\beta$ frame, 73, 366
abc Frame, 71
dq frame, 74, 365

AC-AC conversion
    On-off control, 22
    Phase control, 24
AC-DC conversion, *see* Rectifier
Active power compensator (APC), 174
Active power filter, 235
Auxiliary power supplies, 27

Bode plot, 87, 100, 113, 214, 263, 342
Boost converter, 15, 190
Buck converter, 14, 190
Buck-Boost converter, 17

C-inverters, 152, 165
    parallel operation, 303, 319
Circulating current, 338, 340
Clarke transformation, 73, 365
Concentrated solar power (CSP), 53
Control
    decoupled, 233
    feed-forward, 82, 88, 252
    multiloop, 129
    parallel structure, 219

Controller
    H^∞ current, 221
    cascaded, 109, 129
    deadbeat predictive (DB), 269
    hysteresis, 241
    proportional (P), 150
    proportional integral (PI), 226, 251
    proportional resonant (PR), 136, 259
    sub-optimal, 99
Controller reduction, 87
Cross-coupling terms, 252
Current sensing, 35
Current source inverter, 18

Damping ratio, 262
DC-AC conversion, *see* Inverter
DC-bus voltage regulation, 179
DC-DC conversion, 14
    boost converters, 15
    buck converters, 14
    buck-boost converters, 17
Digital signal processor (DSP), 25, 36
Discrete-time domain, 262
Doubly-fed induction generators (DFIG), 48, 49
Driving circuit, 27
    ground, 29
Droop coefficient, 306
Droop control
    conventional, 301
    harmonic droop control, 351
    limitations, 304
    robust, 309, 319, 326

Control of Power Inverters in Renewable Energy and Smart Grid Integration, First Edition.
Qing-Chang Zhong and Tomas Hornik.
© 2013 John Wiley & Sons, Ltd. Published 2013 by John Wiley & Sons, Ltd.

Filter
    cut-off frequency, 33
    hold, 241
    impact, 162
    LC, 33
    LCL, 34
    loop (LF), 363
    low-pass, 241
    passive, 33
    phase-lead low-pass, 81
    resonant, 168
Firing angle, 7, 362
Four-quadrant PWM converter, 173
Four-wire three-phase distribution network, 187
Frequency
    adaptive mechanism, 109
    droop control, 284
    oscillations, 168

Gradient descent method, 368
Grid-connected mode, 127

Half cycle, 5
Hall effect, 35
Harmonic compensator, 167, 243, 259, 335
    physical interpretation, 167
Harmonic current
    bypassing, 165
Harmonic droop control, 347
Harmonics, 4
Hold filter, 179, 241
Hysteresis control, 12

IGBT, 4, 26
Intelligent power modules (IPM), 27
Internal model, 67, 96
    frequency-adaptive, 109
Inverter
    C-inverters, 152
    current-source inverter, 18
    forced commutated, 18
    grid-friendly, 277
    inversion, 8
    L-inverters, 149
    line commutated, 18
    model, 347
    R-inverters, 150
    single-phase, 20
    three-phase, 21
    voltage-source inverter, 18
Isolation, 25, 35

Kirchhoff's law, 193, 208
    current (KCL), 261
    voltage (KVL), 262

L-inverters, 149, 165, 302
    parallel operation, 326
Lambert W function, 68
Linear extrapolation, 271

MATLAB® function
    c2d, 88
    hinfsyn, 87, 97, 100
Maximum power point tracking (MPPT)
    hill-climb search (HCS) control, 52
    power signal feedback (PSF) control, 52
    solar power, 54
    tip-speed ratio (TSR) control, 52
    wind power, 51
Measurement noise, 222
Measuring resistor, 35
Microgrid, 93, 187
    operation modes, 127
    voltage, 127
    voltage control, 94
Moment of inertia, 280, 383
MOSFET, 4, 26
Motor drives, 233
Multi-level PWM converter, 188

Natural (*abc*) Frame, 71
Negative-sequence current, 173
Neutral leg control
    H^∞ voltage-current, 207
    classical, 193
    parallel PI voltage-H^∞ current, 219
Neutral leg model, 207
Neutral line, 187, 281
    conventional, 189
    independently-controlled neutral leg, 190
    model, 193, 207
    split DC link, 188
Nyquist plot, 102

Output impedance, 66, 166
    capacitive, 152
    inductive, 149

Index

of an inverter, 149
per unit, 307
resistive, 150

Parallel operation
C-inverters, 303, 319
L-inverters, 302, 326
R-inverters, 301, 309
reactive power sharing, 308
real power sharing, 307
voltage regulation, 311
Park transformation, 365
Permanent magnet synchronous generator (PMSG), 48, 50
Phase detection unit (PD), 363
Phase locked loop (PLL), 81, 130, 239, 363
Phase sequence
abc, 71
acb, 76
Phasor diagrams, 71, 178, 238
Photovoltaic (PV) cells, 53
Power angle, 300
Power delivery
to a current source, 349
to a voltage source, 300
Power distribution, 127
Power electronic device, 4
Power factor, 173, 233
Power flow control
current DB control, 269
current PI control, 251
current PR control, 259
current repetitive control, 81
parallel operation, 297, 335
voltage-controlled, 277
Power meter, 43
Power modules, 26
Power processing, 4
AC-AC Conversion, 21
AC-DC conversion, 4
DC-AC Conversion, 18
DC-DC Conversion, 14
Single-phase to three-phase, 233
Power quality, 63, 127, 128, 187
commonly-used terms, 64
current quality, 64
of traction power systems, 173
voltage quality, 65
Power stages, 26

Power systems
centralised generation, 55
distributed generation, 55
Processing delay, 113, 262
Protection, 27, 38
over-current, 127
Pulse Width Modulation (PWM), 18
bipolar, 19, 20
sinusoidal, 18
three-phase, 19
unipolar, 19, 20

R-inverters, 150, 165
parallel operation, 309
Railway static power conditioner (RPC), 174
Reactive power
regulation, 286
sharing, 308
Real power
regulation, 284
sharing, 307
Rectifier, 4
bridge, 5
phase-controlled, 4, 7
PWM-controlled, 4, 11
single-phase, 5
three-phase, 5
uncontrolled, 4
uncontrolled with a boost converter, 8
Reference frame, 71
natural, 71, 81, 109, 128
stationary, 72, 259
synchronously rotating, 74, 251, 364
Repetitive control, 67
basic principles, 67
frequency-adaptive internal model, 112
internal model, 68
Resonance, 150, 165
Resonant frequency, 259, 367
Robust droop control
C-inverters, 319
L-inverters, 326, 335
R-inverters, 309, 336, 353
Root-locus, 262
Root mean square (RMS) value, 5
Rotation matrix, 75, 76, 78

Scalability, 354
Seamless transfer, 128, 136

Second order generalised integrator based PLL (SOGI-PLL), 366
Shoot-through, 32
   deadtime generator, 32
Signal conditioning, 36
   Impedance matching, 36
   level-shifting, 37
   protection, 38
   scaling, 37
Single-phase to three-phase conversion
   conventional topologies, 234
   line-interactive, 235
   neutral leg, 241
   phase leg, 242
   rectifier leg, 241
   with a universal APF, 236
   with an independently-controlled neutral leg, 236
Sinusoid-locked loop, 379
   amplitude tracking, 382
   frequency tracking, 382
   harmornic components, 383
   phase tracking, 382
   structure, 380
Sinusoidal tracking algorithm, 179, 239, 368
slip, 50
Smart grid, 55, 56, 187
   areas, 56
   integration, 58
   layered structure, 56
   scope, 56
Smart grid integration
   anti-islanding, 61
   fault ride-through, 60
   neutral line provision, 60
   power flow control, 59
   power quality control, 59
   synchronisation, 59
Snubber circuit, 29
   lump, 29
   wiring inductance, 30
Solar power, 53
   MPPT, 54
   processing, 54
Space vector, 72, 74, 75, 78
Space vector pulse width modulation (SVPWM), 189
Spatial diagrams, 71
Spatial operator, 75

Spatial order, 72, 74, 77
Squirrel-cage induction generator (SCIG), 48, 49
   excitation, 49
Stabilising compensator, 82, 96, 110
Stability analysis, 151, 169
Stability evaluation, 86, 115
Stand-alone mode, 127
Static power conditioner (SPC), 175
Static var compensators (SVC), 174
Stationary reference ($\alpha\beta$) frame, 72
Switching frequency
   variable, 241
Synchronisation, 81, 88, 127, 239
   close-loop methods, 361, 361
   open-loop methods, 361, 379
Synchronous generator, 277
   model, 278
Synchronous machine, 379
Synchronously rotating reference (dq) frame, 74
Synchronously rotating reference frame PLL (SRF-PLL), 364
Synchronverter, 277, 282
   electronic part, 283
   frequency droop, 284
   power part, 282
   regulation of reactive power, 286
   regulation of real power, 284
   voltage droop, 286, 379

Test equipment, 42
TI
   TMS320F28335 ControlCARD, 42
   TMS320F28335 digital signal controller (DSC), 38
   TMS320F28335 Experimenter Kit, 42
Torque
   electromagnetic, 280
   mechanical, 280
Total harmonic distortion (THD), 64, 81, 127, 149
Traction power systems, 173
   compensation of harmonic currents, 179
   compensation of reactive power, 178
   negative-sequence currents, 175
   topology, 175
Transformers
   Le Blanc, 174
   Scott, 174
   Steinmetz, 174

V/V, 174
Woodbridge, 174

Unbalanced load, 187
Unbalanced voltage, 82, 128

Virtual resistor, 150, 165
Voltage
   droop control, 286
   follower, 37
   quality, 65
   sags, 82, 128
Voltage controlled oscilator (VCO), 363
Voltage drop, 310
Voltage quality
   Degradation mechanisms, 65
Voltage sensing, 35
Voltage source inverter (VSI), 18
   current-controlled, 93, 251, 269
   voltage-controlled, 93

Weighting factor, 222
Weighting function, 97, 135, 214, 221
Weighting parameters, 84, 113, 132
Wind power, 44
   basics, 44
   Betz law, 45
   generators and topologies, 48
   grid integration, 52
Wind turbine
   active stall control, 51
   desired power curve, 52
   pitch control, 51
   power coefficient, 44
   rotor power control, 51
   rotor speed control, 51
   stall control, 51
   tip-speed ratio, 45
   yaw control, 51

Zero-crossing method, 362